INSIDEOut

INSIDEOut

DESIGN PROCEDURES FOR PASSIVE ENVIRONMENTAL TECHNOLOGIES

Second Edition

G. Z. Brown
University of Oregon

Bruce Haglund
University of Idaho

Joel Loveland
University of Washington

John S. Reynolds
University of Oregon

M. Susan Ubbelohde
University of California, San Diego

Illustrations by Amy Dowty

JOHN WILEY & SONS, INC.

New York / Chichester / Brisbane / Toronto / Singapore

This text is printed on recycled paper.

In recognition of the importance of preserving what has been written, it is a policy of John Wiley & Sons, Inc., to have books of enduring value published in the United States printed on acid-free paper, and we exert our best efforts to that end.

Library of Congress Cataloging-in-Publication Data:

Insideout : design procedures for passive environmental technologies / [edited by] G.Z. Brown . . . [et al.]. — 2nd ed.

p. cm.
Rev. ed. of: Insideout / G.Z. Brown. © 1982.
Includes bibliographical references and index.
ISBN 0-471-55461-8 (pbk.)
1. Buildings—Environmental engineering. 2. Solar energy—Passive systems. 3. Buildings—Energy conservation. I. Brown, G. Z. II. Brown, G. Z. Insideout. III. Title: Inside out.

TH6025.I57 1992 91-44625
720'.47—dc20 CIP

Printed in the United States of America

10 9 8 7 6 5 4 3 2 1

Printed and bound by Courier Companies, Inc.

CONTENTS

PREFACE		XIII	
ACKNOWLEDGMENTS		XVII	
1	THERMAL		
A	PRECEDENTS OF THERMAL RESPONSE		
	Introduction	1	A1.0
	Architecture, Climate, and People—Vernacular Precedents	2	A1.1
	Cooling Precedents	3	A1.2
	Heating Precedents	4	A1.3
B	CLIMATE AND SITE ANALYSIS		
	Introduction	5	B1.0
	Bioclimatic Chart Plot	6	B1.1
	Site Wind Flows	8	B1.2
	Sun Peg Shadow Plot	9	B1.3
	Climate Response Matrix	10	B1.4
	Climate Response Matrix Analysis	11	B1.5
	Site Concepts	12	B1.6
C	PROGRAM ANALYSIS		
	Introduction	13	C1.0
	Use Schedules—People, Lights, and Equipment	14	C1.1
	Thermal Zoning	16	C1.2
	Heat Gains	17	C1.3
	Heat Loss	19	C1.4
	Balance Point Temperature Estimate	20	C1.5
	Energy Conservation	24	C1.6
D	SCHEMATIC DESIGN		
	Introduction	25	D1.0
	Design	27	D1.1
	Bioclimatic Chart Thermal Strategies	29	D1.2
	Solar Envelope	31	D1.3
	Site Microclimate	32	D1.4
	Ventilation Aperture	33	D1.5
	Ventilation Air Flow	35	D1.6
	Solar-Glazing Aperture	36	D1.7
	Solar Glazing and Thermal Mass	37	D1.8
	Window Shading	38	D1.9
	Design Review	39	D1.10
E	DESIGN DEVELOPMENT		
	Introduction	41	E1.0
	Design	43	E1.1
	Component-Scale Shading	44	E1.2
	Heat Gains	48	E1.3
	Heat Gain Schedule	54	E1.4

CONTENTS

1 Thermal (continued)

E Design Development (continued)

Heat Gain Peaks	55	E1.5
Envelope Thermal Resistance	57	E1.6
Heat Loss Calculation	62	E1.7
Calculation of Zonal Balance Point Temperatures	66	E1.8
Desiccant Cooling	69	E1.9
Roof-Pond Cooling	72	E1.10
Evaporative Cooling	76	E1.11
Earthtube Cooling	77	E1.12
Design Review	79	E1.13

F Thermal Integration

Introduction	81	F1.0
Design	82	F1.1
Backup Heating System Sizing	83	F1.2
Backup Cooling System Sizing	85	F1.3
HVAC System Choice	86	F1.4
HVAC System Layout	89	F1.5

G Building Evaluation

Introduction	91	G1.0
Thermal Design Narrative	92	G1.1
Heat Loss Criteria	93	G1.2
Solar Heating	95	G1.3
Interior Temperature Swing	97	G1.4
Computer Simulation of Thermal Performance	103	G1.5

2 Lighting

A Precedents of Lighting Response

Introduction	105	A2.0
Architecture, Climate, and People—Vernacular Precedents	106	A2.1
Lighting Precedents	107	A2.2

B Climate and Site Analysis

Introduction	109	B2.0
Sky Conditions	110	B2.1
Available Illumination	111	B2.2
Site Concepts	114	B2.3

C Program Analysis

Introduction	115	C2.0
Recommended Footcandle Levels	116	C2.1
Recommended Lighting Energy Budget	117	C2.2
Required Exterior Daylight Levels	119	C2.3
Illumination Potential	121	C2.4
Daylighting Strategies	122	C2.5

CONTENTS

2 Lighting (continued)

D Schematic Design

Introduction	123	D2.0
Design	125	D2.1
Light Distribution Rendering	127	D2.2
Luminance Evaluation	128	D2.3
Recommended Daylight Factors	129	D2.4
Daylighting Aperture Sizing	131	D2.5
Solar and Ventilation Aperture	133	D2.6
Design Review	134	D2.7

E Design Development

Introduction	135	E2.0
Design	137	E2.1
"Raw" Illuminance Footprints	139	E2.2
"Corrected" Illuminance Footprints	144	E2.3
GDDM Renderings	150	E2.4
Daylight Model	151	E2.5
Design Review	154	E2.6

F Electric Lighting Integration

Introduction	155	F2.0
Design	156	F2.1
100% Electric Lighting	157	F2.2
Backup Daytime Lighting	159	F2.3
Electric Lighting Performance	162	F2.4
Savings from Daylighting	163	F2.5
Design Review	164	F2.6

3 Acoustics

A Precedents of Acoustic Response

Introduction	165	A3.0
Acoustic Precedents	166	A3.1

B Climate and Site Analysis

Introduction	167	B3.0
Site Sound Contours	168	B3.1
Site Concepts	170	B3.2

C Program Analysis

Introduction	171	C3.0
Noise Criteria	172	C3.1
Acoustic Zoning	173	C3.2

CONTENTS

3 ACOUSTICS (CONTINUED)

D SCHEMATIC DESIGN

Introduction	175	D3.0
Design	176	D3.1
Site Sound Rendering	177	D3.2
Thin-Wall Barrier	178	D3.3
Building Sound Rendering	179	D3.4
Design Review	180	D3.5

E DESIGN DEVELOPMENT

Introduction	181	E3.0
Design	182	E3.1
Room Absorbency	184	E3.2
Reverberation Time	186	E3.3
Room Sound Level	188	E3.4
STC Ratings	190	E3.5
Outdoor Sound and Ventilation	192	E3.6
Design Review	193	E3.7

4 WATER AND WASTE

A PRECEDENTS OF WATER AND WASTE RESPONSE

Introduction	195	A4.0
Water and Waste Precedents	197	A4.1

B CLIMATE AND SITE ANALYSIS

Introduction	199	B4.0
Annual Rainfall Comparisons	200	B4.1
Soil Characteristics	201	B4.2
Site Concepts	202	B4.3

C PROGRAM ANALYSIS

Introduction	203	C4.0
Minimum Required Plumbing Fixtures	204	C4.1
Conventional Fixture and Water Use Estimate	205	C4.2
Solid-Waste Resource Estimate	206	C4.3
Water and Waste Strategies	207	C4.4

D SCHEMATIC DESIGN

Introduction	209	D4.0
Design	210	D4.1
Conventional Water System	211	D4.2
Water-Conserving System	213	D4.3
Solar Collector Sizing Estimate	215	D4.4
Cistern Sizing Estimate	216	D4.5
Septic Tank and Drainfield Sizing I	218	D4.6
Septic Tank and Drainfield Sizing II	219	D4.7
Composting Toilets	220	D4.8
Solid-Waste Center	221	D4.9
Design Review	222	D4.10

CONTENTS

4 WATER AND WASTE (CONTINUED)

E DESIGN DEVELOPMENT

Introduction	223	E4.0
Design	224	E4.1
Toilet Room	225	E4.2
Supply Pipe System	226	E4.3
Waste Pipe System	228	E4.4
Drainfield	229	E4.5
Domestic Hot Water (DHW) System Requirements	230	E4.6
Solar DHW System Sizing	232	E4.7
Cistern Sizing	235	E4.8
Design Review	236	E4.9

APPENDICES 237

A BUILDING PROGRAMS 239

B SITE PLANS 263

C NOAA CLIMATOLOGICAL DATA 273

D SUN PEG SHADOW PLOTS 291

E LOF SUN CHARTS 295

F GDDM FOOTPRINTS 303

BIBLIOGRAPHY 325

INDEX 329

PREFACE

> *Architecture occurs at the meeting of interior and exterior forces of use and space. These interior and environmental forces are both general and particular, generic and circumstantial. Architecture as the wall between the inside and the outside becomes the spatial record of this resolution and its drama.*
>
> *—Robert Venturi, Complexity and Contradiction in Architecture*

As teachers of architectural design, we have observed the difficulty students experience as they attempt to resolve the forces that affect the skin of the building. *Insideout* has been written to foster design for thermal, luminous, acoustic, and aqueous environments that is well-integrated, technically sound, and aesthetically pleasing through a framework of exercises that emphasizes response to precedent, environment, and building program. Following this framework clarifies the architect's responsibility and ability to produce buildings that are sensitive to environmental forces as well as occupants' needs. Even in this era of specialization, the abdication of this responsibility leads to design decisions that are made without regard to the environment, are potentially devastating, and cannot be corrected by technical fixes offered later by design consultants. We cannot stress strongly enough the importance of the architect's role in addressing these concerns throughout the design process.

This second edition embraces the theory and intentions of its predecessor, which was published in 1982. In the interim, much has occurred that suggests the need for a new edition. The companion reference, *Mechanical and Electrical Equipment for Buildings (MEEB)*, has undergone two revisions, and a fine, new reference for schematic design procedures, *Sun, Wind, and Light (SWL)*, has been written. On a more general level, the definition of the role of building design in energy conservation has been refined and, after a period of complacency, energy and environment have reemerged as crucial issues as a result of such crises as the Chernobyl disaster, the Exxon Valdez oil spill, the nationwide landfill shortage, Superfund toxic waste cleanup sites, the drought in California, the Iraqi-Kuwaiti war, and global warming. This listing of symptoms points to a larger illness. The adoption of preventive measures has begun in the form of an emerging national building energy code, the ASHRAE 90 series, that virtually eliminates extremely wasteful, new building designs; recognition by many utility companies that conservation in both new and existing buildings is the most cost-effective energy resource for heating, cooling, and lighting; and demonstrations of viable, alternate energy sources, such as solar-thermal power plants, wind farms, and solar automobiles. This workbook attempts to address means to solve these problems in the realm of architectural design.

Insideout takes a broad view of the problem of creating buildings responsive to environmental forces by investigating problems in different contexts and at different scales—those of culture, climate, site, building cluster, building, building zone, and building component—through stages of the design process—analysis of precedent, site and climate, and building program; schematic design; design development; systems integration; and evaluation. This approach helps the architect avoid the dilemma of designing the elegant element that responds beautifully and efficiently yet neglects gross inefficiencies at another scale, thereby obviating its usefulness. It also helps prevent the design of an inefficient building shell that houses a superbly efficient mechanical system. Emphasis is placed on the architect's role in integrating environmental, technical, and aesthetic concerns.

PREFACE

Workbook Format and Use

This workbook was written as a text for an upper-division course in environmental technology and as a companion to *MEEB* and *SWL*. However, we believe it can also be used effectively either in its entirety or as topical units in a range of classes from introductory environmental issues courses, to design studios that address the effects of environmental forces, to advanced seminars in environmental technologies. It is also appropriate as a reference for practicing architects.

There are numerous references to the eighth edition of *Mechanical and Electrical Equipment for Buildings* (Stein and Reynolds 1991) and to *Sun, Wind, and Light* (Brown 1985). For the sake of convenience and space we have used the acronym for each book—*MEEB* and *SWL*.

There are three aspects of the workbook that require further explanation:

1. modular format
2. diversity of climate, program, and site
3. scales of design.

Modular Format

The format of *Insideout* is best described by a matrix of learning units [A1–E4] formed by environmental forces (thermal, lighting, acoustics, and water and waste) and design stages (analysis, synthesis, and evaluation).

DESIGN STAGES		ENVIRONMENTAL FORCES			
		Thermal	Lighting	Acoustics	Water & Waste
Analysis	Precedents	A1	A2	A3	A4
	Climate/Site	B1	B2	B3	B4
	Building Program	C1	C2	C3	C4
Synthesis and Evaluation	Schematic Design	D1	D2	D3	D4
	Design Development	E1	E2	E3	E4
	Systems Integration	F1	F2	--	--
	Evaluation	G1	--	--	--

Insideout has been organized to be used in various ways. Lecture courses typically deal with one environmental force at a time (e.g., the accompanying *Insideout* thermal sequence would be A1, B1, C1, D1, E1, F1, G1), while design studios typically cycle through the stages of design (e.g., the accompanying *Insideout* design sequence would be A1, A2, A3, A4, B1, B2, and so forth). Of course, any class organization that combines these approaches can also be easily accommodated.

The modular organization continues within each cell in the matrix. A single page is the smallest, discrete unit—only one procedure is presented on a single page and each procedure is identified by its place in the matrix (e.g., C1.5 Balance Point Temperature Estimate). Single procedures can be used, deleted, or modified by the instructor. The matrix is like a road map, showing alternate routes, offering some that you may choose not to follow. None of the authors uses *Insideout* in its entirety or without modifications. Completed exercises from the workbook are easily identified by section number (e.g., C1.5) to help cross-referencing and grading.

Although there is no fully illustrated example as in the first edition, all procedures that require complex data manipulation offer "suggested formats" with sample data, in italics, given to demonstrate the appropriate scale and scope of the expected data. We encourage students to generate their own data tables specific to their assigned building programs based on these "suggested formats." Moreover, many illustrations are of a hypothetical linen supply designed as an *Insideout* project for the Charleston, South Carolina, climate. Master copies of graphs and site matrices are also included. We strongly recommend that these be photocopied for multiple use. Overall, students are guided through the creation of diaries, which can be kept in three-ring binders, of the environmental design of assigned buildings.

Diversity of Climate, Program, and Site

Data for a range of temperate climates typical of continental U.S. conditions are presented in Appendix C. Any other location could be used by the student or instructor, but climatic data gathering would be required.

CITY	SUMMER CLIMATE	WINTER CLIMATE
Charleston, SC	hot, humid	cool, humid
Dodge City, KS	hot, humid	cool, dry
Madison, WI	hot, humid	cold, humid
Phoenix, AZ	hot, arid	cool, dry
Spokane, WA	hot, arid	cold, humid

Twenty-two residential, commercial, and institutional building programs are included in Appendix A. These programs represent the range of small buildings (except single-family dwellings) encountered in architectural practice. We have chosen these building types to emphasize the point that all building types can benefit from energy-conscious design. Four sites that are appropriate in scale for these programs are given in Appendix B. Exercises are written so the student or instructor may easily use any other building program or site with this text.

For our environmental technology lecture courses we use all five climates, four of the building programs, and one or two of the sites each year. We organize these variables to form a matrix for assigning projects to student teams. This mode of assigning teams facilitates coverage of the whole spectrum of issues related to differences due to building program and climate. When large classes are assigned *Insideout* exercises, we find it beneficial to divide the class into discussion groups of less than twenty. Teaching assistants lead these discussions and grade the exercises.

PREFACE

SITE A	BUS TERMINAL	LIBRARY	HOMELESS SHELTER	BOWLING ALLEY
Charleston	*Ken and Ryan*		*Esther and Jose*	
Dodge City		*Cisco and Pancho*		*Cleo and Thalia*
Madison	*George and Gracie*		*Askia and David*	
Phoenix		*Pam and Rob*		*Chico and Harpo*
Spokane	*Batman and Robin*		*Bud and Lou*	

Scales of Design

The first edition included procedures that addressed four scales of design—site, cluster, building, and component. Concern for addressing these scales has been maintained; procedures for coping with a crucial intermediate scale, that of the building zone (between building and component scales), have been added. Organizing a building into a few zones of similar environmental needs greatly affects both the aesthetics and efficiency of the building. All the exercises are written to recognize the thermal, luminous, acoustic, and aqueous diversity inherent even in small buildings. Generally, larger scales (e.g., the site) are considered earlier in the workbook sequence than are smaller scales (e.g., building components). *Insideout's* schematic design procedures emphasize zoning for environmental forces as a basis for design. This sequencing encourages a holistic approach to the integration of responses to environmental forces and to the integration of environmental, spatial, and aesthetic issues in design. It is feasible to achieve a well-integrated design that responds to all scales and is appropriate for an eight-week design studio problem by following the procedures in the portion of the organizational matrix defined by modules A1 (Thermal Precedents) through D4 (Waste and Water Schematic Design).

We intend this edition to be a text that stresses an approach to design that is conscious of environmental forces—one that helps architectural designers gain confidence and competence in addressing the mounting concern for holistic, environmentally sensitive design.

—BTH and JL

ACKNOWLEDGMENTS

The second edition has evolved slowly over several years and has been nurtured by the support of many people whom we wish to thank. Prominent among them are:

Teaching and Research Assistants—Gary Ard, John Barnes, Dave Blodgett, Scott Caver, Joe Clark, Peter Cohan, Susan Cooper, Brian Corcoran, Marc Corney, Amy Dowty, Ty Harder, Rick Hatcher, Hal Jensen, Julie Kriegh, Chris Laraway, Jeff Myhre, Patrick O'Hare, Mike Porter, Kristin Ragan, Bragdon Shields, Craig Slocum, Jason Springer, Dave Thomas, Ed Worrell, and Roland Zoller.

Production Assistants—Gail Edney, Pamela Egashira, Tisha Egashira, Tony Franklin, Joanne Hanley, Merideth Monserud, Carol Orr, and Cheryl Wheaton.

Colleagues—Dale Brentrup, Virginia Cartwright, Charlie Huizenga, Marietta Millet, and Paul Weir.

We would also like to thank numerous unwitting helpers—the students in environmental control systems classes at the Universities of Idaho, Oregon, Minnesota, and Washington—who have aided the authors' experiments in content and format.

Special thanks for help instrumental in producing the final product to multitalented Amy Dowty who not only served as a teaching assistant but also drafted first versions of many new procedures and crafted the illustrations and to fastidious, persnickety Tisha Egashira, who edited the text and produced camera-ready copy with an unerring eye for detail and a passion for quality.

A1 THERMAL

PRECEDENTS OF THERMAL RESPONSE

INTRODUCTION

A1

A1.0

GOAL

Assessing technologies in well understood ethnological and climatic contexts can yield principles of design technology that can be generalized and applied to similar contexts. Thus, through the study of less energy-intensive building technologies in either traditional or industrialized cultures, you will gain insights for solving energy-related architectural problems.

DISCUSSION

Bruno Zevi has posed simplistic questions (*right*) that challenge the heart and reason for the study of architectural precedents and their applications in the design of buildings today. Does reference to design precedent reflect a "decadent boredom" or merely quixotic stylistic preference? Such a black-and-white assessment grossly over-simplifies the reasons for incorporating architectural precedents. Rather, the questions to ask are: what should we study, why should we study it, and what can we learn from it that will improve our design skills?

> *Modern history has centered its attention on anonymous architecture . . . vernacular architecture has become a subject of intense research. What is the cause of this interest? Is it simply social or psychological curiosity, a decadent boredom with official architecture, or a romantic preference for the primitive or exotic?*
>
> *—Bruno Zevi*

The study of architectural precedents should focus on the processes of cultural and physical geographic differentiation through which people create and inhabit various systems of settlement. The study should not be of buildings alone, but of the values, activities, traditions, and physical constraints that influenced the design and the construction methods used in response to climate and comfort (Rapoport 1990).

If we only imitate architectural precedents, the resulting buildings can be out of context and, at the worst, decadent and romantic. Architectural precedents must be studied both as processes and final products of contextual decision-making. This investigation can lead to a clearer understanding of the vernacular form from which design principles can be delineated and later applied.

It is the application of these principles in their appropriate contexts that gives the study of architectural precedents importance. Rather than looking at decisions based on high finance or influenced by the vagaries of the style of the moment, focus on architectural decisions based on intuitive, analytical assessments of the problems faced in contemporary design. Or study examples of the more slowly evolving, tested craft of owner-built traditional architecture (Hubka 1984). These principles are important lessons for not only the designer, but anyone attempting to understand the richness of cultural and physical adaptation in architecture as a whole.

> *In developing theories, generalization is of great importance . . . and can only be made if based on the broadest possible evidence. . . . this evidence must include the full range of what has been built: preliterate, vernacular, popular, and spontaneous (squatter) environments as well as the more familiar high-style settings; . . . the relationships among different environments; all cultures, so that the evidence must be cross-cultural; and the full time span of built environments . . .*
>
> *—Amos Rapoport*

IN THIS SECTION YOU WILL:

1. Investigate precedents of climatic response in buildings.
2. Generalize principles of climatic response in climatic and cultural settings.

A1 THERMAL

PRECEDENTS OF THERMAL RESPONSE

ARCHITECTURE, CLIMATE, AND PEOPLE—VERNACULAR PRECEDENTS

A1

A1.1

PROCEDURE

Assess how a particular culture's vernacular architecture, the indigenous and traditional architecture of the region, balances culture and climate. Observe how cultural, technical, and economic forces were integrated with those of climate and thermal comfort to influence building form.

Choose a vernacular building or cluster of buildings in a climate zone similar to your site's. Consider the climate of the locale and the living patterns of its residents. Look for evidence of how such patterns are affected by climate and thermal comfort. Analyze the heating and cooling strategies used in terms of their cultural and physical contexts.

> ***. . . vernacular [architecture] could best be studied from a specific point of view, raising specific questions about topics or themes, rather than chronologically, the way traditional architectural history is studied.***
>
> —*Amos Rapoport*

DOCUMENT YOUR CHOICE AS FOLLOWS:

1. Identify the location of your vernacular building.
2. Describe the climatic characteristics of its locale.
3. Draw a building response diagram to illustrate how seasonal or diurnal changes in climate affect the living patterns.
4. Document your findings using vignettes, photocopies, photographs, etc. with concise and clearly written annotations. Include any uncertainties, questions, or ambiguities that would require additional research. Your analysis should be no longer than two 8½" x 11" pages.

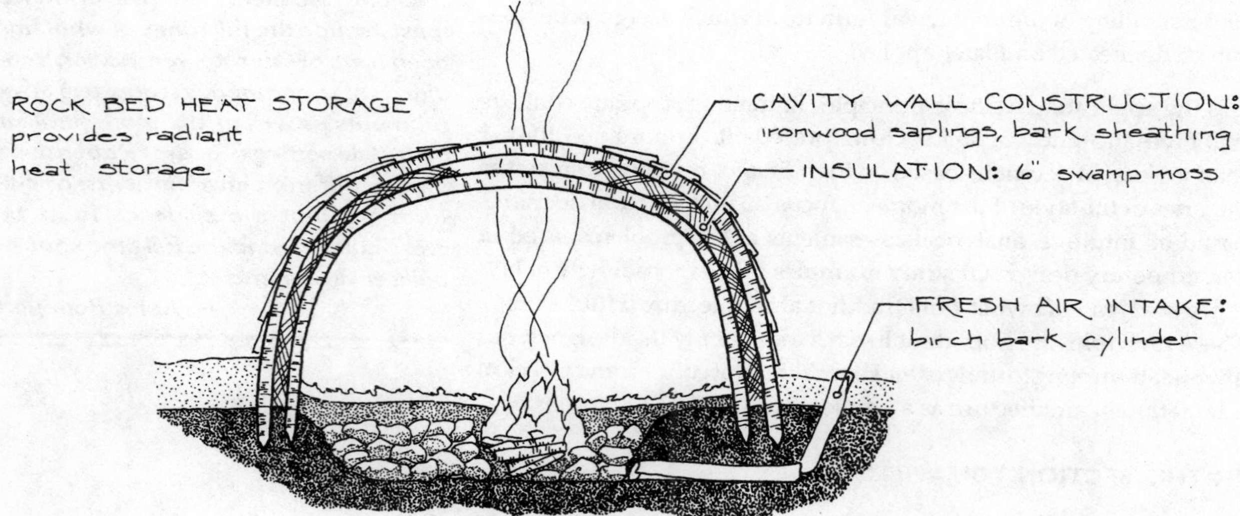

Figure A1.1.1 Modern Great Lakes Wigwam. Adapted from *Native American Architecture* by Peter Nabokov and Robert Easton, 73 (after E. St. Germaine, n.d.). Copyright © 1989 by Peter Nabokov and Robert Easton. By permission of Oxford University Press, Inc.

A1 THERMAL

PRECEDENTS OF THERMAL RESPONSE

A1

COOLING PRECEDENTS

A1.2

Cooling Design Strategies

Site-Scale Strategies

- Choose cool places for summer spaces.
- Use the wind for cooling.
- Use heat sinks for cooling. (A heat sink is any cool mass that is available for the absorption of excess heat, including water bodies, the ground, and massive building materials.)

Cluster-Scale Strategies

- Arrange the buildings to shade each other and their outdoor spaces.
- Preserve each building's access to cooling breezes during overheated periods.
- Use arcaded courtyards for shading and night-sky radiant cooling.

Building-Scale Strategies

- Orient buildings toward cooling breezes.
- Couple buildings to heat sinks.

Component-Scale Strategies

- Provide openings for ventilation that can be separate from those used for view, sunlight, or daylight.
- Use water features for evaporative cooling.

Figure A1.2.1 The Déjà-Vu House. Starkville, Mississippi, Robert Ford (architect), 1983.

Procedure

Choose an existing building or site that has: (1) a building program, summer climate, or both that is similar to your assigned building program or climate and (2) a clear, conceptual approach that incorporates some of the design strategies above.

Document your choice as follows:

1. Identify the location, program, architect (if known), and source of your information.
2. Include photocopies or drawings (whichever is quick and easy for you) to illustrate the design.
3. Evaluate the building or site design with a building response diagram and short annotations that explain how this design is organized for cooling.

HEATING PRECEDENTS

Heating Design Strategies

Site-Scale Strategy

- Choose sunny places for winter spaces.

Cluster-Scale Strategies

- Make winter places warmer by arranging buildings to form sun traps.
- Use earth berms and vegetation to block winter winds.

Building-Scale Strategies

- Provide windows for solar gain that are different from windows for light, view, or ventilation. Large, south-facing windows contribute the most to solar gain.
- Use greater wall mass for less severe indoor temperature swings.
- Couple buildings to heat sinks which tend to be warmer than outdoor winter temperatures.
- Minimize north-facing building skin and glazing areas to reduce heat loss.

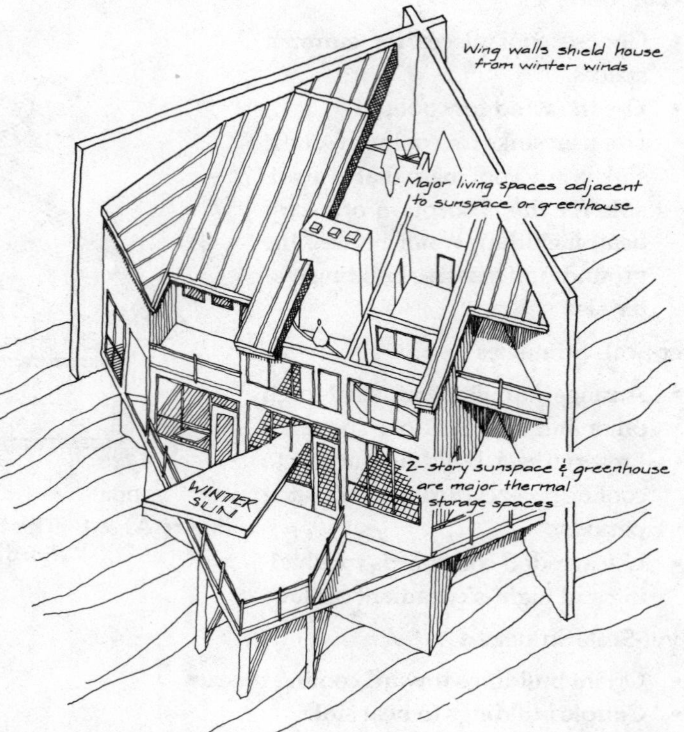

Figure A1.3.1 Davis House. In the Great Smokey Mountains, Virginia, A. J. Davis (architect), 1986.

Procedure

Choose an existing building or site that has: (1) a building program, winter climate, or both that is similar to your assigned building program or climate and (2) a clear, conceptual approach that incorporates some of the design strategies above.

Document your choice as follows:

1. Identify the location, program, architect (if known), and source of your information.
2. Include photocopies or drawings (whichever is quick and easy for you) to illustrate the design.
3. Evaluate the building or site design with a building response diagram and short annotations that explain how this design is organized for heating.

B1 THERMAL

CLIMATE AND SITE ANALYSIS

B1

INTRODUCTION

B1.0

Goal

Determine when your climate is considered comfortable for people and what potential benefits or problems may arise from the sun and wind on your site. Propose a conceptual design responsive to your climate and site.

Discussion

People tend to be comfortable within a fairly narrow range of temperature and relative humidity called the comfort zone. This zone can be extended by providing wind and shade when temperatures are higher, by providing sunshine and blocking wind when temperatures are lower, and by adding moisture when the humidity is low. It is possible, by analyzing your site's seasonal sun and wind patterns, to place the building and outdoor spaces to take advantage of the site's climatic conditions.

In this section you will:

1. Analyze your climatic data to determine which months' conditions fall within the comfort zone and which months' conditions need modifications for comfort.
2. Analyze the wind and sun patterns on your site.
3. Synthesize this information in a schematic design that takes advantage of your site's wind and sun patterns to provide the maximum amount of seasonal comfort.

B1 THERMAL

CLIMATE AND SITE ANALYSIS

BIOCLIMATIC CHART PLOT

B1

B1.1

DISCUSSION

The bioclimatic chart graphically illustrates the human comfort zone in terms of temperature and relative humidity. It relates comfort information directly to outdoor spaces because it bases the comfort zone on studies of adults at rest, outdoors, and wearing moderate clothing (shirt and trousers). Plotted climate data will indicate when your climate is comfortable and when it may be made comfortable by blocking or admitting the wind or sun. Knowing these potentials allows you to determine when it is possible to comfortably use outdoor spaces on your site and what the design has to accomplish in order to make the best use of the sun and wind.

FOR MORE INFORMATION

Olgyay, *Design with Climate*, Chapter 2, pp.14–23 (discussion of the bioclimatic chart).

PROCEDURE

For your location:

1. Record mean minimum and maximum temperatures and relative humidities (RH) for each month [Appendix C, Climatological Data].

2. On the bioclimatic chart [Figure B1.1.1] plot two points for each month—High Temperature/Low RH and Low Temperature/High RH. The higher temperature generally coincides with the lower humidity of the day and vice versa. Connect the points with a straight line. Label and/or color code each month's plot.

3. For a worst-case summer condition, find the summer dry-bulb temperature [*MEEB*, Table A.1, pp.1429–1437], and extend your hottest month's line to that temperature on the chart.

> NOTE: The slope of the lines you have just plotted represents a constant moisture content. Extending the line to the summer high temperature reflects the probable humidity at that temperature.

4. Use the results of your bioclimatic chart plot to discuss the natural potential for seasonal comfort, particularly in regard to outdoor spaces (decks, courtyards, etc.).

CLIMATE AND SITE ANALYSIS

BIOCLIMATIC CHART PLOT (continued)

Average Monthly Temperature (°F) and Relative Humidity (%) **[suggested format]**

	Temp (HI)	RH (LO)	Temp (LO)	RH (HI)		Temp (HI)	RH (LO)	Temp (LO)	RH (HI)
Jan	60.2	64	40	85	Jul				
Feb					Aug				
Mar					Sep				
Apr					Oct				
May					Nov				
Jun					Dec				

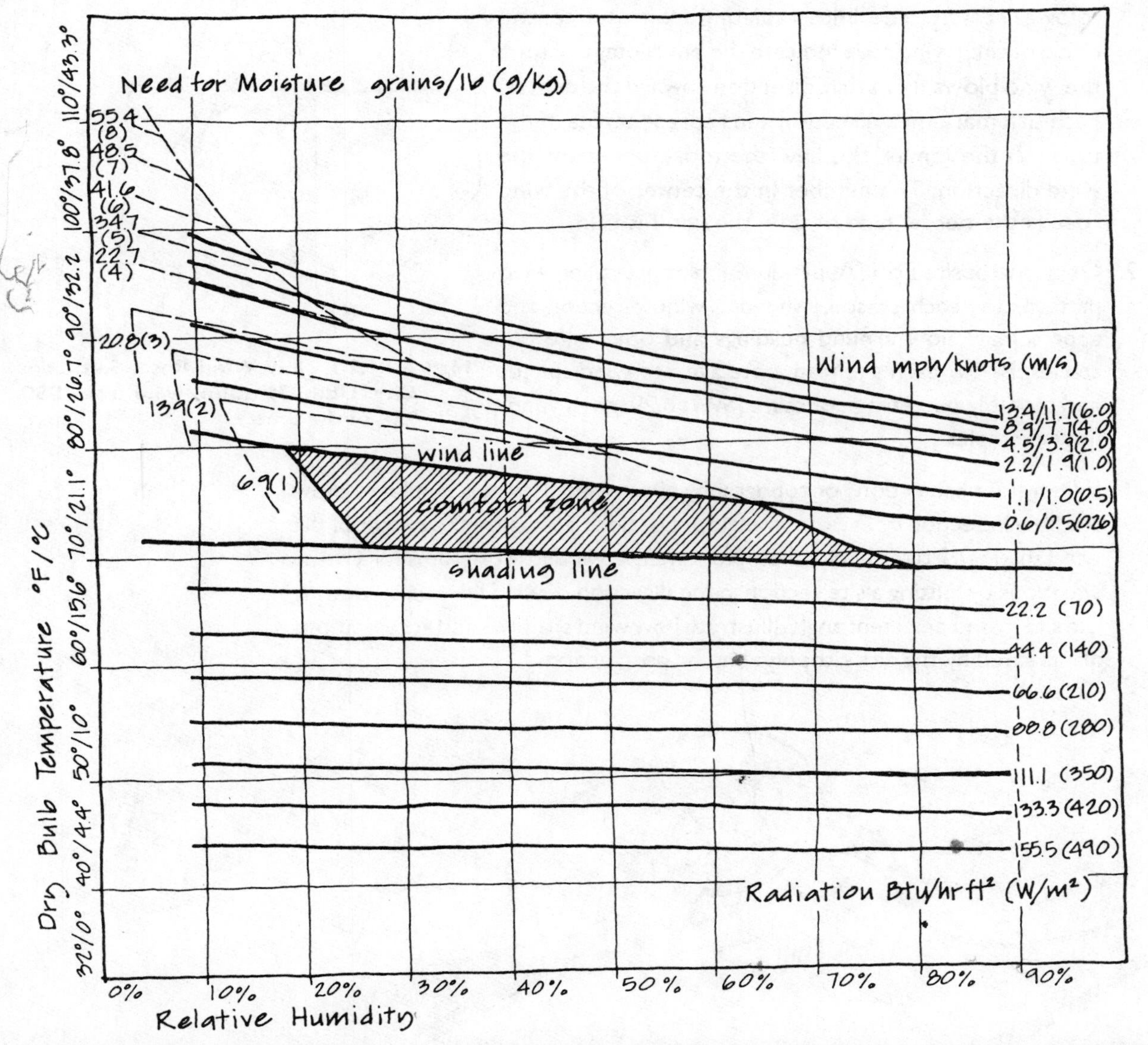

Figure B1.1.1 **Bioclimatic Chart with Design Strategies. Reprinted, by permission, from Brown, *Sun, Wind, and Light*, 35, copyright © 1985, John Wiley & Sons, Inc.**

B1 THERMAL

CLIMATE AND SITE ANALYSIS

SITE WIND FLOWS

Discussion

Wind speed and direction often vary seasonally and are affected by site constraints, topography, and vegetation. Additionally, during periods of calm, microclimatic thermal winds caused by valley effects, city effects, or bodies of water may occur. Site windflow mapping of each season can help you orient buildings and outdoor spaces to block wind during cooler months and admit a free flow of wind during warmer months.

Procedure

1. For each season, determine the prevailing wind direction, percentage of calm time [Appendix C, Wind Roses], and average wind speeds [Appendix C, Climatological Data]. The lines radiating from the central circle of each wind rose indicate the percentage of time the wind blows from that direction toward the center. Each tick mark on a direction line represents 5% of the time, so the longer the line, the more prevalent the wind direction. The number in the center of the wind rose is the percentage of time the wind is calm.

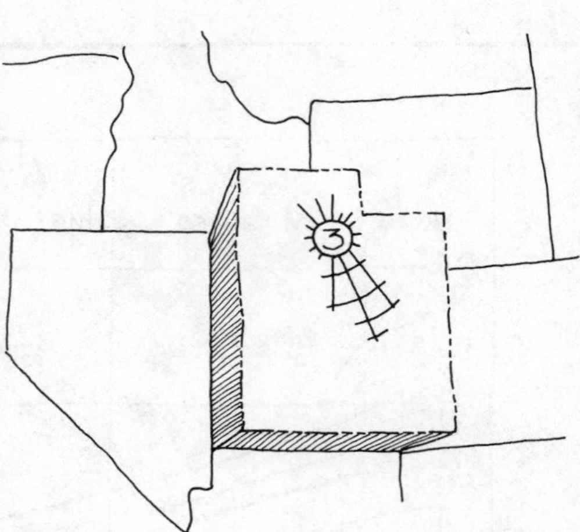

Figure B1.2.1 July Wind Rose, Salt Lake City, Utah. 3% calm, wind from SSE 22% of the time.

2. On a matrix site plan [Appendix B], plot prevailing wind patterns for each season, showing wind direction and approximate flow around buildings and other site obstacles. Below each site plan, note average wind speeds and monthly percentage of calm. [*SWL*, p.29, gives windflow examples.]

3. When site obstructions or topography affect wind flow, it is helpful to draw the flow in section or in axonometric projection to visualize the extent of the wind shadow. Make sectional or axonometric windflow diagrams for critical situations by cutting a site section in the direction of wind flow or by viewing the site from a pertinent angle. Illustrate how wind shadows and accelerations are created by the site's topography or obstructions.

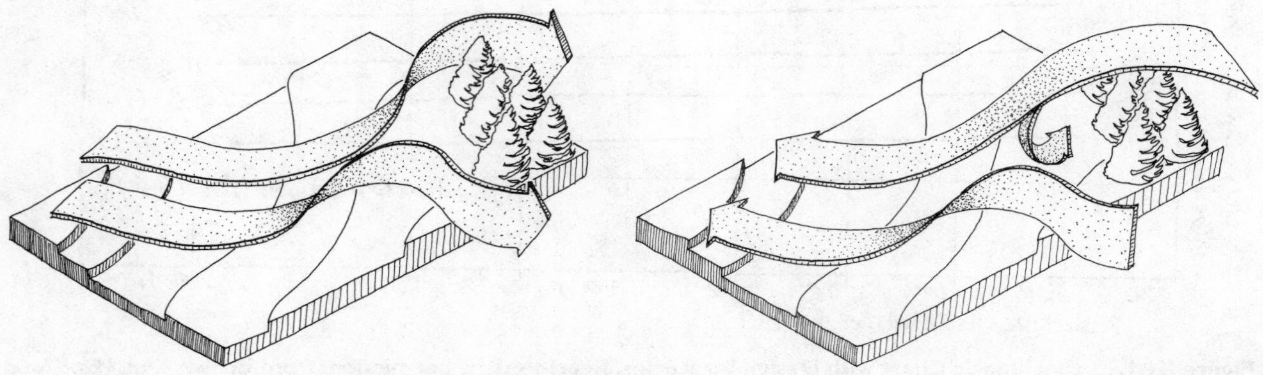

Figure B1.2.2 Summer and Winter Windflow Diagrams

SUN PEG SHADOW PLOT

Discussion

This technique shows you how the sun and your site will interact daily and seasonally to form sunny and shady places. This sun peg study requires a site model and a sunny day. Plan to finish your model early so you do not fall victim to extended cloudy weather.

Procedure

1. Use your site context map to make a 1" = 40' scale contour model of your site. Mark off the surface into a 10' grid and model all vertical obstructions, such as existing buildings, vegetation, and land forms. Leave about 4" of horizontal space across the southern end of your model for mounting the sun peg shadow plot chart [Appendix D].
2. Select the sun peg shadow plot chart for your latitude, and mount it horizontally on your model base so the north arrow points due north. At the location indicated by the cross just above true south, mount a vertical peg of the proper length (as indicated on the chart).
3. Determine the sun-shadow patterns for a typical month of each season at 9 a.m., noon, and 3 p.m.
   a. Set up the model in the full sun or in a sun simulator. **Ordinary indoor lamps don't work!**
   b. Tilt the model until the tip of the peg's shadow falls on the desired time–date intersection on the shadow plot.
   c. Trace and label the resultant shadows on your model to indicate the actual shading condition for that time of day and year.
4. Record the patterns on a copy of the matrix site plan [Appendix B], one matrix site plan for each season.

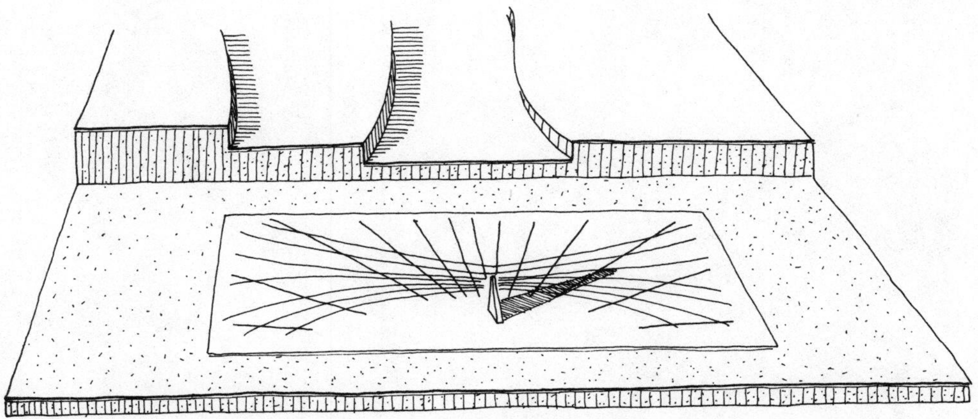

Figure B1.3.1 Sun Peg Shadow Plot Mounted on Model Base

B1 THERMAL

CLIMATE AND SITE ANALYSIS

CLIMATE RESPONSE MATRIX

DISCUSSION

With the bioclimatic chart plot you identified requisite climate conditions for comfortable spaces. During hot months block the sun while admitting the wind. During cold months block the wind while admitting the sun. These wind–sun conditions may already exist on your site. You can combine your windflow and sun pattern information with your bioclimatic chart plot information to identify where and when these conditions occur.

PROCEDURE

1. The matrix below represents the four possible combinations of wind and sun access. Color code it to represent the relative warmth of each condition. You will use this rendered matrix as a legend for the site plans to be developed in step 2.

	Sun	Shade
Windward		
Leeward		

2. Determine the wind–sun conditions that naturally occur on the site at 9 a.m., noon, and 3 p.m. in each season.

   a. Synthesize the wind flow and sun shadow pattern plots from B1.2 and B1.3.

   b. Choose a format for your data presentation. Use three pages (9 a.m., noon, and 3 p.m.) with all four seasons on each page, or use four pages (winter, spring, summer, and fall) with all three times of day on each page. Either choice is equally advantageous; they are just different ways of organizing your data.

   c. Using the legend created in step 1, color each 50' x 50' square of each site plan matrix [Appendix B] to indicate the naturally occurring wind–sun condition.

B1

CLIMATE RESPONSE MATRIX ANALYSIS

B1.5

Discussion

The climate response matrix gives you a graphic representation of the naturally occurring climatic conditions on your site. You can look at the twelve matrices that you have made, and make an educated guess as to the most suitable building sites. However, this method is subjective and is done more skillfully with site selection experience. A more analytical alternative is to assign to the squares on your matrices a numerical value based on its viability (after McHarg [1969] 1971).

Procedure

1. Refer to your bioclimatic chart plot [B1.1] to ascertain whether it is necessary to add heat, provide cooling, or do neither to maintain comfort at 9 a.m., noon, and 3 p.m. in each season.
2. For each square of each climate response matrix [B1.4], assign a value representing its climatic suitability for the particular time of day and year it represents. Use the following scale:

   +1 = suitable (has both required wind–sun conditions)

   0 = neutral (has one required condition)

   –1 = unsuitable (has neither required condition).
3. On a separate site plan, record the totals for each square of your matrix. Render the site plan to indicate desirable building locations (positive scores), neutral locations (zero scores), and undesirable locations (negative scores).
4. Examine the rendered site plan. Do you think it's a valid representation of the microclimatic conditions? If not, propose a more effective way of interpreting the climatic data. (For example, if your building has no late afternoon use, the 3 p.m. data should be disregarded or assigned a value less than the 9 a.m. or noon data.)

B1 THERMAL

CLIMATE AND SITE ANALYSIS

SITE CONCEPTS

Discussion

The rendered site plan you developed in B1.5 identifies climatically favorable locations on your site. For each favorable area, propose a conceptual design using very simple, annotated diagrams, the most appropriate level of detail for this stage of the design process. The concept should address the specific thermal needs of your building—what time of day and year outdoor spaces will be used, and what spaces in the building should be exposed to the sun for winter heat or to the wind for summer ventilation.

Procedure

1. Diagram each site design.
2. Annotate the diagrams to explain thermal considerations.
3. Show how you would alter site vegetation and topography to further improve the livability of your site.

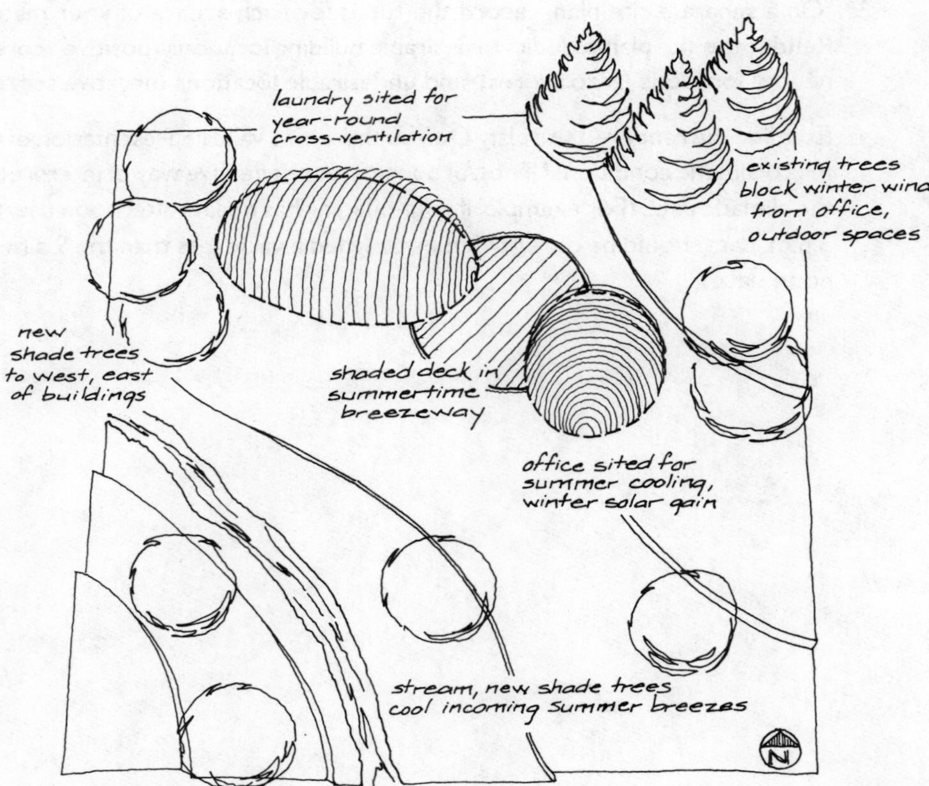

Figure B1.6.1 Site Design Proposal. Linen supply, Charleston, South Carolina.

C1 THERMAL

PROGRAM ANALYSIS

INTRODUCTION

C1

C1.0

Goal

Determine whether the building has an internally dominated load (IDL) or a skin dominated load (SDL) in order to establish appropriate thermal design strategies.

Discussion

The load on a building refers to the heat imbalance between the inside and the outside that makes the building skin a necessary barrier between the two. Buildings with a heating load need to be heated; buildings with a cooling load need to be cooled. Your building may require both heating and cooling over the course of the day or year, and it may have zones with heating loads, while the rest of the building requires cooling. The customary unit for measuring thermal loads (in English units) is British thermal units per hour or Btuh (a few references abbreviate Btuh as Btu/hr or Btu/h). When comparing the intensity of thermal loads in similar buildings or in zones of a single building, thermal load density (Btuh/ft^2) is used.

For an IDL building the major thermal problem over the year is caused by thermal forces generated within the building—heat from people, lights, and equipment. IDL buildings have a cooling load most of the year because they generate heat faster than they can dissipate it to their surroundings.

For an SDL building the major thermal problem over the year is caused by thermal forces acting on the skin of the building—convective, conductive, and radiant heat flows through the building skin as well as infiltration of outside air. SDL buildings are more strongly affected by these external loads than by their internal loads: They need to be cooled when it's hot and heated when it's cool.

In this section you will:

1. Analyze the programmed use of each space in your building.
2. Divide the building into a few zones according to anticipated thermal performance.
3. Estimate the heat gains and heat losses for each zone.
4. Calculate balance point temperatures for each zone to establish heating and cooling requirements.
5. Adopt appropriate, energy-conservation strategies for each zone.

C1 THERMAL

C1

PROGRAM ANALYSIS

USE SCHEDULES—PEOPLE, LIGHTS, AND EQUIPMENT

C1.1

Discussion

Your building program identifies a number of discrete spaces with characteristics that change over the course of a day. The number of occupants will vary from space to space, and each space may be heavily occupied at certain times (peak hours) and sparsely occupied at others (off-peak hours). Each space may have different types of equipment in operation and different lighting requirements. Since people, lights, and equipment are heat sources, the internal heat gains will vary from space to space, and each space will have different thermal needs. Analyzing the programmatic use of each space is a first step toward a design that will address those needs.

Procedure

For each space identified in your building program:

1. Calculate the approximate area of the space according to occupancy [Table C1.1.1] or other design considerations (such as your design intuition or *Architectural Graphic Standards*).
2. Determine when the space is in use, then identify the hours of heavy use or higher occupancy (peak hours) and the hours of light use or lower occupancy (off-peak hours). You may want to add a third use category, unoccupied, to distinguish between sparsely occupied and completely unoccupied times.
3. Record the number of people during peak and off-peak hours.
4. Determine whether the lighting use level is high, moderate, or low at peak and off-peak hours. Display spaces and visually demanding tasks such as sewing, drafting, or hazardous work require high light levels. Circulation spaces require only low light levels. Most program spaces require moderate light levels. [For a detailed list of lighting levels by task, see *MEEB*, Table K.2, Parts A, C–E, pp.1568–1570.]
5. Determine whether equipment use in the space is high, moderate, or low during peak and off-peak hours. For example, circulation spaces have minimal equipment and low use rates, office spaces generally have moderate use rates, and commercial kitchens have high use rates.

Use Schedule [suggested format]

Space	Area (ft^2)	Peak Hours	Off-Peak Hours	Number of Occupants	Light Level	Equipment Use
kitchen	*240*	*6a.m.–1p.m., 5p.m.–8p.m.*		*6*	*moderate*	*high*
	240		*1p.m.–5p.m., 8p.m.–midnight*	*1*	*moderate*	*moderate*

Table C1.1.1 Occupancy

	Use	Floor Area per Occupant (ft^2)
1.	Aircraft Hangars (no repair)	500
2.	Auction Rooms	7
3.	Assembly Areas, Concentrated Use	7
	(without fixed seats)	
	auditoria	
	churches and chapels	
	dance floors	
	lodge rooms	
	reviewing stands	
	stadia	
4.	Assembly Areas, Less Concentrated Use	15
	conference rooms	
	dining rooms	
	drinking establishments	
	exhibit rooms	
	gymnasia	
	lounges	
5.	Bowling Alleys	5 people per alley
6.	Children's Homes and Homes for the Elderly	80
7.	Classrooms	20
8.	Congregate Residences	
	10 or fewer occupants, less than 3,000 ft^2	300
	10 or more occupants, more than 3,000 ft^2	200
9.	Dormitories	50
10.	Dwellings	300
11.	Exercise Rooms	50
12.	Garages, Parking	200
13.	Hospitals and Sanitaria, Nursing Homes	80
14.	Hotels and Apartments	200
15.	Kitchens (commercial)	200
16.	Library Reading Rooms	50
17.	Locker Rooms	50
18.	Manufacturing Areas	200
19.	Mechanical Equipment Rooms	300
20.	Nurseries for Children (daycare)	35
21.	Offices	100
22.	School Shops and Vocational Rooms	50
23.	Storage and Stock Rooms	300
24.	Stores, Retail Sales	30
25.	Warehouses	500
26.	All Others	100

Reproduced from the 1991 edition of the *Uniform Building Code*, © 1991, with the permission of the publishers, the International Conference of Building Officials, 659–660.

C1 THERMAL

PROGRAM ANALYSIS

THERMAL ZONING

Discussion

If thermal efficiency were your sole design criterion, it would be simpler to meet thermal needs if you grouped the building spaces into two or three major thermal zones, rather than dealing with each space individually. Each thermal zone should contain spaces with similar thermal needs. By identifying similar or compensating use levels (similar hours of use and density of internal loads from people, lights, and equipment), you can easily group spaces into thermal zones. When this grouping is done, an appropriate thermal design strategy can be developed for each zone.

Procedure

1. Evaluate the data you compiled in C1.1. Group those spaces with similar peak and off-peak hours. Then group the spaces with similar or compensating levels of use by people, lights, and equipment.

2. Work with both lists until you have identified two or three zones with relatively consistent thermal characteristics. (More than three zones should not be necessary for your small building.)

3. Develop a schematic design drawing indicating the spatial layout of each zone and the placement of zones relative to each other on your site. Annotate this drawing to explain your rationale. Label the zones A, B, or C in the order of thermal load density (e.g., from lowest [A] to moderate [B], to highest [C]).

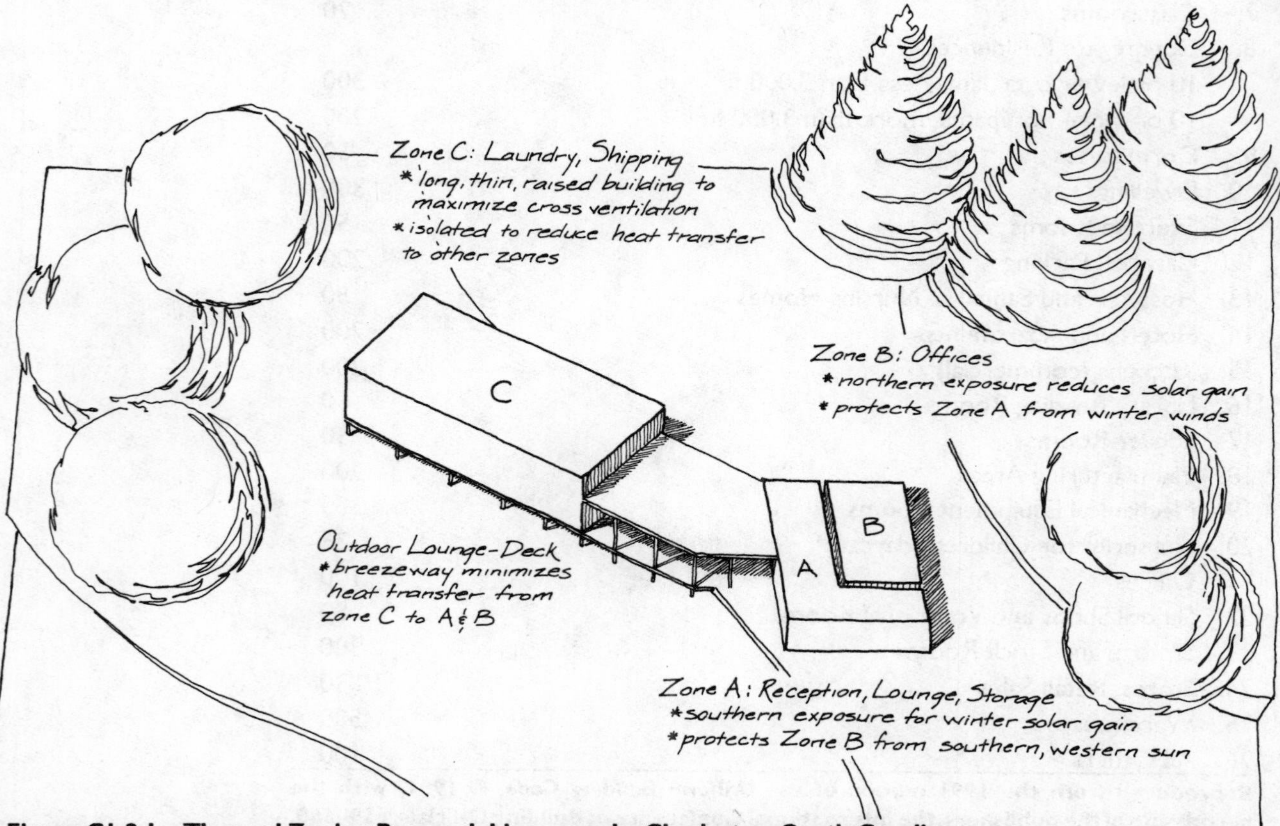

Figure C1.2.1 Thermal Zoning Proposal. Linen supply, Charleston, South Carolina.

C1 THERMAL

PROGRAM ANALYSIS

HEAT GAINS

Discussion

There are two basic components of your building's heat gain: The gain from internal thermal forces (people, lights, and equipment) and the gain from external thermal forces (primarily solar gain). You can estimate the people–lights–equipment gains of each thermal zone according to its characteristic functions. Solar gains can be roughly estimated within a narrow seasonal range. More precise heat gain estimates will be developed in E1.3.

Procedure

For each zone:

1. List the peak and off-peak hours.

2. Select the function that is most representative [*MEEB*, Table 5.8, A and B, p.214], and use the corresponding heat gain from people, lights, and equipment. If your zone is a composite of two or more functions, calculate or estimate the average intensity of heat gain. Since the *MEEB* table represents heat gains for moderate to high use, you may choose a different function or simply use a reduced heat gain when estimating off-peak gains. For example, a space may act as an activity center during peak hours and a lobby during off-peak hours. In most cases you will have to make an educated guess, and the results will be only as accurate as your estimate. This estimation **is** precise enough for this stage of the design process.

Heat Gains: People, Lights, Equipment [suggested format]

Zone	Peak Hours	Off-Peak Hours	Function	Heat Gain (Btuh/ft^2)
A	*8a.m.–6p.m.*		*office*	*11*
A		*6p.m.–8a.m.*	*office*	*6*
B	*6a.m.–8a.m., noon–2p.m.*		*cafeteria*	*19.5*
B		*8a.m.–noon*	*cafeteria*	*6.5*

3. Use the solar heat gain table [*SWL*, p.54] to approximate your daily and seasonal solar gains.

Heat Gains: Solar (for each zone) [suggested format]

Zone ________		Winter Heat Gain (Btuh/ft^2)	Spring Heat Gain (Btuh/ft^2)	Summer Heat Gain (Btuh/ft^2)	Fall Heat Gain (Btuh/ft^2)
Time:	6a.m.				
	9a.m.				
	noon				
	3p.m.				
	6p.m.				

HEAT GAINS (continued)

4. Make a total heat gain schedule for your building by combining the heat gains from people, lights, and equipment with the solar heat gains. Include enough data to represent the operations of your building during peak and off-peak periods. You will need these schedules to estimate the balance point temperature in C1.5.

Heat Gain Schedule (for each zone) [suggested format]

Zone *A* Month/Season *September/Fall*

Time	People, Lights, Equipment (Btuh/ft^2)	Solar (Btuh/ft^2)	**TOTAL** (Btuh/ft^2)
6a.m.	*7*	*0*	*7*

C1 THERMAL

PROGRAM ANALYSIS

C1

HEAT LOSS

C1.4

Discussion

Building heat gains vary daily and seasonally according to operating schedule and sun intensity. Heat loss—through the building skin and ventilation or infiltration—is defined solely by the temperature difference between inside and outside. In fact, heat loss is constant for each degree of temperature difference (in other words, Btuh/ft^2°F is constant). Both types of heat loss can be estimated according to function.

Procedure

For each zone, select the most representative function in both heat loss tables [*SWL*, p.55]. Use the corresponding heat loss to represent the zone. (You'll need to know the approximate area of each zone to use the tables.)

Heat Loss Schedule [suggested format]

Zone	Area (ft^2)	Heat Loss through Skin (Btuh/ft^2°F)	Heat Loss by Ventilation/Infiltration (Btuh/ft^2°F)	TOTAL (Btuh/ft^2°F)
A	*1,200*	*0.7*	*1.0*	*1.7*

BALANCE POINT TEMPERATURE ESTIMATE

C1.5

Discussion

By comparing the external temperature data for your climate with the heat gain and heat loss data for your thermal zones, you can determine the daily and seasonal heating and cooling requirements of your building. The balance point temperature is the outdoor temperature at which a zone loses heat at the same rate it generates heat. For example, a balance point temperature of 40°F indicates cooling is required when the external temperature is above 40°F, and heating is required when it is below 40°F. Since your heat gain and loss rates change daily and seasonally, your balance point temperatures will also fluctuate.

For more information

SWL, Sections 18–19, pp.52–62.

Procedure

For each thermal zone:

1. Assign minimum and maximum seasonal temperatures for your climate [Appendix C, Climatological Data] to the times of day on the chart below. (These are the typical times when minima and maxima occur—just before sunrise and halfway between noon and sunset.) You may substitute data from months that better represent your climate's seasons.

Month	Time	Min Temp	Time	Max Temp
January	6a.m.		2p.m.	
March	5a.m.		3p.m.	
July	4a.m.		4p.m.	
September	5a.m.		3p.m.	

2. Plot the time–temperature points for each season on the corresponding balance point temperature chart [Figure C1.5.3], and connect the points with a straight line. Then make a copy of the chart for each zone.

3. Calculate the temperature differential (ΔT) between the inside and the balance point temperature using the nomograph [*SWL*, p.53] when heat gain is greater than 30 Btuh/ft^2. If heat gain is less than 30 Btuh/ft^2, use Figure C1.5.1. For the rate of heat gain, use the data from C1.3; for the heat loss rate, use the data from C1.4. For each season you will need to find ΔT for peak and off-peak hours throughout the day. Alternately, you may use the formula below to calculate ΔT.

   ΔT = (rate of heat gain) / (rate of heat loss)

BALANCE POINT TEMPERATURE ESTIMATE (continued)

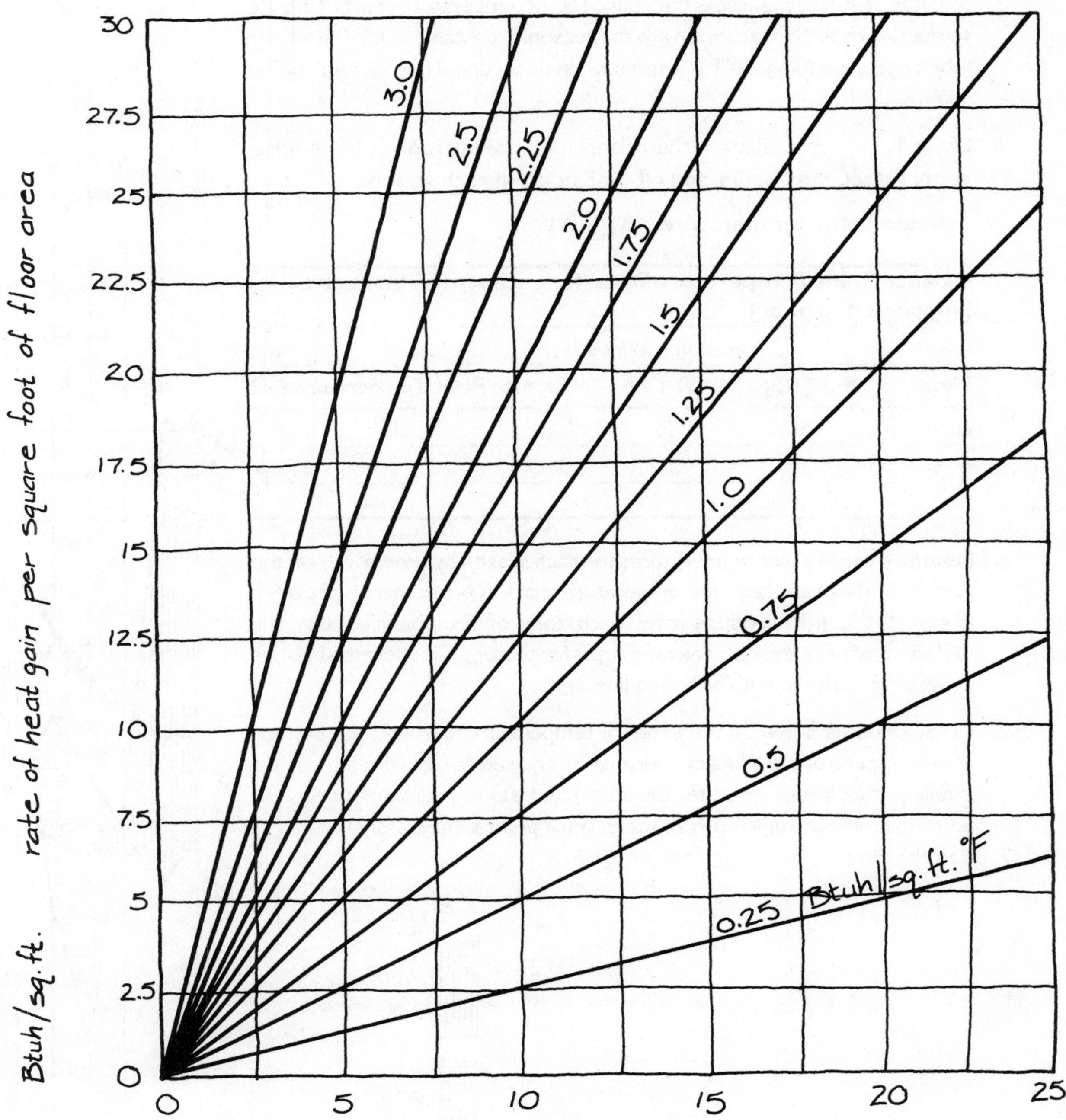

Figure C1.5.1 Balance Point Temperature Nomograph for Low Rates of Heat Gain. Adapted, by permission, from Brown, *Sun, Wind, and Light*, 53, copyright © 1985, John Wiley & Sons, Inc.

C1 THERMAL

PROGRAM ANALYSIS

BALANCE POINT TEMPERATURE ESTIMATE (continued)

4. Set an internal temperature, T_{in}, or thermostat setting for each season, and plot it as a straight line across the balance point temperature charts. Choose temperature settings according to the season. For example, 65°F in winter when occupied (day), 50°F in winter when unoccupied (night), and 80°F in summer.

5. Using T_{in} and ΔT, calculate the balance point temperatures for sunrise, noon, sunset, peak hours, and off-peak hours in each season.

balance point temperature = $T_{in} - \Delta T$

Balance Point Temperature Data (for each season for each zone) [suggested format]

Zone *A* Season *Winter*

Time	T_{in} (°F)	ΔT (°F)	Balance Point Temperature (°F)
6a.m.	*50*	*15*	*35*

6. Plot the balance point temperatures for each season by time of day on the corresponding balance point temperature chart. A blank chart is provided [Figure C1.5.3]. Reproduce it for each zone of your building. Use the midpoints of peak and off-peak time spans for plotting. Connect the balance point temperatures with a dashed line.

7. Color the area between the external temperature level and the balance point temperature level **RED** if heating is required (external temperature is below the balance point temperature) and **BLUE** if cooling is required (external temperature is above the balance point temperature).

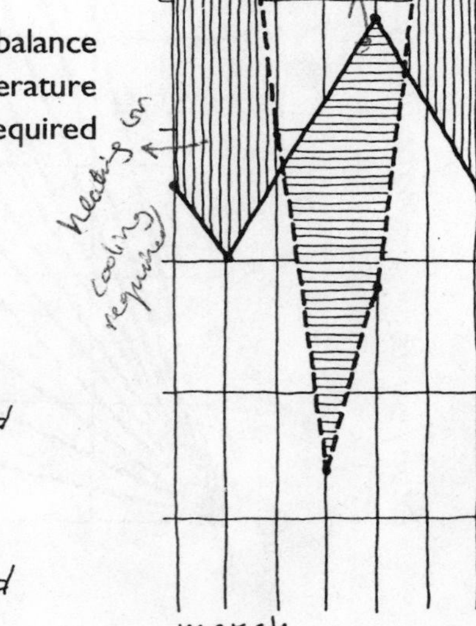

Figure C1.5.2 March Balance Point Temperature Plot. For 1,500 ft² building in Minneapolis, Minnesota. Reprinted, by permission, from Brown, *Sun, Wind, and Light*, 60, copyright © 1985, John Wiley & Sons, Inc.

PROGRAM ANALYSIS

BALANCE POINT TEMPERATURE ESTIMATE (continued)

C1.5

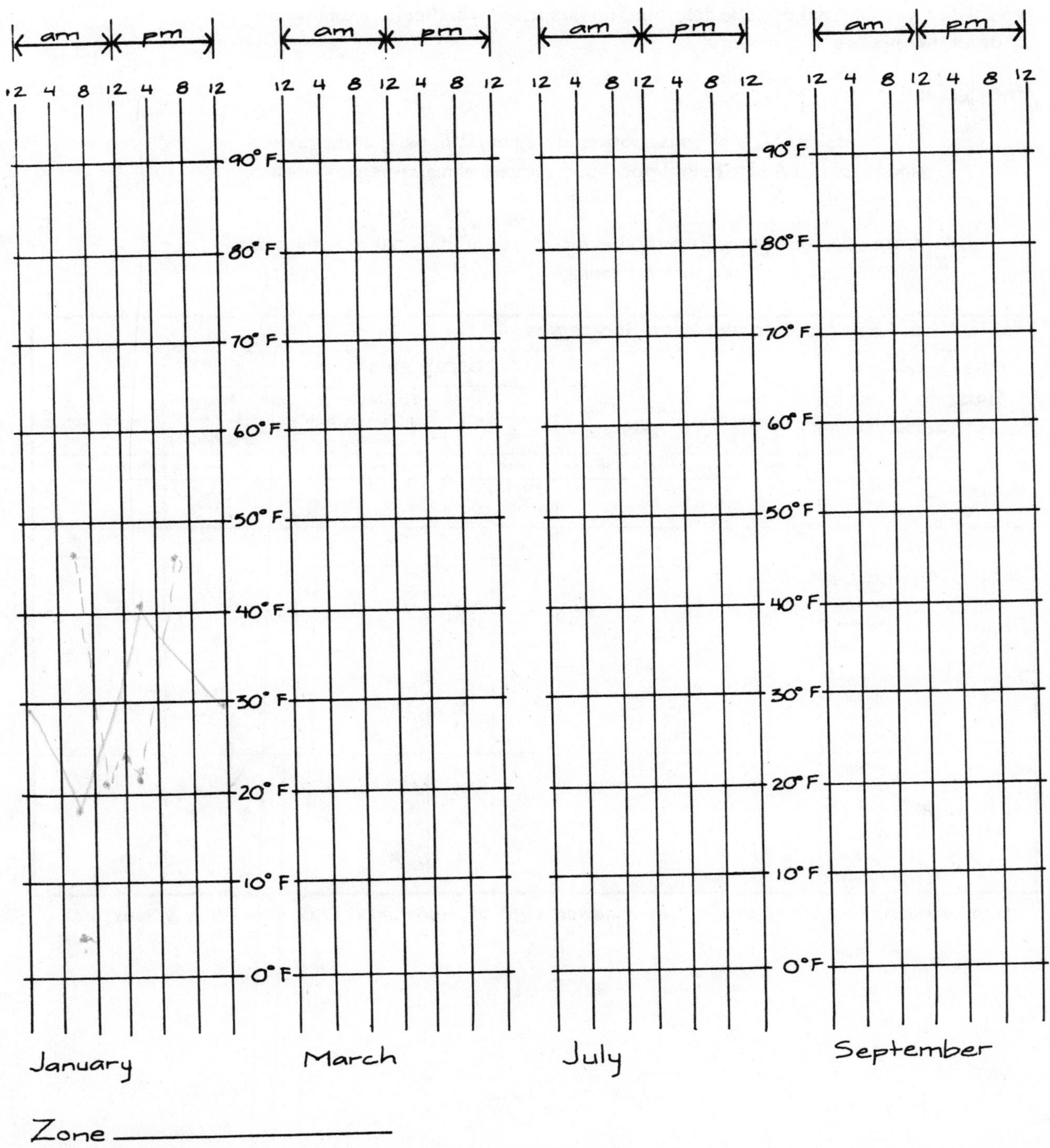

Zone ____________

Figure C1.5.3 Balance Point Temperature Charts

C1 THERMAL

PROGRAM ANALYSIS

C1

ENERGY CONSERVATION

C1.6

Discussion

The balance point temperatures indicate the thermal needs of your zones. IDL zones require cooling year-round, while SDL zones require summer cooling and winter heating. Since the balance point temperature charts show the thermal profile of a typical day, you can make subtle determinations, such as whether morning warm-up or all-day heating is necessary.

Procedure

1. Consult the table of energy conservation strategies [*SWL*, p.61] or the revised version [Table C1.6.1] to help choose appropriate thermal strategies for each zone.

2. Illustrate your energy-conservation choices with an annotated, schematic diagram of your proposed thermal zoning.

Table C1.6.1 Energy Conservation Strategies

Characteristic Balance Point Temperature Chart		**Strategies**											
		interior temperature setting		internal heat generation		solar heat gain		rate of heat gain or loss through envelope		rate of heat gain or loss by ventilation		thermal storage	
		increase	decrease	increase	decrease	increase	decrease	increase	decrease	increase	decrease	store heat	store cool
heating always required			X	X		X			X		X		
heating and cooling required	heating		X	X		X			X		X		X
	cooling--outside warmer than inside	X			X		X		X		X	X	
	cooling--outside cooler than inside	X			X		X	X		X		X	
cooling always required	outside warmer than inside	X			X		X		X		X		
	outside cooler than inside	X			X		X	X		X			X

Adapted, by permission, from Brown, *Sun, Wind, and Light*, 61, copyright © 1985, John Wiley & Sons, Inc.

D1 THERMAL

D1

SCHEMATIC DESIGN

INTRODUCTION

D1.0

Goal

Design a building that synthesizes the information you have acquired through the study of precedents; the analyses of your site, climate, and program requirements; and the application of thermal design strategies for heating and cooling.

Thermal Schematic Design Goals

A. The thermal strategies chosen will work in your climate.
B. Buildings are located so shadows are not cast off the site.
C. The buildings are clustered and sited to create optimal microclimates for heating when you need heat and for cooling when you need to cool.
D. Sufficient aperture is provided to ventilate the building.
E. Ventilation removes heat without leaving hot, dead-air areas.
F. The building is exposed to the south in proportion to the heat needed.
G. The winter sun can reach all the south-facing apertures provided for the collection of heat.
H. Glazing is shaded from the sun during overheated months.

Discussion

Sections D1, E1, F1, and G1 are concerned with design for heating and cooling. We have devoted a large portion of this workbook to thermal considerations because of their significant impacts on building form and energy use.

It is helpful to make an analogy between the ways organisms and buildings respond to their thermal environments (Knowles 1974). Organisms respond in three ways—migration, form, and metabolism. When migrating, they move from an environment that is too cold or too hot to one that is just right. This movement may happen seasonally or diurnally. The ratio of an animal's skin or surface area to its volume determines its ability to lose heat to the environment. For example, the form of an elephant's ear is primarily described by its large surface area that maximizes the elephant's potential to cool. Metabolism refers to the internal chemical conversion of food energy to heat. Those animals with a high heat loss must eat large amounts of food to produce heat within their bodies to balance their heat loss (Thompson 1952 and Stevens 1974).

These three forms of thermal response have their analogies in buildings. Migration in the built world is as commonplace as migration in the natural world. Occupants may move from one area of a building to another or the building may be moved to a thermally different area. For example, to better cope with hot, summer evenings, residents move from bedrooms to sleeping porches. Form—which can include size, shape, skin area, orientation, proportion, volume, openings, articulation—can also be dependent on thermal response. An igloo's compact form (an example of the smallest possible skin area to enclosed volume ratio) minimizes the rate of heat loss and offers a small profile to heat-robbing winds. Metabolism, the fuel-sustained component of the building, seeks to

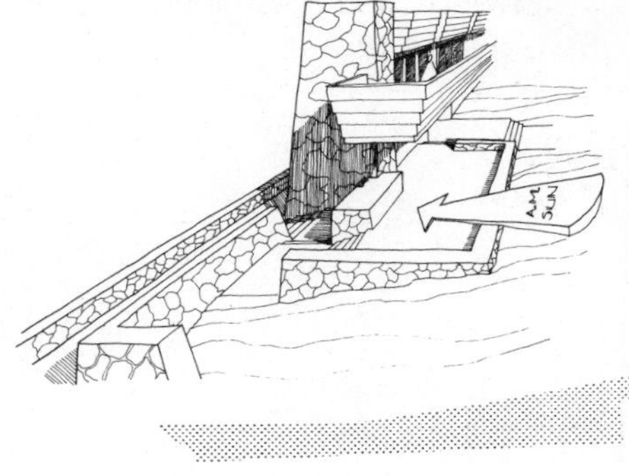

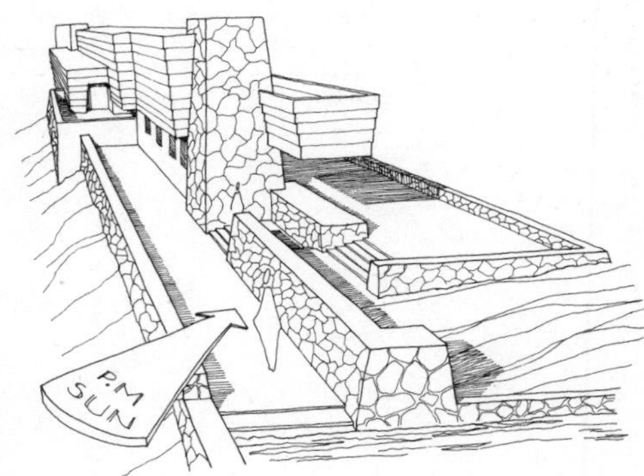

Figure D1.0.1 ***Above,*** **Morning Terrace;** ***below,*** **Evening Terrace. Pauson House, Phoenix, Arizona, Frank Lloyd Wright (architect), 1941.**

maintain thermal equilibrium. The more heat-producing people, lights, and equipment that occupy a building, the more cooling is required.

Current design practice usually separates form from metabolism in order to simplify preliminary design development. They are not recombined until later in the process when a building's form must be used as the basis for determining its metabolic rate. Even though form and metabolism are dependent on different sets of variables, they are intrinsically linked. If thermal needs are not taken into consideration when determining form, the cost required for the building to obtain metabolic equilibrium can be enormous.

IN THIS SECTION YOU WILL:

1. Employ appropriate thermal design strategies.
2. Complete your thermal schematic design.
3. Use a variety of analysis techniques to test the performance of your schematic design and to alter it, where appropriate.
4. Review and critique your schematic design.

DESIGN

Thermal Schematic Design Strategies

Site-Scale Strategies

- Schedule building use periods to avoid the hottest times of the day.
- Use wind breaks to protect building clusters without shading them in winter.
- Use trees for shading during the building's overheated months, particularly on the east and west building façades, and for outdoor spaces during the months when it is hot outside.
- Pre-cool ventilation air by allowing the wind to pass over damp vegetation, through shaded areas, or over bodies of water.
- Design for wind—wind tends to keep moving in one direction and flows from high pressure to low pressure areas; hot air rises while cool air falls.

Cluster-Scale Strategies

- Arrange buildings so the winter wind is blocked from sunny exterior spaces.
- Provide courtyards that are open to the night sky for summer cooling.
- Shape the building so that it does not block the neighbor's access to the sun for heating and/or daylighting.

Building-Scale Strategies

- Place the zones with higher internal gains on the cooler side of the building.
- Use spaces that can tolerate greater temperature variations as buffer areas.
- Provide alternative cool places for activities during the hottest periods of the day or the hottest season.
- Use apertures for natural ventilation. Building codes generally allow commercial buildings to be built without mechanical ventilation systems if at least 5% of the floor area is provided as fully operable window area. The more you depend on natural ventilation for cooling, the larger these window openings should be.
- Keep a clear path through buildings for unobstructed ventilation.
- Adopt a "closed," "open," or "mixed" building ventilation strategy.
  - ❖ "Open" buildings have large openings that provide continuous ventilation. These buildings have little thermal mass to increase the effect of the cooling exterior air.
  - ❖ "Closed" buildings use high mass without simultaneous ventilation. These may use evaporative cooling, desiccant cooling, roof ponds, and/or mechanical refrigeration.
  - ❖ "Mixed" buildings can be "closed" during very hot hours, then "opened" for night ventilation to remove stored heat.
- Orient the solar glazing for heating between 20°E and 32°W of South.
- Exclude direct sun when using daylight instead of electric light.

Component-Scale Strategies

- Minimize or fully shade from direct sun glass areas in the hottest summer exposures (east- and west-facing walls and roofs).
- Provide clerestories with protection from direct sun. They are sources of daylight as well as effective ventilators, especially if they open on the leeward side of the building.

DESIGN (continued)

PROCEDURE

Propose a schematic design for your building based on analyses of precedent [A1], site and climate [B1], and program [C1]. Use the appropriate thermal design strategies for your design.

DOCUMENT YOUR DESIGN AS FOLLOWS:

1. Site plan, including parking and access drives (scale: 1" = 100').
2. Cluster plan, including outdoor spaces (scale: 1" = 40').
3. Floor plans (scale: 1" = 20').
4. Roof plan and elevations, or axonometrics, illustrating all building sides and roof (scale: 1" = 20').
5. Sections (scale: 1/8" = 1'-0").
6. Design diagram, annotated to identify design strategies and thermal zones.

D1 THERMAL

SCHEMATIC DESIGN

BIOCLIMATIC CHART THERMAL STRATEGIES

D1

D1.2

Discussion

The bioclimatic chart of thermal strategies [Figure D1.2.1] can be used to identify appropriate thermal responses for your climate. It separates cooling strategies into "open" (cross-hatched pochet) and "closed" (diagonal pochet). In the comfort zone (no pochet), both strategies are applicable. Below the comfort zone (vertical pochet), passive solar heating strategies are applicable. In extremely dry climates, only evaporative cooling is effective. In hot, dry southern climates (less than 30% relative humidity and south of 36°N latitude), roof ponds can be used for cooling and heating. In warm, wet climates (greater than 50% relative humidity), desiccant radiators can be used for cooling. In climates that have characteristically moist soil, earthtube cooling can be effective. Further discussion of desiccant, roof-pond, evaporative, and earthtube cooling can be found in E1.9–E1.12. Programmatic considerations may suggest different strategies for different thermal zones in your building.

Procedure

Evaluate how well your design attains Thermal Schematic Design Goal A—**The thermal strategies chosen will work in your climate.**

1. For each thermal zone use the balance point temperature estimates [C1.5] to determine which seasons require heating, cooling, or both.
2. Make a separate bioclimatic plot for each zone. Plot the monthly climatic data on the bioclimatic chart [Figure D1.2.1]. Use the data from B1.1. Represent months that need cooling with a **BLUE** line, months that need heating with a **RED** line, and months that need both cooling and heating with a **BLUE** line that turns **RED**.
3. Identify the appropriate thermal strategies for your climate. Note that when your building requires cooling and outdoor temperatures are below the comfort zone, ventilation is the most effective cooling strategy.
4. Explain how your design employs the appropriate strategies or how you must change your design to meet Goal A.

BIOCLIMATIC CHART THERMAL STRATEGIES (continued)

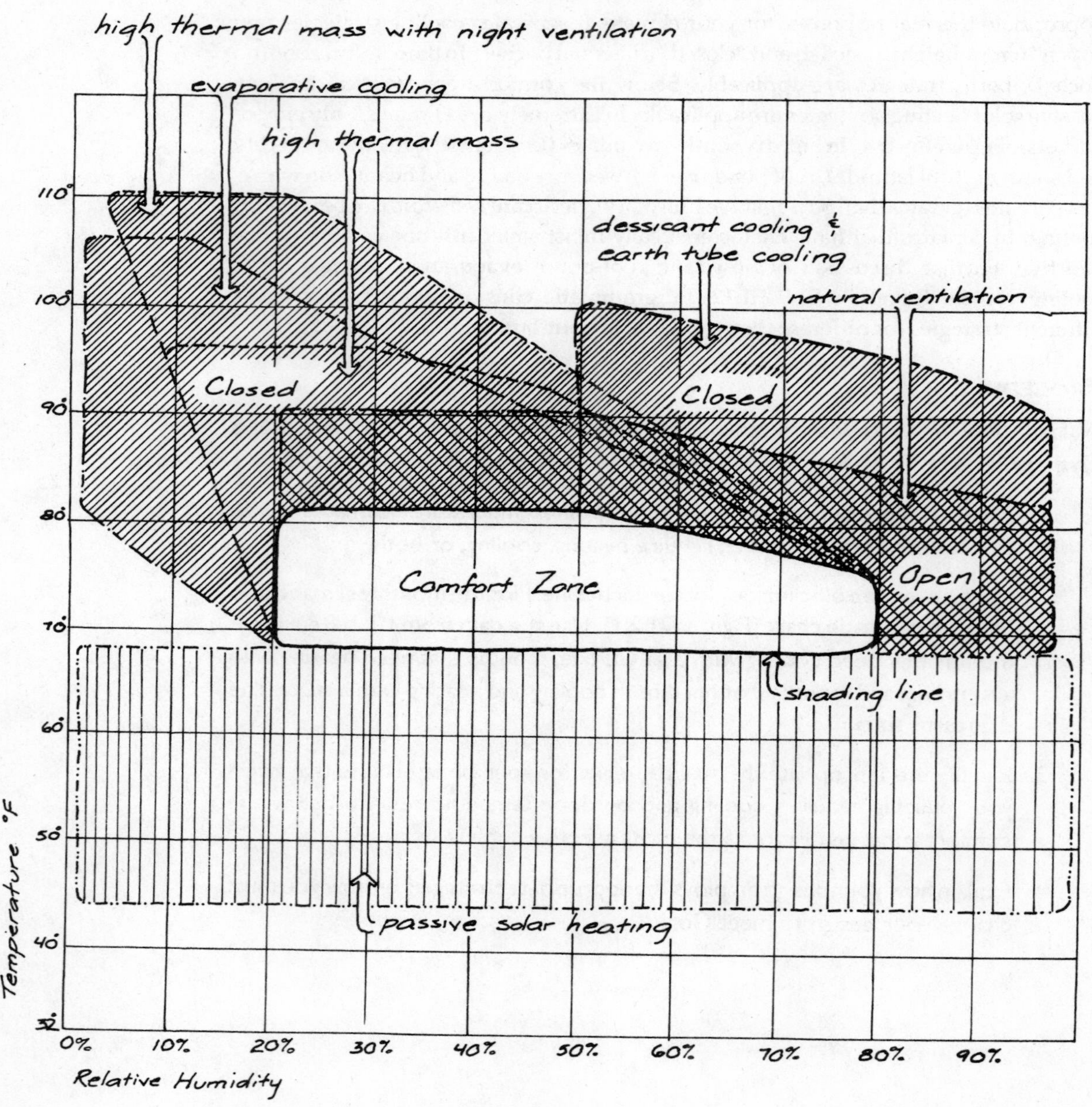

Figure D1.2.1 Bioclimatic Chart of Thermal Strategies

D1 THERMAL

SCHEMATIC DESIGN

SOLAR ENVELOPE

D1
D1.3

Discussion

To be a good solar neighbor, to ensure that your building and the buildings on adjacent sites have full access to the sun at the peak of the heating season, your building must be placed within the solar envelope of the site [*MEEB*, Section 3.4(b), pp.74–76]. The solar envelope is a "roof" over your site that is based on the sun's position at 9 a.m. and 3 p.m. on December 21, the lowest sun path during the heating season in the Northern Hemisphere. To fit within the solar envelope, your building must not penetrate the planes of this "roof" (Knowles 1974).

For more information

SWL, Section 20, pp.68–71 (this discussion is based on a flat site).

Procedure

Evaluate how well your design attains Thermal Schematic Design Goal B—**Buildings are located so shadows are not cast off the site.**

1. Make a clay or cardboard model of your building and site improvements (scale: 1" = 40'). Place this model on your site model [B1.3]. The building model must be accurate in size and massing as well as in its location on the site.

   > NOTE: You may want to draw your proposed windows, which you will need for D1.8, on the model before you assemble it.

2. Construct a wire-frame or cardboard solar envelope for your site.
   a. Find the sun's azimuth and altitude for 9 a.m. and 3 p.m. on December 21 [*MEEB*, Tables B.11–B.13, pp.1450–1455, or use the Libbey-Owens-Ford sun angle calculator].
   b. Build "roof envelope hips" from the site's NE corner (at the 3 p.m. azimuth and altitude) and the NW corner (at the 9 a.m. azimuth and altitude). If these "hips" cross the south property line before they intersect, construct an E–W "ridge" above the south property line. If these "hips" intersect on the site, construct a N–S "ridge" from the point of intersection southward. If the site is sloped and the two "hips" intersect at different elevations, build the N–S "ridge" at the lower elevation.

3. Discuss whether or not your building and site improvements are within the solar envelope and meet Goal B. Describe how your design must be altered if the goal has not been met.

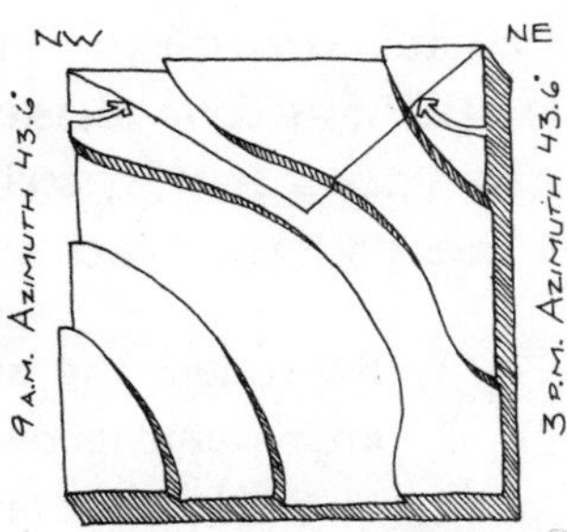

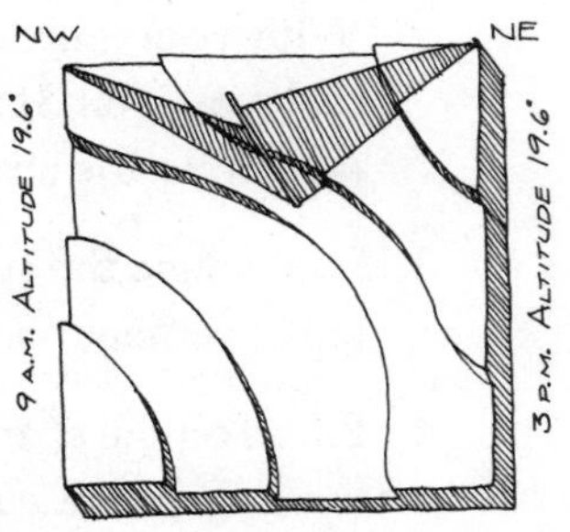

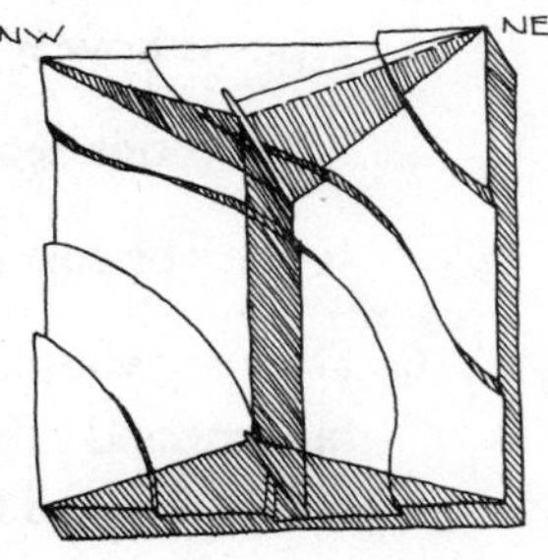

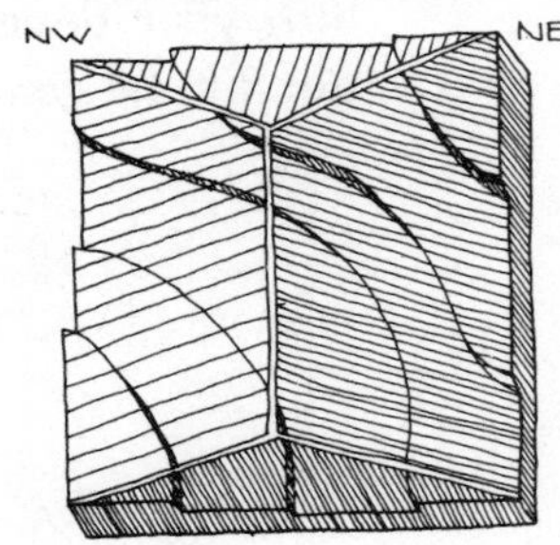

Figure D1.3.1 Typical Solar Envelope Construction. *Top, down:* Draw azimuths; construct altitude angle; construct north-south ridge; build roof planes.

D1 THERMAL

SCHEMATIC DESIGN
SITE MICROCLIMATE

D1

D1.4

PROCEDURE

Evaluate how well your design attains Thermal Schematic Design Goal C—**The buildings are clustered and sited to create optimal microclimates for heating when you need heat and for cooling when you need to cool.**

1. Reproduce the windflow diagram [B1.2], and draw your building and site improvements on the site. Show seasonal changes in wind patterns that are caused by your proposed building placement and site development.
2. With your building and site model [D1.3], use the sun peg shadow plot technique [B1.3] to study the yearly cycle of shadows. On a copy of your site plan draw the shadows cast at 9 a.m., noon, and 3 p.m.
3. Reproduce the sun peg shadow plot [B1.3], and draw your building and site improvements on the site plan, including the new shadow patterns.
4. Based on the changes in wind and shadow patterns caused by your proposed building, plot a new set of climatic response matrices [B1.4].
5. Discuss how well you have manipulated the microclimates created for:

   a. outdoor spaces

   b. indoor spaces.
6. Discuss whether or not your building and site improvements have met Goal C, and describe how your design must be altered if the goal has not been met.

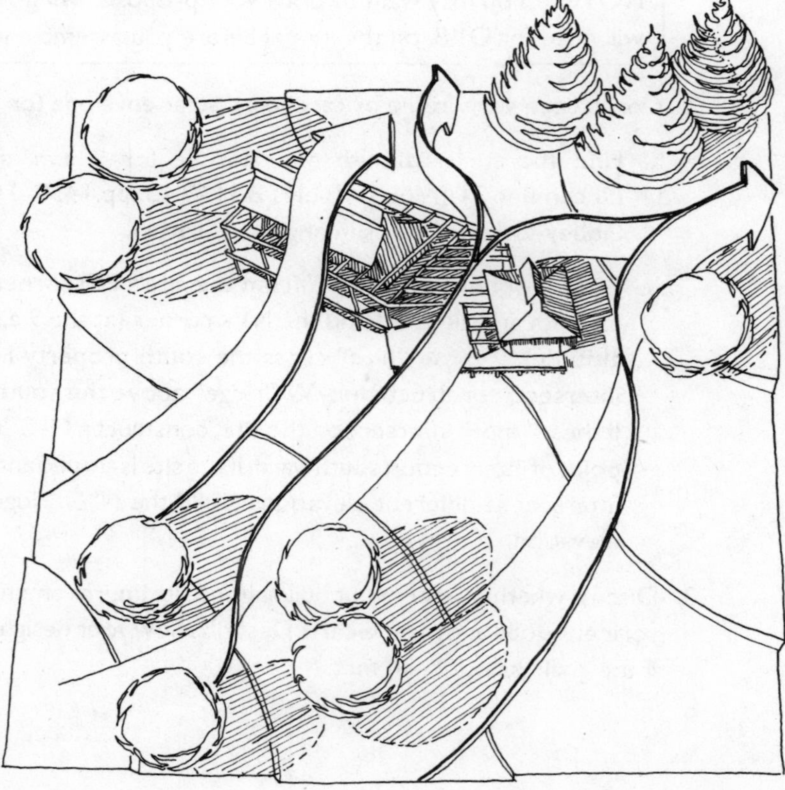

Figure D1.4.1 Site Wind Flow and Shadows. Linen supply complex, Charleston, South Carolina.

D1 THERMAL

SCHEMATIC DESIGN

VENTILATION APERTURE

Discussion

Explore the use of either an "open" or "mixed" ventilation scheme for your building, even if you have selected a "closed" scheme. In climates where a "closed" scheme is appropriate for the overheated months, a "mixed" scheme is often appropriate for the cooler months. This "mixed" scheme is especially appropriate for an internally dominated load building.

"Open" buildings use primarily natural ventilation and depend on thermally or wind-induced internal air flow for cooling. This cooling strategy works only if the inside air temperature is higher than the outside temperature. For some climates, natural ventilation can be used even in the hottest month. In drier climates, cooling can be accomplished near the upper limit of the comfort zone by moving air as warm as 87°F. This temperature has to be significantly lower in more humid climates.

"Closed" buildings use high mass without simultaneous ventilation. These buildings may also use evaporative cooling, desiccant cooling, earthtube cooling, roof ponds, and/or mechanical refrigeration.

"Mixed" buildings can be "closed" during very hot hours, then "opened" for night ventilation to remove stored heat.

The significant difference between "open" and "mixed" ventilation strategies is that "open" ventilation removes heat as it is generated, while "mixed" strategies remove the entire day's heat gain during a short ventilation period.

There are two basic ways to passively ventilate a building.

- Cross-ventilation (wind-induced) depends on the force of the wind to expel hot air from the leeward side of the building to be replaced by cooler air forced in the windward side.
- Stack ventilation (thermally induced) depends on hot air rising to expel heat from openings high in the building. This hot air is then replaced by cooler air drawn in through much lower openings in the building envelope. This strategy is particularly useful if you have calm wind conditions during overheated months. It can also be effective with earthtube cooling.

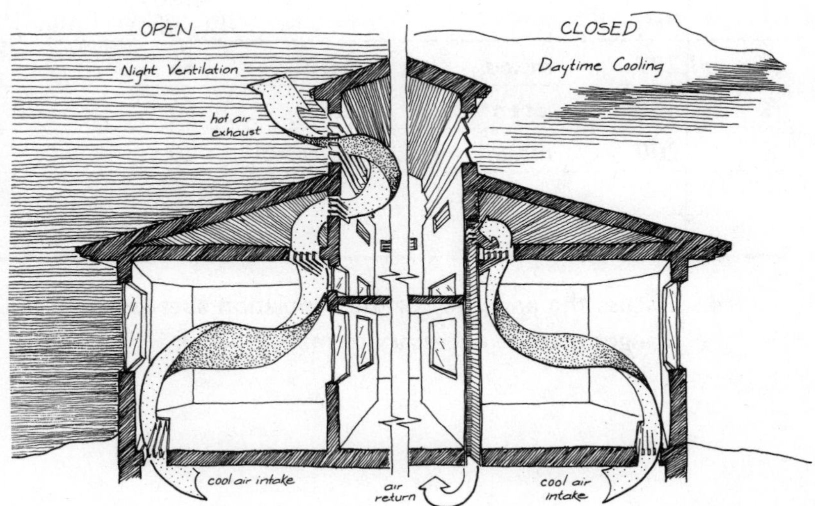

Figure D1.5.1 **"Mixed" Cooling Strategy. Linen supply office, Charleston, South Carolina.**

SCHEMATIC DESIGN

VENTILATION APERTURE (continued)

PROCEDURE

Evaluate how well your design attains Thermal Schematic Design Goal D—**Sufficient aperture is provided to ventilate the building.**

1. Calculate the internal heat gain density (Btuh/ft^2) for each thermal zone using data from C1.3.

   If the zone uses a "mixed" strategy, the heat for an entire day must be removed during a short period. Estimate the heat to be removed during this ventilation period by doubling the Btuh/ft^2 calculated in C1.3.

2. Tabulate the following information about your ventilation strategy:

   a. strategy used

   b. ventilation method (cross-, stack, or both)

   c. internal heat density

   d. wind speed [Appendix C, Climatological Data] or stack height (the measurement from the horizontal centerline of the upper window to the horizontal centerline of the lower window)

   e. recommended ventilation aperture [*SWL*, p.91 (cross-ventilation) and p.99 (stack ventilation)].

3. If you are using earthtube cooling, estimate the length and diameter of the tube required [*MEEB*, Figure 5.12, p.223].

Ventilation Aperture [suggested format]

Zone	Area (ft^2)	Ventilation Strategy	Ventilation Method	Internal Heat Density (Btuh/ft^2)	Wind Speed (mph)	Stack Height (ft)	Tube Size (diam,length)	Recommended Ventilation Aperture (ft^2)
A	*1,200*	*mixed*	*cross-*	*65*	*6*			*96*
B	*400*	*open*	*stack*	*50*		*25*		*80*
C	*200*	*mixed*	*earthtube(forced)*	*20*			*2@12"ϕ, 50'*	

4. Discuss the adequacy of the ventilation aperture, and indicate any design changes that are necessary to meet Goal D.

D1 THERMAL

SCHEMATIC DESIGN

VENTILATION AIR FLOW

D1.6

PROCEDURE

Evaluate how well your design attains Thermal Schematic Design Goal E—**Ventilation removes heat without leaving hot, dead-air areas.**

1. For "open" and "mixed" buildings, show the airflow ventilation path.

   On your site plan, floor plans, and sections draw the flow of ventilating air around and through your building for the months you need to cool. In drawing these flows, remember that wind flows from positive to negative pressure areas, hot air rises, and cool air falls.

2. For "closed" and "mixed" buildings, show the radiation or airflow paths.

   a. On your floor plans and sections indicate the locations of thermal mass. In order for mass cooling strategies to be successful, the mass must have a large amount of surface area as compared to its volume.

   b. If the mass will be cooled primarily by radiation, draw the path of radiant transfer from thermal mass to heat sink, and specify the type of heat sink (night sky, body of water, or the earth).

   c. If your building relies on night ventilation, where the mass is cooled by conduction/convection, draw the path of air currents over the surface of the mass.

3. Discuss the effectiveness of your ventilation or radiation strategies. Indicate any design changes that are necessary to accomplish Goal E.

Figure D1.6.1 Cross-Ventilation Flow. Linen supply laundry, Charleston, South Carolina.

D1 THERMAL

SCHEMATIC DESIGN

SOLAR-GLAZING APERTURE

D1
D1.7

Discussion

Adequate solar aperture is determined by two factors, the recommended solar savings fraction (SSF) and an adjustment for internal heat gains. SSF is the percent of energy a solar-heated building can be expected to save when compared to a conventionally heated, energy-conserving building. The SSFs from the map (*right*) can be used with tables from *SWL* to determine solar aperture for a residential, SDL building. This aperture must be adjusted to account for internal heat gains that reduce the need for solar heating. The adjustment is based on a building's or thermal zone's balance point temperature during the daytime in the winter.

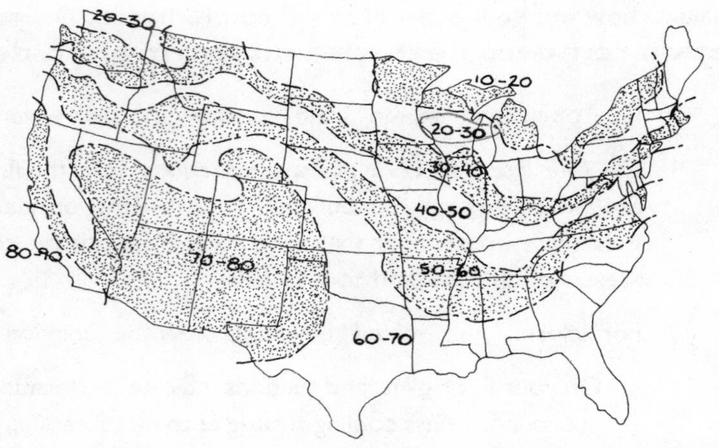

Figure D1.7.1 Recommended SSF Map. Reprinted from Balcomb et al., *Passive Solar Design Handbook*, 3:18.

Procedure

Evaluate how well your design attains Thermal Schematic Design Goal F—**The building is exposed to the south in proportion to the heat needed.**

1. Determine the recommended SSF for your climate [Figure D1.7.1].
2. Determine the amount (or range) of solar glazing required to heat an SDL building [*SWL*, pp.131–133]. Use the city and the passive heating system that is most appropriate for your project. You can compare the heating degree days [*MEEB*, Table B.15, column D65, pp.1459–1463] to match climate locations.
3. Adjust the solar-glazing area required for each zone of your building based on its internal heat gains.

$$A_{adj} = (A)\ [(T_{BP} - T_{Jan})\ /\ (60°F - T_{Jan})]$$

where:

A_{adj} = adjusted aperture (ft^2)
A = area of solar aperture (ft^2)
T_{BP} = average winter daytime balance point temperature (°F)
T_{Jan} = average January temperature (°F)

Solar Aperture Calculation [suggested format]

Zone	Floor Area	Type of Strategy	Recommended Aperture Area	Balance Point Temperature	Avg Jan Temperature	Adjusted Solar Aperture
A	*1,200 ft²*	*trombe wall w/ night insulation*	*(.30) (1,200) = 360 ft²*	*45°F*	*32°F*	*167 ft²*

4. Discuss the adequacy of the solar heating aperture and any design changes you deem necessary to meet Goal F.

SOLAR GLAZING AND THERMAL MASS

DISCUSSION

Solar apertures must be coupled to thermal mass to provide effective passive solar heating. A very large ratio of thermal mass surface area to volume is required. This surface area should be directly exposed to the sun's rays. The thermal mass must be located within about 20 feet of the occupants to effectively re-radiate heat to them. Remember, the total volume of the mass is less significant than the total surface area that is exposed to the sun.

PROCEDURE

Evaluate how well your design attains Thermal Schematic Design Goal G—**The winter sun can reach all the south-facing apertures you have provided for the collection of heat.**

1. Determine which months to use the sun to heat [B1.1 and C1.5]:
   a. outdoor spaces
   b. building interiors in each zone.
2. Clearly identify the areas of solar aperture on your 1" = 40' scale model. Remember, the roof pond is an effective heating device and should be identified as a solar aperture for heating.
3. Use the sun peg and model to see if the sun reaches these areas between 10 a.m. and 2 p.m. during the months when your building needs heat.
4. Draw the shadows on your matrix site plan [Appendix B] and your south elevation, or include photographs of the study. Indicate the time and month shown.
5. Draw a heat flow diagram to indicate the sun's radiation to and re-radiation from the thermal mass during the periods when your building requires heat.
   a. In a sectional perspective, show the location of the thermal mass designated for heating (scale: 1/8" = 1'-0").
   b. Indicate the paths of direct solar radiation through solar apertures to the heat sink of the thermal mass [step 3]. Also illustrate the re-radiation patterns.
6. Discuss the effectiveness of your design and any changes necessary to meet Goal G.

WINDOW SHADING

Discussion

Your building must be cooled when the outdoor temperature exceeds the balance point temperature [C1.5]. When outdoor conditions fall below 68°F—indicated by the shading line on the bioclimatic chart of thermal strategies [Figure D1.2.1]—cooling can easily be accomplished solely through ventilation. When conditions as plotted on this bioclimatic chart are above the shading line, it is also necessary to shade glazing on all façades of your building for effective cooling.

Procedure

Evaluate how well your design attains Thermal Schematic Design Goal H—**Glazing is shaded from the sun during overheated months.**

1. Record the months that your building requires shading [Figure D1.2.1]. These months occur when the average daily high outdoor temperature is plotted above the shading line.
2. Use the sun peg with your 1" = 40' scale model to evaluate the building aperture shading at site and cluster scales. (Component-scale shading will be evaluated in E1.2.)
   a. Evaluate whether all glazed apertures are shaded at 9 a.m., noon, and 3 p.m. during the building's overheated months.
   b. Document this shading using either photos of your model or drawings (elevations, plans, or axonometrics).
3. If you intend to use desiccant cooling or a roof pond, verify that the roof is exposed to the sky and not shaded by overhanging trees or buildings.
4. Discuss the success of your shading strategies and any design changes needed to accomplish Goal H.

DESIGN REVIEW

D1.10

Discussion

You have formulated a schematic design for your building and site, chosen appropriate thermal strategies, and ensured access to appropriate environmental forces. Each step has been accomplished independently and conflicting decisions may have been made. This design review affords you the opportunity to synthesize your cumulative design decisions.

Procedure

Review how well your design accomplishes Thermal Schematic Design Goals A–H.

1. Discuss the tradeoffs between the heating and cooling strategies that your design required.
2. Discuss situations where your heating and cooling strategies worked effectively together.
3. Make a schematic design sketch that combines all your thermal design strategies. Indicate how and why your design has evolved in response to meeting D1.0 goals.

E1 THERMAL

DESIGN DEVELOPMENT

E1

INTRODUCTION

E1.0

GOAL

Refine and further develop your building's design. Use your schematic design as the foundation for your design development.

THERMAL DESIGN DEVELOPMENT GOALS

A. All glazed apertures are shaded during overheated periods.
B. Daily heat gain peaks are minimized during overheated months.
C. The building envelope meets or exceeds the current ASHRAE minimum thermal resistance standards for its building type and climate.
D. The building's cooling system effectively removes heat and humidity.

DISCUSSION

Energy conservation is the most cost-effective way to save energy. Eighty percent of the energy that can be saved by building design is saved during the conceptual design phase. You have already designed a building that can save a substantial amount of energy. However, in order to fully realize your building's potential for energy savings, a thorough job of analyzing its thermal needs and developing appropriate design details must ensue. You have just realized another general rule-of-life: 80% of the job can be attained with 20% of the effort; the remaining 20% of the job takes the final 80% of the effort. Attaining that final 20% of potential savings is essential. The results of expending this additional effort show not only prudence, but elegance.

> *Selection of the proper design concept in terms of how energy is used can have a much greater impact on energy use than the efficiency of the equipment used.*
>
> *—Lawrence Spielvogel*

A design that reaches only part of its potential, in any aspect, leaves the user unsatisfied. To design a building that responds seamlessly to the changing thermal demands of its environment and occupants requires a comprehensive understanding of those needs and the design of supportive details. When the details of a building not only work efficiently, but add to the elegance of the design concept, the end result is beauty in the energy-saving detail (to paraphrase Mies van der Rohe).

The eminent architectural historian James Marston Fitch corroborates this point-of-view in an article in the winter 1988 *Journal of Architectural Education*:

> *My fundamental thesis is that the ultimate task of architecture is to act in favor of humankind: to interpose itself between us and the natural environment in which we find ourselves, in such a way as to remove the gross environmental load from our shoulders. The central function of architecture is thus to lighten the very stress of life. Its purpose is to maximize our capacities by permitting us to focus our limited energies upon those tasks and activities which are the essence of the human experience.*

> *The central function of architecture is . . . to lighten the very stress of life. . . .*
>
> *—James Marston Fitch*

> *This successful interposition between humanity and its natural environment furnishes the material basis of all great architecture. To wrest the objective conditions for optimal development and well-being from a Nature which only seldom provides it, to satisfy our physiological and psychological requirements at optimal levels—this beyond question is the objective basis of any architecture which is both beautiful and good. To design such a building, as Gropius once so movingly put it, is an authentic act of love.*

E1 THERMAL

DESIGN DEVELOPMENT

INTRODUCTION (continued)

IN THIS SECTION YOU WILL:

1. Complete the thermal design development.
2. Design shading devices for each of the windows of the building.
3. Calculate the internal heat gain from people, lights, and equipment.
4. Calculate the total solar heat gain.
5. Use the energy code requirements to guide the design of the building's skin.
6. Calculate the heat loss through the building's skin.
7. Calculate the balance point temperature of the building's thermal zones.
8. Explore energy-conserving cooling systems for your building.
9. Critique your building's thermal design.

E1 THERMAL

DESIGN DEVELOPMENT

E1

DESIGN

E1.1

Thermal Design Development Strategies

Component-Scale Strategies

- Use effective external shading devices to block the sun during the cooling season.
- Use movable internal shading devices when a device that must be operable is appropriate.
- Place thermal insulation on the exterior of masonry walls.
- Place thermal mass so it "sees" the sun during the heating season and "sees" the occupants during the cooling season.
- Use sufficient levels of insulation in the building's skin.
- Use air-to-air heat exchangers to ensure fresh air in a well-caulked and weatherstripped building.
- When possible, schedule building use to avoid simultaneous heat gain peaks from different sources.
- Use desiccant cooling for heat and humidity reduction.

Procedure

Refine your final schematic design [D1] to include component design considerations. Use the appropriate strategies to aid your thermal design development. Modify your schematic design drawings to include the components you have developed.

Document your design as follows:

1. Site plan, including parking and access drives (scale: 1" = 100').
2. Cluster plan, including outdoor spaces (scale: 1" = 40').
3. Floor plans (scale: 1" = 20').
4. Roof plan and elevations, or axonometrics, illustrating all building sides and roof (scale: 1" = 20').
5. Sections (scale: 1/8" = 1'-0").
6. Design diagram, annotated to identify design strategies and thermal zones.
7. Details of thermal components (e.g., typical roof and wall construction, fenestration including shading devices, and details of thermal mass).

E1 THERMAL

E1

DESIGN DEVELOPMENT

COMPONENT-SCALE SHADING

E1.2

Discussion

It is CRUCIAL to fully shade the windows of your building when the outdoor temperature, as plotted on the bioclimatic chart, is above the shading line. It is also ADVISABLE to shade the windows when the outdoor temperature exceeds the balance point temperature. For buildings that use an "open" cooling strategy, use the occupied hour balance point temperature; for those that use a "mixed" or "closed" cooling strategy, use the 24-hour average balance point temperature. Although you can achieve adequate shading by other means, we recommend that you provide: (a.) ADVISABLE shading with movable devices (either exterior [better] or interior [worse]) which may be adjusted to allow full sun or provide full shade and (b.) CRUCIAL shading with fixed exterior devices.

There are three types of fixed exterior shading devices—horizontal, vertical, and egg-crate (a combination of horizontal and vertical). Generally speaking, the best generic, fixed shading device varies with the window orientation.

N:	vertical	S:	horizontal or egg-crate
NE:	vertical	SW:	horizontal or egg-crate
E:	egg-crate	W:	egg-crate
SE:	horizontal or egg-crate	NW:	vertical

Each device provides shade described by either a profile angle (for horizontal), bearing angle (for vertical), or both (egg-crate). If the device exists, the profile angle, ∠ABC (*right, above*), is in the vertical plane perpendicular to the window. The bearing angle, ∠XYZ (*right, below*), is in the horizontal plane perpendicular to the window.

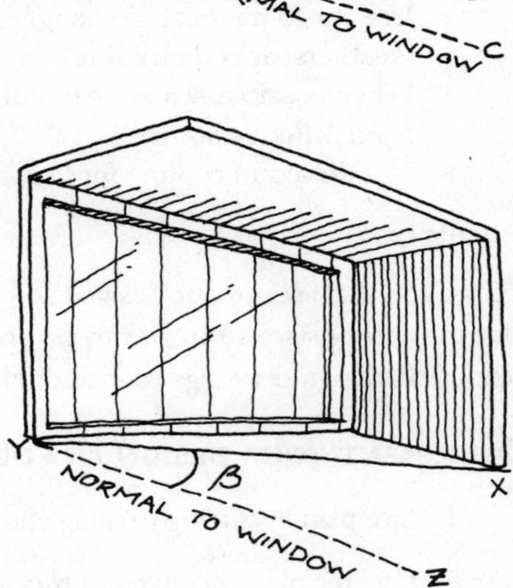

Figure E1.2.1 Determining Angles. ***Above*, profile angle; *below*, bearing angle.**

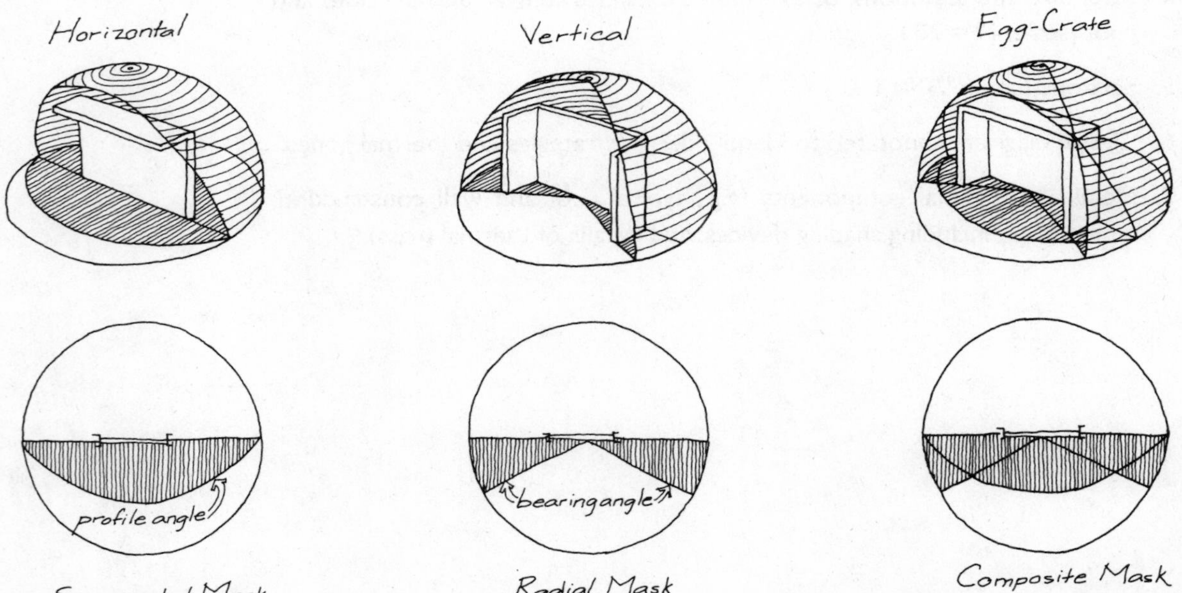

Figure E1.2.2 Shading Device Projection to Sun Chart and Resultant Shading Mask

E1 THERMAL

DESIGN DEVELOPMENT

COMPONENT-SCALE SHADING (continued)

To design shading devices with the LOF sun angle calculator use the sun chart and the profile angle overlay to determine the sizes of the profile and bearing angles [Figure E1.2.2]. The mask of an effective, fixed, exterior shading device will "shade" the sun path (on the sun chart) during CRUCIAL shading periods and won't "shade" the sun path when solar heat is desirable. For example, the egg-crate device mask in Figure E1.2.3 indicates the portion of the sky that is obscured by the shading device. The intersection of the mask and the periods that need shading indicate that the device is only effective until 3 p.m. in the summer. Moreover, it blocks winter afternoon sun. Obviously, this shading device is not optimal.

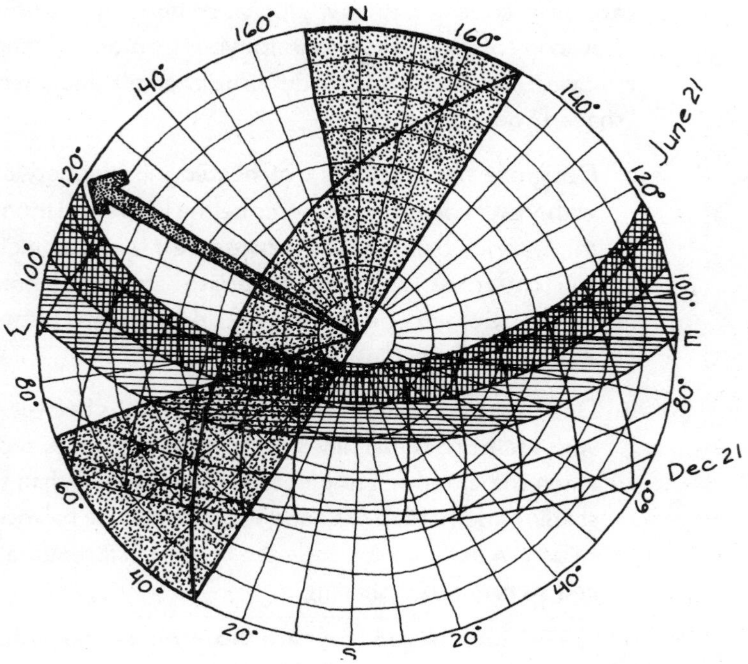

Figure E1.2.3 **Shading Mask Design for 32°N Latitude. Shading mask (stippled) for egg-crate device facing 30° north of West with CRUCIAL (cross-hatched) and ADVISABLE (horizontal lines) shading periods indicated.**

For more information

MEEB, Appendix D, pp.1495–1507.

Olgyay, *Design with Climate*, Section 7, pp.63–83.

Procedure

Evaluate how well your design attains Thermal Design Development Goal A—**All glazed apertures are shaded during overheated periods.**

For all apertures:

1. List your building's overheated months for each thermal zone in terms of:
   a. the months when the temperature is plotted above the shading line
   b. the months when the temperature is plotted above the "occupied hour" or "24-hour average" balance point temperatures you have calculated for your design [C1.5].

Overheated Months [suggested format]		
Zone *A*	Months when the outside temperature is above the:	
Shading Line	Occupied Hour Balance Point Temperature	24-Hour Average Balance Point Temperature
Jun, Jul, Aug, Sep	*Apr–Nov @ 50°F*	*May–Oct @ 60°F*

COMPONENT-SCALE SHADING (continued)

2. Based on the appropriate balance point temperature and shading line for your particular cooling strategy, plot your building's overheated months on a photocopy of the LOF sun chart for your latitude [Appendix E]. These months indicate the appropriate "shade" time intervals, and, therefore, the area to be "shaded" on the sun chart.

   a. Designate the building's overheated months between January and June with **VERTICAL** lines to describe the hours and months of the year when the exterior temperature is **GREATER** than that indicated by the shading line (**CRUCIAL**) and the appropriate balance point temperature (**ADVISABLE**). Color code the lines to differentiate between **CRUCIAL** and **ADVISABLE** shading.

   b. Designate the building's overheated months between July and December with **HORIZONTAL** lines to describe the hours and months of the year when the exterior temperature is **GREATER** than that indicated by the shading line (**CRUCIAL**) and the appropriate balance point temperature (**ADVISABLE**). Color code the lines to differentiate between **CRUCIAL** and **ADVISABLE** shading.

> NOTE: "Solar-paired" days are those whose sun paths are identical (e.g., March 21 and September 21). When overheating occurs during both "solar-paired" days, this condition (indicated on the sun chart [Figure E1.2.3] by a cross-hatched area) means that fixed, exterior shading devices are appropriate. When overheating occurs during only one of the paired days, operable shading devices are needed.

3. For all apertures not shaded at the site, cluster, and building scales, design an external shading device.

   a. For each window orientation (±5°) in each thermal zone, make a copy of your sun chart on which you have plotted the overheated periods.

   b. Select the appropriate device type for your window's orientation.

4. Draw the mask of the shading device on tracing paper over your sun chart [step 2]. Correlate these masks with their apertures and key them to your elevations or axonometrics. To provide effective shading the mask should cover the cross-hatched, **CRUCIAL**, overheated area on the sun chart. Conversely, the mask should not cover the unmarked area of the sun path, as this represents the time when the sun is desired. In most instances you will have to compromise between these two needs.

COMPONENT-SCALE SHADING (continued)

5. Each of your external shading devices has a shading coefficient (SC). This coefficient indicates what percentage of the window's unshaded solar gain is admitted through or around the shading device. The shading coefficients are listed in Olgyay, ***Design with Climate***, pp.68–71; ***SWL***, p.46; and ***MEEB***, Table 4.22, p.169. Shading coefficients for trees are given in Lechner, ***Heating, Cooling, Lighting***, Table 9.C, p.236. List the shading coefficients for each of your externally shaded apertures.

6. Summarize your component shading strategies in tabular form. Make provisions for recommended internal shading devices or glazing that augment the external devices. Shading coefficients for internal shading devices and glazing are listed in Olgyay, ***Design with Climate***, pp.68-71; ***SWL***, p.46; and ***MEEB***, Tables 4.18–4.24, pp.163–171.

7. Calculate the combined SC for each window.

Summary of Shading Components [suggested format]

Zone	Window Orientation	External Device Type	External SC (SC1)	Internal Device Type	Internal SC (SC2)	Combined SC [(SC1)(SC2)]
A	*South*	*overhang*	*.6*	*louvered blinds*	*.4*	*.24*

8. Discuss the success of your shading strategy, and indicate any design changes you view as necessary to meet Goal A.

E1 THERMAL

DESIGN DEVELOPMENT

HEAT GAINS

E1.3.1: HEAT GAIN FROM PEOPLE

DISCUSSION

This series of evaluation techniques gives a more accurate heat gain estimate than the one you made in C1.3. They will help you check how appropriate your thermal zoning strategies are and aid your design of the thermal components of your building. In this technique, the heat gain from people is calculated according to their activity level and how many occupy the space. Since the occupancy level may change over the course of the day, you should calculate heat gain during peak, off-peak, and, if relevant, unoccupied hours.

PROCEDURE

1. Divide the building into thermal zones based on occupancy levels and use schedules [C1.1].

2. For each zone:

   a. calculate the area of each space

   b. identify peak and off-peak hours according to occupancy levels

   c. state the number of occupants present during peak and off-peak hours

   d. state the rate of heat gain, HG, per occupant [*MEEB*, Table 5.22, "Sensible Heat" figures, p.252]

   e. calculate the rate and density of heat gain generated during peak and off-peak hours in Btuh and Btuh/ft².

   $$HG = (p)\ (HG_{pp})$$

   $$HD = (HG)\ /\ (A)$$

   where:

   HG = hourly heat gain (Btuh)
   p = number of people
   HG_{pp} = heat gain per person (Btuh)
   HD = heat density (Btuh/ft²)
   A = floor area (ft²)

Occupancy Schedule [suggested format]

Space	Area (ft²)	Peak Hours	Off-Peak Hours	Occupancy (# of people)	Heat Gain per Person (Btuh)	Hourly Heat Gain (Btuh)	Heat Density (Btuh/ft²)
office	*900*	*8a.m.–noon, 1p.m.–5p.m.*	*NA*	*6*	*275*	*1,650*	*1.83*
		NA	*noon–1p.m, 5p.m.–8a.m.*	*1*	*275*	*275*	*.31*

E1 THERMAL

DESIGN DEVELOPMENT

HEAT GAINS (continued)

E1.3.2: HEAT GAIN FROM LIGHTS

Discussion

You can estimate the heat generated by electric lighting based on your building's lighting requirements. Since you will be daylighting your building and not relying solely on electric lighting, you can estimate the reduction in heat gain made possible by daylighting.

Procedure

1. Divide the building program into zones according to their lighting requirements. For each zone select the appropriate lighting power density, which is a worst-case estimate of the heat density when you only use electric lighting [*MEEB*, Table K.2, pp.1568–1570].
2. Identify which hours daylighting is and is not possible. Estimate when there is adequate daylight [*MEEB*, Figure 19.17, p.989].
3. Select a recommended daylight factor for each space [*MEEB*, Table 5.3, p.197].
4. Select an appropriate adjusted lighting heat density (Btuh/ft^2) for each space according to its daylight factor [*MEEB*, Table 5.8B p.214].
5. Calculate the hourly heat gain for the space when it is daylighted and when it is not.

   heat gain = (heat density) (area)

Lighting Schedule [suggested format]

		Without Daylighting			With Daylighting			
Space	Area (ft^2)	Hours	Heat Density (Btuh/ft^2)	Hourly Heat Gain (Btuh)	Hours	Daylight Factor (%)	Heat Density (Btuh/ft^2)	Hourly Heat Gain (Btuh)
office	*900*	*5p.m.–8a.m.*	*5.1*	*4,590*	*8a.m.–5p.m.*	*1.5–2.5*	*2.0*	*1,800*

E1.3.3: HEAT GAIN FROM EQUIPMENT

Discussion

The heat gain from equipment depends on the types of machinery in your building, how efficient they are, and how much they are used. Machines may produce both sensible heat (heat that increases air temperature) and latent heat (heat that is present in increased air moisture which has the potential to release this heat by condensing). Here we are concerned only with sensible heat gains.

Procedure

1. Divide the building into task areas, identifying the hours the space is in use and the equipment in each space.

2. Determine the rate of sensible heat gain and heat density from equipment, according to the function [*MEEB*, Table 5.13, p.232 (residential buildings) or Table 5.23, p.253 (commercial buildings)].

   If your space isn't listed in either of the *MEEB* tables, use sensible heat gain figures for each machine, in Btuh (for heat gain in Btuh or watts, see ASHRAE tables, Sweet's catalogs, manufacturer's data, or labels on machines).

   NOTE: To convert watts to Btuh, 1 watt = 3.41 Btuh.

3. Calculate the hourly heat gain and heat gain density for each machine type.

   $$HG_t = (m)\ (HG_m)\ (p)$$

   where:

   HG_t = hourly heat gain (Btuh)

   m = the number of machines

   HG_m = heat gain per machine (Btuh)

   p = percentage of time the machine is on during the hours of occupancy

Equipment Use Schedule [suggested format]

Function	Area (ft²)	Hours	Equipment	HG_m (Btuh/machine)	% Time "On"	HG_t (Btuh)	Heat Density (Btuh/ft²)
copying	*100*	*8a.m.–5p.m.*	*1 photocopier*	*10,000*	*50*	*5,000*	*50*

E1.3.4: HEAT GAIN FROM THE SUN

Discussion

Solar Heat Gains through the Roof and Walls

Opaque skin elements receive and store heat from the sun as well as heat from the air. To get an approximate 24-hour average of the heat transfer from the exterior to the interior, use design equivalent temperature differentials (DETD). The rate of heat transfer in walls and roofs is calculated in a greatly simplified manner from a combination of the effects of a material's U-value multiplied by an assumed indoor/outdoor temperature difference (DETD). This product is called the heat transfer factor (HTF). Significant heat transfer through the opaque skin only occurs when the outdoor temperature exceeds 85°F.

A more precise calculation would use cooling load temperature differentials (CLTD), rather than HTFs. This calculation, though, only applies to nonresidential buildings. You will find that most of the heat gain stems from solar and internal heat gains. Therefore, the use of the simplified procedure is a reasonable approximation.

Solar Heat Gains through the Glazing

Hundreds of calculations would be required for an hour-by-hour depiction of heat transfer through the windows. But you will simplify this procedure by looking only at the worst-case pattern for each glazing orientation. Use *MEEB*, Tables B1–B5, pp.1440–1444, to calculate the maximum heat gain through:

- east-facing windows (morning)
- south-facing windows (noon)
- west-facing windows (afternoon)
- horizontal skylights (morning, noon, and afternoon)
- sloping skylights (morning, noon, and afternoon). (Interpolate between the vertical and horizontal solar gains, according to the skylight's altitude angle and orientation.)

The shading coefficients to be used in calculating solar heat gains should be taken from E1.2.

DESIGN DEVELOPMENT

HEAT GAINS (continued)

E1.3

PROCEDURE

1. Tabulate the summer heat gains through the building's skin for each thermal zone. (In very warm climates where average high temperatures exceed 85°F in fall or spring, tabulate fall and spring heat gains also.) Find the DETD (for cooling, DETD = $T_{out} - T_{in}$). Use *MEEB*, Table 5.19, p.249 and the average high summer temperature for your climate or *MEEB*, Table A.1, "summer design temperature," pp.1429–1437.

   HG_o = (DETD) (U_o) (A)

   where:

   HG_o = solar heat gain through the opaque skin (Btuh)

   U_o = overall U-value of the opaque skin component (Btuh/ft²°F)

   (U-values will be calculated in E1.6. U-values are the inverse of R-values. Use R-30 ceilings and R-20 walls and floors for now. Revise these values later, if necessary.)

   A = area of the skin component (ft²)

> NOTE: If you are using a roof pond, trombe wall, or water wall for cooling, assume there is no heat gained through it, since it is producing net cooling.

Opaque Skin Heat Gain from the Sun [suggested format]

Zone *A* — Season/Month *Summer/July*

Hours	Component	A (ft²)	U_o (Btuh/ft²°F)	DETD (°F)	HG_o (Btuh)
all day	*walls*	*370*	*.05*	*16.3*	*302*
	roof	*400*	*.033*	*36.0*	*475*

HEAT GAINS (continued)

2. For morning, noon, and afternoon, tabulate the year-round (winter, spring, summer, and fall) solar heat gains through the glass. Find values for each window in each thermal zone [*MEEB*, Tables B1–B5, pp.1440–1444].

> NOTE: For morning hours, use the time in the left column and the header at the top of the page; for afternoon hours, use the time in the right column and the header at the bottom of the page. The values are in Btuh/ft^2 of glazing.

$$HG_g = [(HG_{sun})\ (SC)\ (A)] + [(HG_{shade})\ (SC)\ (A)]$$

where:

HG_g = solar heat gain through the glass (Btuh)

HG_{sun} = direct solar heat gain incident on the exterior of the glass (Btuh/ft^2)

SC = total shading coefficient (the product of interior and exterior SCs) [E1.2]

A = area of the window (ft^2)

HG_{shade} = diffuse solar gain incident on the glass (Btuh/ft^2)

Window Heat Gain from the Sun [suggested format]

Zone *A* Season/Month *Summer/July*

Time	Component	A (ft^2)	SC	HG_{sun} (Btuh/ft^2)	HG_{shade} (Btuh/ft^2)	HG_g (Btuh)
9a.m.	*east glass*	*10*	*.3*	*215*	*0*	*645*
	south glass	*20*	*.6*	*0*	*41*	*492*

E1 THERMAL

DESIGN DEVELOPMENT
HEAT GAIN SCHEDULE

E1
E1.4

Procedure

Generate heat gain data for your building's thermal zones to calculate its balance point temperature.

For each zone:

1. Add the heat gains (Btuh), as calculated in E1.3, for all the spaces. These will vary from peak to off-peak hours and from season to season.

2. Calculate and record a schedule of total heat gains for:

   a. peak and off-peak hours (unoccupied hours, if necessary)

   b. each season.

Heat Gain Schedule [suggested format]

Season *Winter*

Zone	Occupancy	Hours	People (Btuh)	Lights (Btuh)	Equipment (Btuh)	Skin (Btuh)	Solar (Btuh)	**TOTAL** (Btuh)
A	peak	*all day*	*345*	*1,276*	*NA*	*1,813*	*20*	*3,454*
	off-peak							
	unoccupied							
B	peak	*8a.m.–11p.m.*	*59,970*	*10,026*	*72,000*	*33,075*	*1,018*	*176,089*
	off-peak	*11p.m.–8a.m.*	*1,490*	*1,002*	*27,000*	*33,075*	*NA*	*32,567*
	unoccupied							
C	peak							
	off-peak							
	unoccupied							
D	peak							
	off-peak							
	unoccupied							

DESIGN DEVELOPMENT

HEAT GAIN PEAKS

PROCEDURE

Evaluate how well your design attains Thermal Design Development Goal B—**Daily heat gain peaks are minimized during overheated months.**

1. Plot the hourly summer heat gains for each zone from the preceding calculation [E1.4]. Use different colors or patterns to distinguish different heat gain sources. Grids for your plots are provided on the next page.
   a. Plot the gain through the building skin (roof and wall) as a constant over the 24-hour period.
   b. **ADDITIVELY** plot the total internal heat gains (from people, lights, and equipment) for the hours of their operation.
   c. **ADDITIVELY** plot each of the solar gains through the glazing for the hours calculated in E1.4. Connect these points of solar gain by plotting "zero" solar gain at sunrise (5 a.m.) and sunset (7 p.m.).

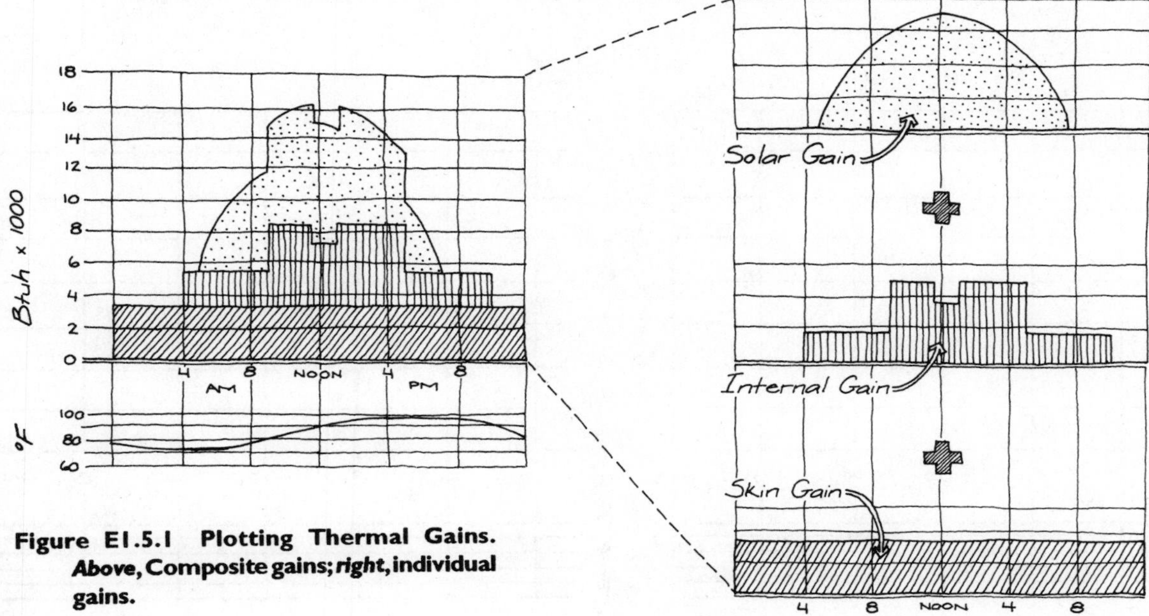

Figure E1.5.1 Plotting Thermal Gains. ***Above***, **Composite gains;** ***right***, **individual gains.**

2. Below the graph of heat gains, plot the 24-hour temperature pattern for August [B1.1]. The high temperature generally occurs at 4 p.m. and the low at 5 a.m.
3. Discuss the success of your design in avoiding simultaneous heat gain peaks from the variety of sources in your building. Indicate any design changes that you view as necessary to avoid these simultaneous peaks and meet Goal B.

HEAT GAIN PEAKS (continued)

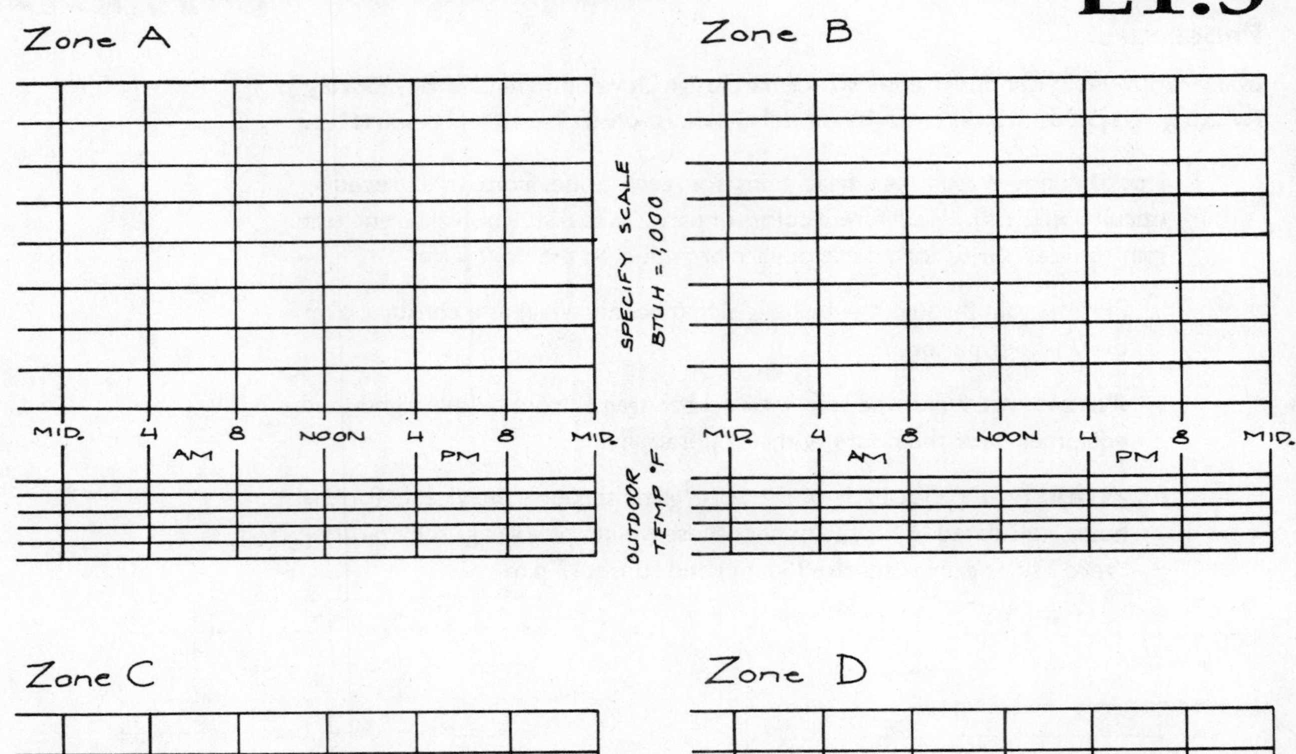

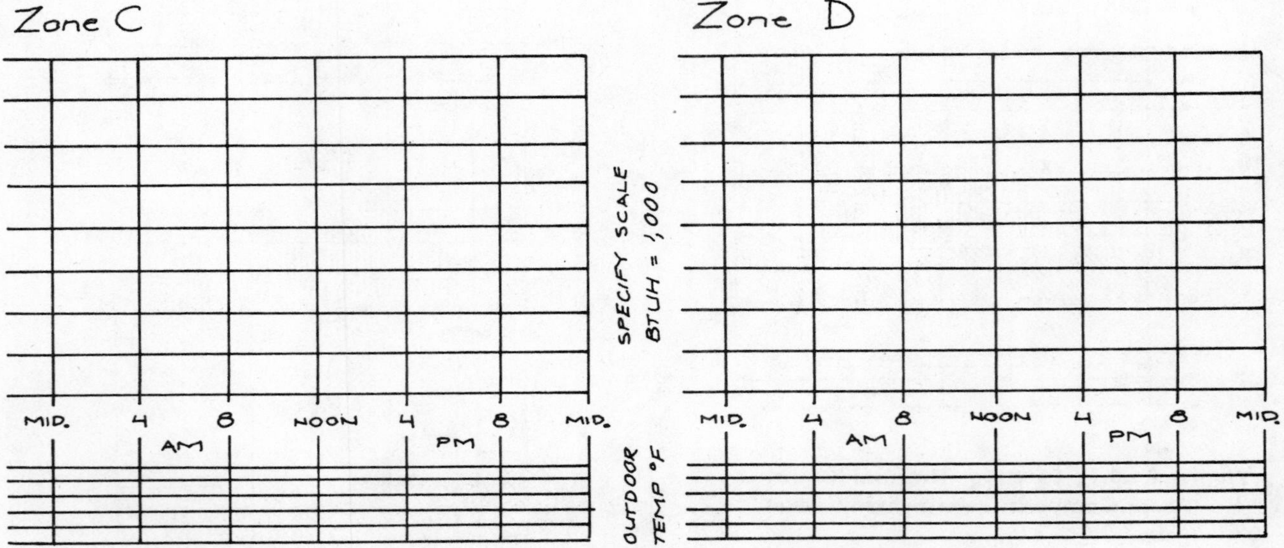

Figure E1.5.2 Peak Heat Gain Charts

ENVELOPE THERMAL RESISTANCE

DISCUSSION

Most locales in the United States have energy code requirements for the building envelope. These can be summarized into three categories:

- prescriptive (e.g., walls must have R-19 insulation)
- component packages (specific combinations of components meet the requirements)
- performance (e.g., the building must not use more than 9.5 Btu/DDft^2, no matter how it is built).

Most energy codes base their standards on the most recent *ASHRAE Standard*. In this procedure, you will use the alternate component package (ACP) approach, outlined in *ASHRAE/IES 90.1–1989*, which rewards integrated thermal and lighting design. By using the ACP table for your climate, you will be able to determine the allowable percentage of fenestration and the maximum U_o for the opaque walls, the roof, the floor, the slab or basement, and the walls adjacent to unconditioned enclosed spaces.

FOR MORE INFORMATION

MEEB, Example 5.1, Part D, pp.201–203 (a completely worked out example using *ASHRAE/IES 90.1*).

PROCEDURE

Evaluate how well your design attains Thermal Design Development Goal C—**The building envelope meets or exceeds the current ASHRAE minimum thermal resistance standards for its building type and climate.**

Based on the code recommendations, you will design the components of your building's skin.

1. Calculate or select the ACP entry data for each zone of your building.

   a. State whether your building is base case (non-daylighted) or perimeter daylighted with automatic lighting controls.

   b. Calculate the internal load density (ILD) in watts/ft^2 during the peak hour.

   $$\mathbf{ILD = (HG_P + HG_L + HG_E) / [(A)\ (3.41\ Btuh/watt)] - (0.6\ watts/ft^2)}$$

   where:

   HG_P = total peak heat gain from people (Btuh) [E1.3.1]

   HG_L = total peak heat gain from lights (Btuh) [E1.3.2]

   HG_E = total peak heat gain from equipment (Btuh) [E1.3.3]

   A = total floor area of conditioned space (ft^2)

NOTE:	–0.6 watts/ft^2 is a correcting factor for the ACP data.

ENVELOPE THERMAL RESISTANCE (continued)

c. Calculate the external shading projection factor (PF) if horizontal or egg-crate devices are used on south, west, and east façades [Figure E1.6.1].

$$PF = P_d / H$$

where:

P_d = depth of overhang (ft)
H = height from sill to bottom of overhang (ft)

d. Calculate the shading coefficient of all shading devices (SC_x), except for overhangs considered in step 1c.

$$SC_x = (SC_I)(SC_E)$$

where:

SC_I = SC of internal devices [E1.2]
SC_E = SC of external devices (except overhangs) [E1.2]

e. Select the glazing type and record its U-value, U_oF, and visual transmittance (VT) [Table E1.6.1 or *MEEB*, Table 4.16, pp.160–162]. If you are using daylighting, VT should be greater than or equal to the solar transmittance.

2. Find the maximum allowable percentage of fenestration (glazing as a percentage of wall area) for your building using the appropriate ASHRAE ACP table from *MEEB*, Appendix K, Table K.12 (Charleston), p.1581; Table K.17 (Phoenix), p.1586; Table K.25 (Spokane), p.1594; Table K.28 (Dodge City), p.1597; or Table K.31 (Madison), p.1600. Use ILD, PF, SC_x, and base case U_oF or daylighted U_oF to index the ACP table [Figure E1.6.2]. Where values exceed the minimum or maximum index in the table, use the minimum or maximum value. Compare the allowable fenestration with the amount you've provided in your design.

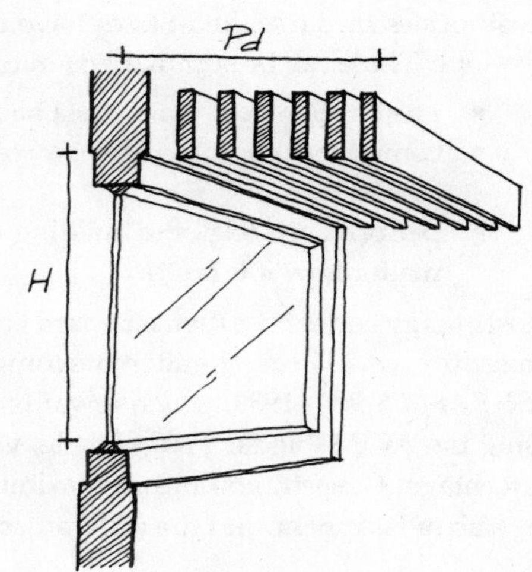

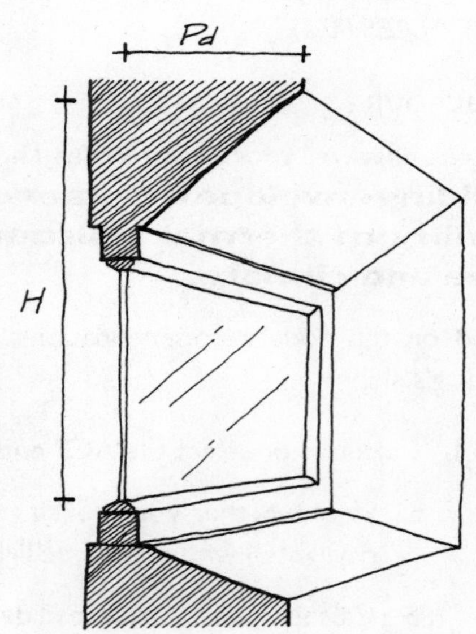

Figure E1.6.1 Determining the External Shading Projection Factor (PF). ***Above,*** **louvered, horizontal device;** ***below,*** **recessed window.**

Maximum Percentage of Fenestration [suggested format]

Base Case or Daylighted	ILD (watts/ft^2)	PF	SC_x	U_oF	VT	Max % Fenestration	Design % Fenestration
daylighted	*1.74*	*1.0*	*.25*	*.5*	*.8*	*66*	*40*

E1 THERMAL

DESIGN DEVELOPMENT

ENVELOPE THERMAL RESISTANCE (continued)

Table E1.6.1 Transmittance of Glazing

Glazing Type		Transmittance			
		Single Pane		Double Pane*	
		Visual (VT)	Solar	Visual (VT)	Solar
clear		.88	.78	.79	.62
blue green		.79	.55	.70	.43
grey		.51	.53	.45	.42
bronze,	light	.59	.55	.53	.43
	medium	.28	.34		
	dark	.06	.14		
solar blue		.58	.65	.51	.50
brown,	light	.55	.55	.49	.44
	medium	.28	.34		
	dark	.09	.16		
translucent white				.58	.45
low emissivity	residential (R = 3.3)			.86	.71
	commercial (R = 7.1)			.57	.36
silica aerogel (R = 7)**				.85	.65
evacuated (R = 12)**				.79	.62
fiberglass insulated glass (R = 5)					
	crystal (ext and int)			.15	.12
	white (ext and int)			.10	.09
	bronze (ext) and white (int)			.04	.07

*inside pane is clear glass except as noted **experimental glazing, not in production

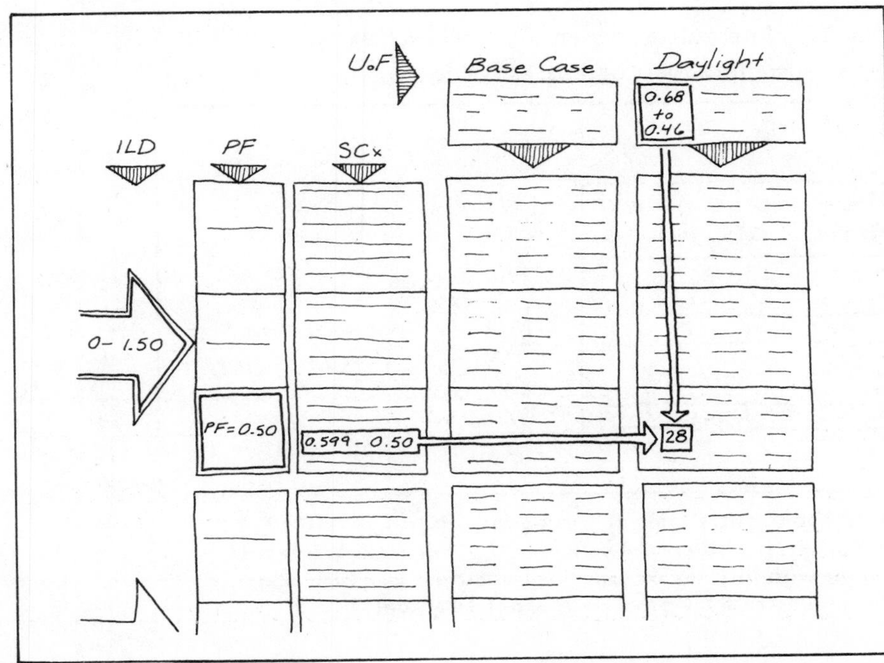

Figure E1.6.2
Procedure for Determining Maximum Allowable Percentage of Fenestration.

Enter ACP table at appropriate internal load density (ILD) range. Select projection factor (PF). Select shading coefficient (SC_x) range. Select U_oF range for your base-case or daylighted building. Use SC_x and U_oF to locate maximum percentage of fenestration value.

ENVELOPE THERMAL RESISTANCE (continued)

3. Determine the maximum U_{ow} for the opaque wall assembly. This establishes the maximum weighted average overall U-value for the non-glazed portion of all exterior walls. If lightweight framing is used, one U_{ow} is given. If walls with significant thermal storage capacity are used, the U_{ow} value must be adjusted for the heat capacity, insulation placement, percentage of fenestration, and internal load density range [Figure E1.6.3].

   a. Determine the heat capacity of your wall assembly using either this formula or the wall assembly table below.

   $HC = (r)(t)(C_p)$

   where:

   HC = heat capacity (Btu/ft²°F)

   r = density (lb/ft³) [*MEEB*, Table 4.2, pp.136–142]

   t = wall thickness (ft)

   C_p = specific heat (Btu/lb°F) [*MEEB*, Table 4.2, pp.136–142]

Wall Assembly	HC(Btu/ft²°F)
Wood or metal stud frame with gypsum board or paneling inside and wood siding or stucco outside	1
2" masonry with gypsum board inside and stucco outside	5
4" masonry	8
8" masonry	16

   b. Use the ACP table to find U_{ow}. Interpolate between PCT FEN values, but **NEVER EXTRAPOLATE**—use the largest or smallest value given.

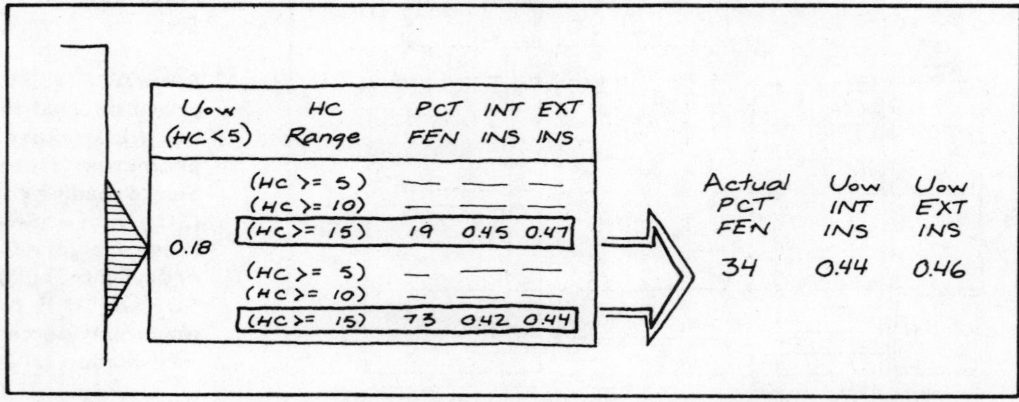

Figure E1.6.3 Detail of ACP Table. Enter table at appropriate HC range (HC < 5, ≥ 5, ≥ 10, ≥ 15). For HC < 5, use listed value for U_{ow}. For HC≥ 5, select U_{ow} values for walls with interior or exterior insulation. Use high and low percentage of fenestration figures to interpolate actual U_{ow}.

DESIGN DEVELOPMENT

ENVELOPE THERMAL RESISTANCE (continued)

E1.6

4. Using the values in the lower right corner of the ACP table, determine other envelope criteria (e.g., R-values of wall below grade or unheated slab-on-grade, U-values of roof, walls adjacent to unconditioned spaces, and floor over unconditioned space).

Envelope Design Criteria [suggested format]

Wall HC (Btu/ft^2°F)	U_{ow}	R-Value Below Grade Wall	R-Value Slab	U_o Roof	U_o Floor	U_o Wall Adjacent to Unconditioned Space
8	*.45*	*NA*	*NA*	*.08*	*.13*	*.29*

5. Based on your analysis of the ASHRAE standard and your decisions on how to design your building envelope to comply with that standard, draw typical sections (scale: 1" = 1'–0"). Indicate the assemblies, materials, and R-values for each of the following connections:

- non-south wall at window opening
- south wall at window opening
- roof at wall
- floor at wall
- thermal mass at floor
- thermal shutters (if you used them) [*MEEB*, Table 6.6, pp.314–315].

> HINT: Show how thermal bridges are avoided and where vapor barriers are installed. See *MEEB*, Tables 4.5–4.13, pp.147–156, for typical construction details.

6. Discuss the success of your thermal resistance strategies, and indicate any design changes that you view as necessary to meet Goal C.

E1 THERMAL

DESIGN DEVELOPMENT

HEAT LOSS CALCULATION

E1

E1.7

Discussion

Your building's heat gains will vary daily and seasonally according to its operating schedule and the intensity of the sun. Heat loss will change with respect to only one variable—temperature difference between the inside and the outside. In fact, heat loss is quite constant for each degree of temperature difference (Btuh/ft^2°F is constant). The major components of heat loss are: (1) loss by conduction through the skin of the building and (2) loss through ventilation or infiltration of outside air. The conductive heat loss can be calculated for the building envelope you designed in E1.6. The ventilation loss can be calculated based on the prescribed ventilation rate. The building's infiltration rate can be estimated based on its method of construction.

The basic formula for calculating conductive heat loss through the skin of the building is

$$HL = \sum [(U)\ (A)\ (\Delta T)],$$

where:

HL = hourly heat loss (Btuh)

U = U-value for each skin component assembly (Btuh/ft^2°F)

A = exterior surface area of each skin component (ft^2)

ΔT = difference between the inside air temperature and the outside air temperature (°F).

To calculate balance point temperature, it is more important to know the HL/°F.

$$HL/°F = \sum (U)\ (A)$$

The basic formula for calculating ventilation or infiltration heat loss is

$$HL = (ACH)\ (V)\ (.018)\ (\Delta T),$$

where:

ACH = number of air changes per hour

V = heated volume of the building (ft^3).

Remember, in order to calculate the balance point temperature, it is more important to know the HL/°F, thus solving the equation in the following manner:

$$HL/°F = (ACH)\ (V)\ (.018).$$

DESIGN DEVELOPMENT

HEAT LOSS CALCULATION (continued)

E1.7

PROCEDURE

1. Calculate the conductive heat loss rate (Btuh/°F) for each zone *[MEEB*, Table 4.16, pp.160–162 (window U-values); Table 4.15, pp.158–159 (door U-values)].

$$HL/°F = U_w A_w + U_r A_r + U_g A_g + U_d A_d + U_f A_f$$

(walls) (roof) (glass) (doors) (floors)

Slab-on-grade floors are a special case. All heat loss is assumed to occur through the exterior perimeter of the slab. The simplified formula for slab-edge heat loss is

$$HL_{se} / °F = (F_2)(L_{se}),$$

where:

HL_{se} = heat loss through slab edge (Btuh)

F_2 = heat flow coefficient per lineal foot of slab edge (Btuh/ft°F) [*MEEB*, Table 4.13, p.156]

L_{se} = length of slab perimeter (ft).

Conductive Heat Loss [suggested format]

Zone *A*				Floor Area *400 ft²*
Component	U-Value	Area	(U) (A)	Annotation
walls	*.022*	*370*	*81.4*	*E1.3*
roof	*.033*	*400*	*13.2*	*E1.3*
glazing–east	*.49*	*10*	*4.9*	*½" double-glazed, wood frame [MEEB, p.160]*
glazing–south	*.98*	*20*	*1.96*	*single-glazed, wood frame [MEEB, p.160]*
doors	*0*	*0*	*0*	
floor	*0*	*0*	*0*	
slab	*.64*	*40LF*	*25.6*	*R-5.4 insulation with 4" concrete slab*
TOTAL (Btuh/°F)			*127.1*	

HEAT LOSS CALCULATION (continued)

E1.7

2. Calculate the infiltration or ventilation heat loss rate (Btuh/°F) for each zone of your building.

   a. If your building type requires forced ventilation, it will be listed in *MEEB*, Table 4.25, pp.172–178. To convert the values in CFM to ACH,

   ACH = (60 / V) (CFM),

   where:

   V = heated volume of the building, zone, or room (ft^3).

   b. You may reduce the effect of forced ventilation during the heating season by using an air-to-air heat exchanger. A fair assumption is that the heat exchanger is 80% efficient—it only loses 20% of the heat when transferring it from outgoing air to incoming air.

   $$ACH_{he} = (ACH)\,(1 - e_{he})$$

   where:

   ACH_{he} = revised value of the ACH

   e_{he} = heat exchanger efficiency (.80 is a good assumption for efficiency)

   c. If your building does not require forced ventilation, infiltration is a function of tightness of construction. Select an ACH value [table below or from *MEEB*, Table 4.26, parts A–C, pp.179–180].

Rule-of-Thumb for Infiltration

To safely achieve lower and lower values of infiltration, a series of strategies is usually used. Little information is available on the costs, but the following scenario is a reasonable possibility for a 1,500 ft^2 house.

ACH	Strategy	Incremental Cost Increase	**TOTAL** Cost
1.5	practice standard construction methods	na	$0
1.0	add plastic vapor barrier	$125	125
0.5	seal joints, foam cracks, and take more care	375	500
0.375	add window heat recovery unit and take more care	250	750
0.3	add another window unit and take more care	250	1,000
0.25	add heat exchanger, 70% efficient	250	1,250
0.214	add better heat exchanger, 80% efficient	250	1,500
0.167	add better heat exchanger, 90% efficient	250	1,750

Reprinted from Balcomb *et al., Passive Solar Design Handbook* 3:23–24.

HEAT LOSS CALCULATION (continued)

Infiltration Heat Loss [suggested format]

Zone *A* Size *(400 ft²) (10 ft high walls)*

Forced Ventilation? (yes/no)	Heat Exchanger? (yes/no)	ACH	V (ft^3)	HL/°F (Btuh/°F)
(yes)@(35CFM)	*(yes)@80% efficiency*	*0.105*	*4,000*	*7.56*

3. Summarize your building's heat loss.

Heat Loss Summary [suggested format]

Zone	Conductive Loss (Btuh/°F)	Infiltration Loss (Btuh/°F)	**TOTAL** (Btuh/°F)
A	*127.1*	*7.6*	*134.7*
B	*75.4*	*61*	*136.4*
C	*24.3*	*32*	*56.3*
TOTAL			***327.4***

CALCULATION OF ZONAL BALANCE POINT TEMPERATURES

E1.8

Discussion

Determine the heating and cooling requirements of each thermal zone, daily and seasonally, to further develop appropriate design strategies. This calculation is much more accurate than the estimate you made in C1.5, and it incorporates the design refinements you have made.

Balance point temperature is the outdoor temperature at which the zone loses heat at the same rate it generates heat. For example, if the balance point temperature is 40°F, the zone must be cooled when the outside temperature is over 40°F and heated when it's under 40°F. The balance point temperature depends on the rates of heat loss and heat gain of the zone, and since these change daily and seasonally, so does the balance point temperature.

For more information

SWL, Sections 18–19, pp.52–62.

Procedure

1. Make a copy of the balance point temperature chart [Figure E1.8.1] for each thermal zone of your building.

2. For the times of day on the chart below, assign minimum and maximum seasonal temperature data. (These are the typical times when minima—just before sunrise—and maxima—halfway between noon and sunset—occur.) Plot these points on the balance point temperature charts for each season, and connect the points with a sine curve.

Month	Time	Minimum Temp	Time	Maximum Temp
January	6a.m.		2p.m.	
March	5a.m.		3p.m.	
July	4a.m.		4p.m.	
September	5a.m.		3p.m.	

3. For each zone calculate the morning, noon, and afternoon balance point temperature difference (ΔT) between inside and outside.

   $$\Delta T = HG / (HL/°F)$$

   where:

   HG = zone's heat gain [E1.4]

   HL/°F = zone's rate of heat loss [E1.7]

DESIGN DEVELOPMENT

CALCULATION OF ZONAL BALANCE POINT TEMPERATURES (continued)

4. Set an internal temperature, T_{in}, or thermostat setting for each season, and plot it on the balance point temperature charts. (We recommend that you use 65°F in the winter when the building is occupied, 50°F in the winter when unoccupied, and 80°F in the summer at all times.)

5. Calculate each zone's balance point temperatures for peak and off-peak hours and for each season. Record your findings on the chart below.

   balance point temperature = $T_{in} - \Delta T$

Balance Point Temperature Data (for each zone for each season) [suggested format]

Zone	Season	Hours	T_{in}	ΔT	Balance Point Temperature
	winter				
A	spring/fall				
	summer				
	winter				
B	spring/fall				
	summer				
	winter				
C	spring/fall				
	summer				

6. Plot the balance point temperatures at the midpoint of each peak or off-peak time span. Connect the balance point temperatures with dashed lines.

7. Color the area between the external temperature curve and the balance point temperature curve **RED** if heating is required and **BLUE** if cooling is required.

8. Compare your calculated balance point temperatures with your estimated balance point temperatures [C1.5]. Comment on how and why they differ.

CALCULATION OF ZONAL BALANCE POINT TEMPERATURES (continued)

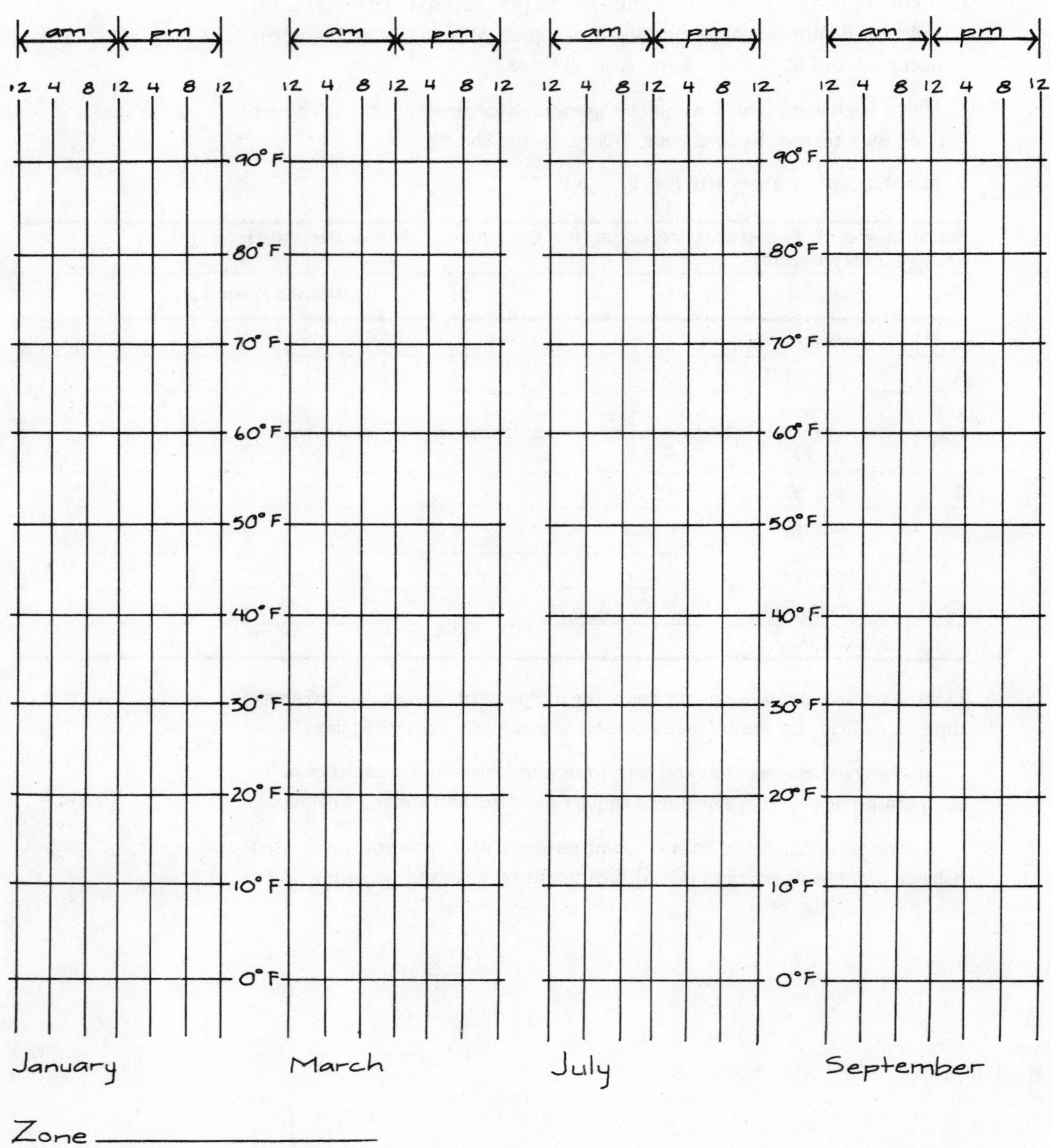

Zone ____________________

Figure E1.8.1 Balance Point Temperature Charts

DESIGN DEVELOPMENT
DESICCANT COOLING

Discussion

The desiccant enhanced nocturnal radiator is a passive cooling system for hot, humid climates and is currently being researched and developed at the Florida Solar Energy Center. The system is capable of providing both cooling and dehumidification. It will accommodate a cooling load of at least 70% greater than the sensible cooling load to account for the removal of water vapor (latent heat) from the air. The system requires the building to be "closed" both day and night, while the nocturnal radiator is coupled to the building for cooling at night and uncoupled for regeneration during the day. The radiator should have an unobstructed view of the sun and sky during the cooling season (Fairey, Vieira, and Keresteciogiu 1985).

Desiccant Desorption Mode

The desiccant desorption mode occurs during the daylight hours. The roof radiator is thermally uncoupled from the building. For about six hours, outside air is passed over the desiccant in the radiator to remove water from it. The temperature in the radiator reaches about 140°F. It is necessary to provide a radiant barrier between the radiator and the building, as well as a thermal barrier to isolate this high temperature space. For the next four hours the radiator is allowed to cool to about 80°F while outside air flow is cut off. This cooling will occur in late afternoon and early evening when radiation incident on the radiator is low.

During these 10 hours, the inside air is circulated through a hygroscopic, thermal storage space to remove heat and humidity. The Florida Solar Energy Center suggests the hollow cores of concrete masonry units as an efficient storage space. These masonry units could be in interior or exterior walls as long as they are within the insulated skin of the building.

Desiccant Adsorption Mode

The desiccant adsorption mode occurs during the night. The roof radiator is coupled to the heat and moisture storage space and to the building's interior. Air is circulated over the desiccant to remove moisture. The working temperature for this process is 82–85°F which is maintained by the heat of adsorption. The dry air is further cooled to 70–76°F in the lower half of the radiator, which has no desiccant. The cool, dry air is then passed through the thermal storage where it gains heat and moisture to complete the cycle. The resultant interior conditions should be about 70–72°F and 50% RH.

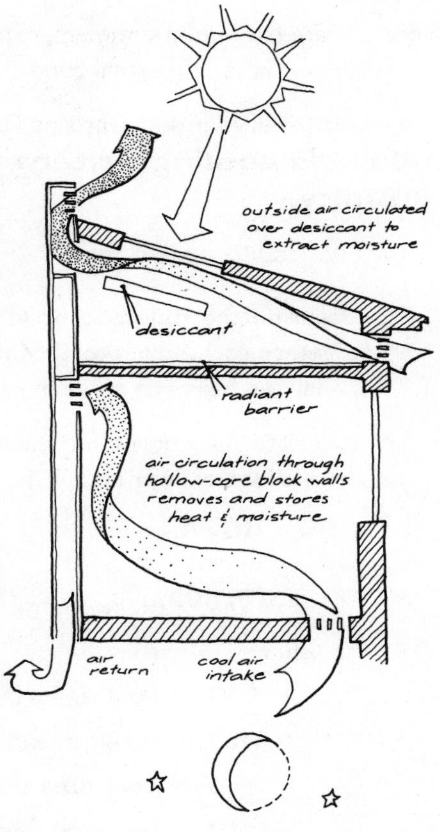

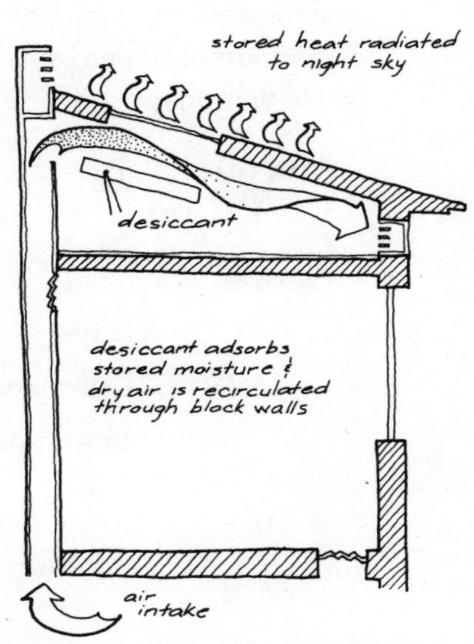

Figure E1.9.1 Desiccant System Schematic. ***Above,*** **desorption mode (daytime);** ***below,*** **adsorption mode (nighttime). Linen supply office building, Charleston, South Carolina.**

DESICCANT COOLING (continued)

Procedure

If your climate has humid summers, explore using the desiccant enhanced nocturnal radiator in at least one thermal zone.

Evaluate how well your design attains Thermal Design Development Goal D—**The building's cooling system effectively removes heat and humidity.**

For each desiccantly cooled zone:

1. Calculate the daily cooling load for the worst cooling month by adding together each hour's total heat gain. (This total is represented by the area under the heat gain curve in E1.5.)

2. Calculate the cooling load density and required amount of desiccant per square foot of floor area.

   $$HGD = HG / A$$

   $$D = (HGD)(.021)$$

   where:

   HGD = cooling load density (Btu/dayft^2)
   HG = cooling load (Btu/day) [step 1]
   A = floor area (ft^2)
   D = required desiccant (lb/ft^2)

3. Size the collector and the depth of desiccant. For proper operation, the radiator must be sloped with the upper half layered with desiccant and the lower half without desiccant. The desiccant layer thickness must not exceed 2".

   $$T = \frac{(D)(A)(12''/ft)}{(DD)(\tfrac{1}{2}RA)}$$

   where:

   T = desiccant thickness (inches)
   DD = desiccant density (45 lb/ft^3)
   RA = total radiator area (ft^2)

DESIGN DEVELOPMENT

DESICCANT COOLING (continued)

Desiccant Enhanced Nocturnal Radiator Sizing [suggested format]

Zone	Area (ft^2)	Cooling Load (Btu/day)	Load Density (Btu/day ft^2)	Required Desiccant (lb/ft^2)	Radiator Area (ft^3)	Desiccant Thickness (inches)
A	*900*	*38,400*	*43*	*0.90*	*1,100*	*0.4*

4. Discuss the success of your proposed desiccant cooling scheme. If it has met Goal D, draw a section through the building showing the air paths through the thermal storage and the connections between the radiator and the building.

ROOF-POND COOLING

Discussion

A roof pond is usually not a pond in the true sense: the water is contained in storage units (waterbed mattresses work well) on the roof. The pond must have a movable insulating cover to protect it during the day and must be in thermal contact with the space to be cooled (a metal roof deck between the pond and the space to be cooled works best). Roof ponds can only cool the space directly below them, which implies that they are useful only in one-story buildings. At latitudes above 36°N, a pitched roof with a solar aperture is needed over the roof pond. For space heating, the insulating cover can be removed in the daytime and replaced at night.

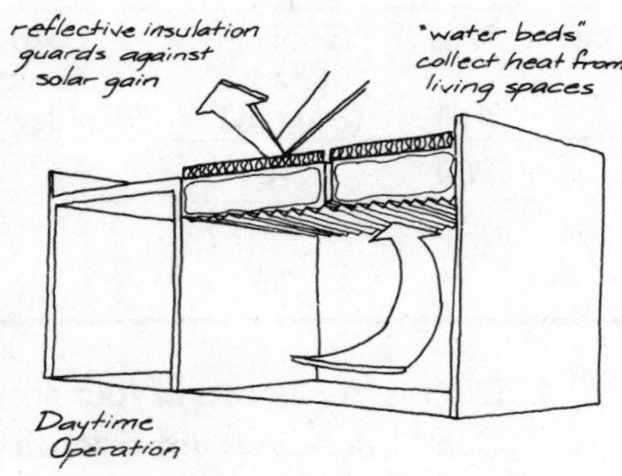

For more information

MEEB, Section 5.9(c), pp.271–274 (a complete example).

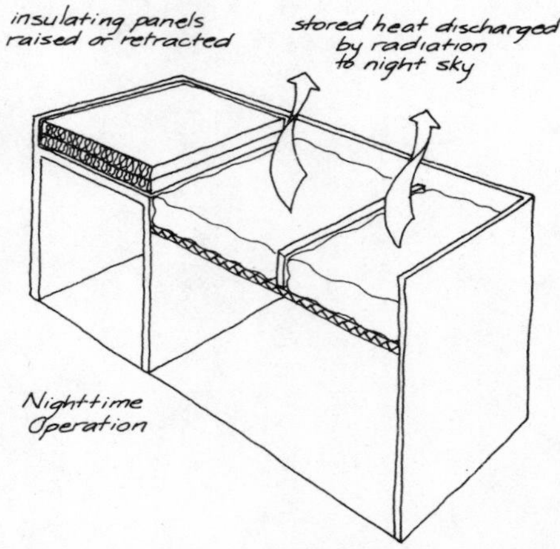

Figure E1.10.1 Roof-Pond Cooling System Schematic. *Above,* daytime; *below,* nighttime.

Procedure

If your climate has arid summers, explore using a roof pond cooling system in at least one thermal zone.

Evaluate how well your design attains Thermal Design Development Goal D—**The building's cooling system effectively removes heat and humidity.**

For each zone using roof-pond cooling:

1. Assemble climate data:

   a. maximum dry bulb temperature [*MEEB*, Table A.1, pp.1429–1437]

   b. mean daily summer temperature range [*MEEB*, Table A.1, pp.1429–1437]

   c. minimum dry bulb temperature [(maximum dry bulb temperature) – (mean daily range)]

   d. highest average relative humidity for July (usually at 6 a.m.) [Appendix C, Climatological Data]

   e. July average temperature [Appendix C, Climatological Data or *MEEB*, Table B.15, "TA," pp.1459–1463]

   f. minimum wet bulb temperature [*MEEB*, Figure 5.23 or Figure 4.8(a), Psychometric Chart, p.271 or 130]

   g. desired summer interior temperature, T_{op} [T_{in} in E1.8].

Climate Data [suggested format]

Max DB Temp (°F)	Mean Daily Temp Range (°F)	Min DB Temp (°F)	July RH (%)	July Avg Temp (°F)	Min WB Temp (°F)	Interior Temp (T_{op}) (°F)
91	*27*	*64*	*55*	*79*	*59*	*80*

2. Calculate the daily external heat, Q_E (Btu), to be stored in the roof pond.

   $$Q_E = [(HG_p)(N)] / (31 \text{ days})$$

   where:

   HG_p = heat gain peak from glass and skin in July (Btuh) [E1.4]

   N = number of average July operating hours for residential air conditioners [*MEEB*, Figure 5.24, p.271]

3. Calculate the daily internal heat, Q_I (Btu), to be stored in the roof pond. This value is the daily total gain from people, lights, and equipment [E1.4].

   $$Q_I = (\text{hourly internal heat gain})(\text{daily hours of occupancy})$$

   NOTE: Q_I is more accurately calculated as the area under the summer internal heat gain curve during the hours of occupancy [E1.5].

4. Calculate the daily heat gain through the insulated cover of the roof pond, Q_P (Btu). The formula assumes an R-16 cover with a white upper surface and foil-faced lower surface.

   $$Q_P = (0.4)(A_c)[4(DB_{max}) - DB_{min} - 200]$$

   where:

   A_c = surface area of the pond (ft^2)

   DB_{max} = maximum dry bulb temperature (°F)

   DB_{min} = minimum dry bulb temperature (°F)

5. Determine the maximum internal temperature, T_{imax}, based on climate and the use/non-use of a fan. Choose the value of F that corresponds to the climate whose temperature range approximates your climate's temperature range.

$$T_{imax} = T_{op} + F$$

where:

T_{op} = desired summer interior temperature [step 1g]
F = temperature differential [*MEEB*, Values of F, p.272]

6. Determine the heat transfer coefficient, h.

	Air Velocity			
	0 fpm	44 fpm	73 fpm	115 fpm
h	1.25	1.47	1.53	1.70

7. Calculate temperature swings. (If neither ΔT_{dry} nor ΔT_{wet} is greater than zero, a roof pond cannot cool effectively.)

   a. For maximum pond temperature:

   max pond T = HG_{peak} / [(h) (A_c)],

   where:

   HG_{peak} = peak hourly heat gain (external + internal) (Btuh) [E1.5]
   h = heat transfer coefficient
   A_c = pond surface area (ft²).

   b. For a pond that does not use external water to cool itself evaporatively:

   min pond T_{dry} = (DB_{min}) + (1.5) ± (corrections)

   and

   ΔT_{dry} = max pond T – min pond T_{dry},

   where:

   corrections = corrections for temperature values [*MEEB*, p.272 (bottom of page)].

Roof Pond Sizing: Part One [suggested format]

Zone	Pond Area (ft²)	Q_E (Btu)	Q_I (Btu)	Q_P (Btu)	T_{imax} (°F)	F (°F)	h	HG_{peak} (Btuh)	Corrections (°F)
A	*2,000*	*24,000*	*36,100*	*80,000*	*82*	*2*	*1.25*	*37,000*	*0*

ROOF-POND COOLING (continued)

c. For a pond that uses external water:

$$\textbf{min pond } T_{wet} = DB_{min} - [(DB_{min} - WB_{min}) / 2]$$

and

$$\Delta T_{wet} = \textbf{max pond } T - \textbf{min pond } T_{wet}.$$

8. Determine the required pond depth, D.

$$D = [(0.19)(Q_E + Q_I + Q_P)] / [(\Delta T)(A_c)]$$

> NOTE: D = 4" is optimal. If D < 2", the pond needs to be smaller, and if D > 10", the pond needs to be larger.

Roof Pond Sizing: Part Two [suggested format]

Max Pond T (°F)	Min Pond T (dry or wet, °F)	ΔT (dry or wet, °F)	Depth (inches)
67	*(dry) 65.5*	*(dry) 1.5*	*19.45*

9. Discuss the success of your proposed roof pond. If it has met Goal D, draw a section of the building that shows how the roof pond is integrated with the design and how it operates.

E1 THERMAL

E1

DESIGN DEVELOPMENT

EVAPORATIVE COOLING

E1.11

Discussion

Evaporative cooling is not a passive cooling strategy. Rather, it is a low-energy cooling strategy for hot, arid climates that offers the advantage of not using ozone-attacking coolants. Evaporative cooling is done adiabatically—the total sensible heat plus latent heat remains constant—by absorbing sensible heat through the evaporation of water. Dry air at a high temperature is transformed into moderately humid air at a lower temperature. Roof-mounted evaporative coolers (also called swamp coolers) require both a blower to force chilled air into the building and ventilation to allow warm exhaust air to escape.

Procedure

If your climate has arid summers, explore using evaporative cooling in at least one thermal zone.

Evaluate how well your design attains Thermal Design Development Goal D—**The building's cooling system effectively removes heat and humidity.**

For each evaporatively cooled zone:

1. Record the worst-case (peak) heat gain density [E1.4].

   HD = HG / area

2. Find the summer dry bulb (DB) and wet bulb (WB) temperatures for your location [*MEEB*, Table A.1, pp.1429–1437].

3. Calculate the potential cooling provided by evaporative cooling [*MEEB*, Figure 5.11, p.221].

Evaporative Cooling Check [suggested format]

Zone	Peak Heat GainDensity (Btuh/ft²)	Summer DB Temp (°F)	Summer WB Temp (°F)	Potential Cooling (Btuh/ft²)
A	*32*	*95*	*65*	*23*

4. Compare the cooling load to the potential cooling provided. Explain why evaporative cooling is or is not appropriate for your building. If evaporative cooling is an appropriate strategy to meet Goal D, draw a building section that shows how the cooling works and how it is integrated with the building.

EARTHTUBE COOLING

Discussion

Earthtube cooling can provide passive or hybrid (using a blower) cooling. It uses moist earth as a heat sink for cooling incoming fresh air. Three variables—the length, depth, and diameter of the tube—determine the effectiveness of the system. In general, several small diameter tubes are more effective than one large tube. Moist soil conducts heat away from the tube. In a very efficient system the air temperature in the tube can get to within 4°F of the ground temperature. Earthtubes can easily be integrated with earth sheltering and berming. As with all ventilation strategies, precooling intake air will enhance cooling. The calculations in this procedure assume an airflow rate of 500 feet per minute.

Procedure

If your climate has humid summers or moist earth in summer, explore using earthtube cooling in at least one thermal zone.

Evaluate how well your design attains Thermal Design Development Goal D—**The building's cooling system effectively removes heat and humidity.**

For each earthtube cooled zone:

1. Calculate the ambient summer ground temperature at the average depth of the earthtubes.

   $$T_{SG} = (T_{GW}) + (T_{AMP})\,[1 - (D / CF)]$$

   where:

   T_{SG} = ambient summer ground temperature (°F)

   T_{GW} = ground water temperature (°F) [*MEEB*, Figure 5.25, p.275]

   T_{AMP} = temperature amplitude (°F) [*MEEB*, Figure 4.4, p.122]

   D = average depth of earthtube (ft)

   CF = correction factor for soil type

   - 14 for dry soil
   - 18 for average soil
   - 22 for moist soil

   NOTE: When $D \geq CF$, $T_{SG} = T_{GW}$.

EARTHTUBE COOLING (continued)

2. Calculate the difference between the intake air and the ambient ground temperatures. Assume the worst case—use the summer design dry bulb temperature [*MEEB*, Table A.1, pp.1429–1437].

$\Delta T = T_{SD} - T_{SG}$

where:

T_{SD} = summer design dry bulb temperature (°F)

3. Calculate the amount of cooling performed by the earthtubes.

$C = (1.3)\ (\Delta T)\ (L)\ (D / 8)$

where:

L = total length of all tubes (ft)
D = tube diameter (inches)

4. Find the worst-case cooling requirements. Use the peak hourly heat gain (Btuh) [E1.4].

Earthtube Cooling Schedule [suggested format]

Zone	T_{GW}	T_{AMP}	Tube Dimensions			CF	T_{SG}	T_{SD}	ΔT	Cooling	Cooling
			Depth	Diam	Length					Provided	Required
	(°F)	(°F)	(ft)	(inches)	(ft)		(°F)	(°F)	(°F)	(Btuh)	(Btuh)
A	*52*	*19*	*12*	*12*	*4 @ 60*	*22*	*60.6*	*90*	*29.4*	*13,759*	*3,454*

5. Compare the cooling provided to the cooling required. Discuss how the use of earthtubes should be implemented in your design. If earthtube cooling has met Goal D, draw a section of the building that shows how the earthtubes are integrated with the design and how they operate.

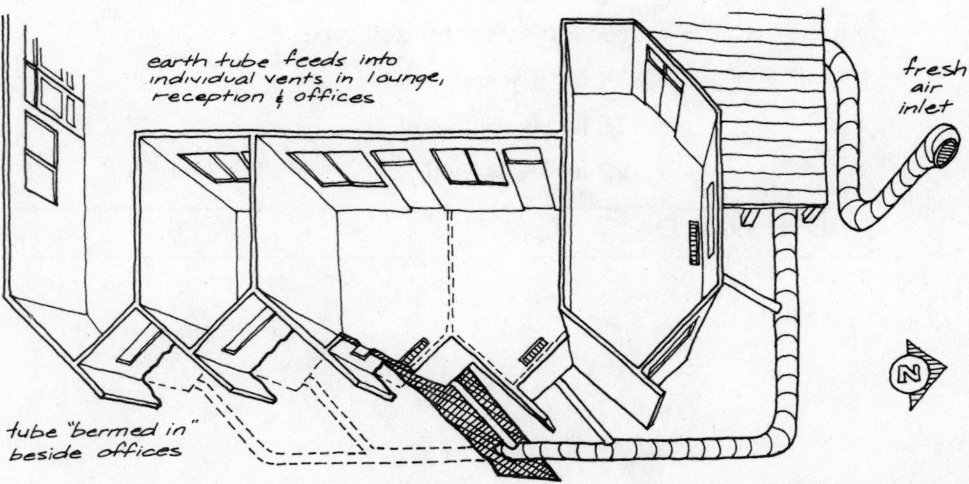

Figure E1.12.1 Earthtube Cooling Scheme. Linen supply office, Charleston, South Carolina.

DESIGN REVIEW

E1.13

Discussion

You have just completed design development of the thermal components of your building. Even though the design decisions have been based on an evolving body of information about the building's thermal needs, conflicting decisions may have been made. This design review affords you the opportunity to view your thermal design holistically and to address the conflicts and tradeoffs you have encountered.

Procedure

1. Compare your developed design with your schematic design [D1]. Discuss how it has evolved and why.
2. You have explored the use of external and internal shading devices, thermal zoning, design of the building's skin, and various cooling strategies. Comment on what tradeoffs have been made and what strategies worked well together.
3. Illustrate your discussion of points (1) and (2) with an annotated, schematic diagram of your building and site.

F1 THERMAL

THERMAL INTEGRATION

F1

INTRODUCTION

F1.0

GOAL

Select auxiliary thermal systems that are sensitive to the thermal and aesthetic needs of your building.

THERMAL INTEGRATION GOALS

A. The backup heating and cooling systems are sized to accommodate the thermal needs of the building.

B. The backup heating and cooling systems are appropriate for the auxiliary thermal needs.

C. Backup thermal systems are efficiently integrated with the structural and spatial schemes of the building.

DISCUSSION

All buildings designed for passive heating and cooling will inevitably require mechanical systems to supplement their heating or cooling. Care must be taken to integrate both passive and mechanical systems to prevent inefficiency and thermal conflict.

Several design questions are at issue in the choice of backup thermal systems:

- Is the system sized efficiently to complement the passive systems?
- Is the system energy- and cost-efficient?
- Does the system thermally complement the passive systems, use similar components, or provide thermal energy where it is most needed?
- Is the system integrated with the spatial and structural framework of the building?

IN THIS SECTION YOU WILL:

1. Address issues of systems integration at both the small and large scales.
2. Calculate backup heating and cooling loads for your building.
3. Specify backup thermal systems for your building.
4. Demonstrate how the backup thermal systems are integrated with your design.

THERMAL INTEGRATION DESIGN

Procedure

Refine your thermal design development [E1] to include auxiliary heating and cooling systems. Generate schematic drawings illustrating your design.

Document your design as follows:

1. Floor plans (and roof plans, if necessary), illustrating heating and cooling equipment placement and distribution trees (scale: 1" = 20').
2. Sections, illustrating heating and cooling equipment placement and distribution trees (scale: 1/8" = 1'-0").
3. Details of system components (e.g., air grilles, fan-coil units, or other typical in-space components).
4. Design diagram, annotated to identify thermal zones and HVAC system design strategies.

THERMAL INTEGRATION

BACKUP HEATING SYSTEM SIZING

Discussion

The "conservative" approach to backup heating system sizing is based on maximum heat loss figures and assumes no contributions from either solar or internal gains. This approach ensures ample supplies of auxiliary heat under extremely cold and cloudy conditions, but the heating units are oversized and may never be used to full capacity.

The "reasonable risk" approach sizes heating systems based on smaller heat loss figures. This approach assumes that buildings with reliable internal gains (people, lights, and equipment that consistently supply some of the required space heating) can be fitted with smaller heating units.

Procedure

Evaluate how well your design attains Thermal Integration Goal A—**The backup heating and cooling systems are sized to accommodate the thermal needs of the building.**

For each thermal zone:

1. Find the winter design dry bulb temperature for your location [*MEEB*, Table A.1, pp.1429–1437], and calculate ΔT.

   $\Delta T = T_{INT} - DB$

   where:

   T_{INT} = desired winter internal temperature (°F) [E1.8, step 5]

   DB = winter design dry bulb temperature (°F)

2. Calculate maximum hourly heat loss, HL_{max}.

   $HL_{max} = (HLR)\ (\Delta T)$

   where:

   HLR = total heat loss rate (Btuh/°F) [E1.7, step 3]

3. Find the total hourly internal heat gain (from people, lights, and equipment) [E1.4]. Calculate average internal heat gain, HG_{AI}, by averaging the total internal heat gain over 24 hours.

BACKUP HEATING SYSTEM SIZING (continued)

4. Calculate reasonable risk hourly heat loss, HL_{RR}.

$$HL_{RR} = HL_{max} - HG_{AI}$$

Heating Loads [suggested format]

Zone	Heat Loss Rate (Btuh/°F)	ΔT (°F)	HL_{max} (Btuh)	HG_{AI} (Btuh)	HL_{RR} (Btuh)
A	*521*	*45*	*23,445*	*17,505*	*5,940*

5. Choose a backup heating system capacity based on either conservative or reasonable risk heat loss figures. Justify your choice of strategy, and discuss how you have met Goal A.

THERMAL INTEGRATION

F1

BACKUP COOLING SYSTEM SIZING

F1.3

Discussion

A mechanical cooling system should be sized to handle peak internal, solar, and building-skin heat gains. It also should remove enough moisture from the air (the latent heat) to prevent an uncomfortable increase in humidity at the lowered temperature. Removing latent heat increases the cooling load. For typical office or commercial buildings, the latent heat gain may be estimated as 30% of the sensible heat gain; food preparation and other moisture-generating activities will cause higher latent heat gain.

Procedure

Evaluate how well your design attains Thermal Integration Goal A—**The backup heating and cooling systems are sized to accommodate the thermal needs of the building.**

For each thermal zone:

1. Find maximum hourly heat gain, HG_{max}, as indicated by the heat gain plots [E1.5]. Indicate the time of day and year the peak occurs.
2. Calculate the total cooling load (sensible plus latent heat gains).

   $$\textbf{cooling load} = HG_{max} + [(0.3)\ (HG_{max})]$$

   NOTE: Where high moisture-generating activities occur, triple the latent heat gain estimate or use *MEEB*, Table 5.24, p.254.

3. Determine the required refrigeration capacity.

   1 ton refrigeration capacity = 12,000 Btuh cooling load

Cooling Loads [suggested format]

Zone	Maximum Heat Gain (Btuh)	Peak Time (hr/mo)	Latent Heat Gain (Btuh)	Cooling Load (Btuh)	Refrigeration Capacity (tons)
A	*24,570*	*4p.m./Aug*	*12,285*	*36,855*	*3.07*

THERMAL INTEGRATION

HVAC SYSTEM CHOICE

Discussion

Analysis of the building's auxiliary heating and cooling loads and use schedules will help ascertain whether a central or local backup thermal system is appropriate for each zone. Specific systems can then be chosen to suit the building's thermal and spatial characteristics.

"Central" systems provide heating and cooling for all thermal zones by means of a single equipment package. These systems require large mechanical spaces or rooms (typically at basement or roof level) and large distribution trees of air ducts and/or water pipes. Centralized equipment is easy to maintain and simplifies control of machine heat and noise. It also presents many opportunities for energy conservation (e.g., waste heat may be recycled from one zone to another or within equipment or distribution trees). However, since these systems operate most efficiently at or near peak capacity, it is expensive and inefficient to use central systems for thermal zones with very different heating and cooling schedules.

"Local" systems have heating and cooling equipment adjacent to the zone or task area served. These systems offer substantial first-cost savings by minimizing distribution trees. Energy efficiency and convenience may be enhanced by local control, however, there are fewer opportunities for recycling waste heat. Controlling maintenance, machine heat, and noise are all more difficult, but it may be easier to accommodate smaller, local units than to provide large mechanical rooms.

For more information

Allen and Iano, *The Architect's Studio Companion*, pp.189–218 (concise guidelines for choosing backup systems for small buildings).

MEEB, Section 6.3, pp.283–301 (extensive discussion of system design issues).

MEEB, Section 6.7, pp.341–381 (discussion of alternatives in small buildings).

MEEB, Table 7.1, pp.390–391 (typical backup system choices based on building type).

HVAC SYSTEM CHOICE (continued) F1.4

Table F1.4.1 HVAC Pros and Cons			
Type	**Pros**	**Cons**	**MEEB Reference**
Centralized Systems			
All Air	Best control of air quality	Largest distribution trees	Section 7.1(b), 384-386 Section 7.4, 429-444
Air-Water	Smaller distribution trees than "All Air" Central control of air quality	Several distribution trees required Larger distribution trees than "All Water"	Section 7.1(c), 386-388 Section 7.5, 444-452
All Water	Smallest distribution trees	Local fresh-air supply required No central control of air quality	Section 7.1(d), 388-389 Section 7.6, 452-459
Localized Systems			
Electrical Resistance (heating)	Low first cost	High-grade energy for low-grade task	344
Heat Pump (heating and cooling)	May incorporate solar assistance for increased efficiency	Poor heating performance at low, winter and high, summer temperatures	370-375
Air-Conditioning Units (cooling)	Low first cost	Poor efficiency; high noise output	375-377
Direct Evaporative Cooler (cooling)	Low operating cost	Limited to hot, arid climates	376-379
Make-Up Air Units (heating and cooling)	Appropriate for spaces with high exhaust air rates	Low efficiency	Section 7.1(a), 384 Section 7.2, 389-401

HVAC SYSTEM CHOICE (continued) F1.4

PROCEDURE

Evaluate how well your design attains Thermal Integration Goal B—**The backup heating and cooling systems are appropriate for the auxiliary thermal needs.**

1. Determine the probable auxiliary heating and cooling schedule for each thermal zone.

   a. Identify the months when internal gains may be insufficient to heat each zone [E1.8 and E1.5]. Remember, you are designing for the worst case (cold, cloudy days) when your passive solar system will be inoperative.

   b. Identify the months when outdoor temperatures are too high [T_{out} > (T_{in} – 3°F)] to allow successful cooling by ventilation [Appendix C, Climatological Data or D1.2].

   c. Record the peak auxiliary heat required [F1.2] and the auxiliary cooling required [F1.3].

Auxiliary Heating Schedule [suggested format)]

Zone	Heating Season	Peak Auxiliary Heat Required
A	*Oct–Mar*	*5,940 Btuh max*

Auxiliary Cooling Schedule [suggested format)]

Zone	Cooling Season	Auxiliary Cooling Required
A	*Aug*	*3.07 tons*

2. Identify the backup heating and cooling systems you have chosen to meet Goal B, and discuss how they are particularly suited to your thermal requirements. Describe any energy-conservation measures incorporated into your systems. Illustrate your discussion with schematic drawings.

THERMAL INTEGRATION

HVAC SYSTEM LAYOUT

Discussion

Once you've identified the appropriate mechanical heating and cooling systems for your building, select and integrate system components. Heating and cooling units should be chosen (from the manufacturer's data or Sweet's catalog) whose net ratings meet or exceed your required system capacities. Ratings for boilers, furnaces, and heating units are listed in Btuh or in square feet of radiation (1 ft^2 radiation = 240 Btuh). Ratings for cooling equipment are given in Btuh or tons. This equipment should be located in your building to minimize the size of the distribution system, facilitate maintenance, control noise, and easily vent waste heat. Access to outside air must be provided for combustion equipment, and storage areas must be provided for fuel oil or wood burning systems.

There are many options for distribution tree placement and design [Figure F1.5.1]. The benefits and applications of each option are discussed in *MEEB*, pp.293–298. Pipes and ductwork may be concealed to minimize noise and maximize control of the appearance of interior or exterior spaces. Exposing system components facilitates maintenance or system redesign and may provide visual interest. Here are some guidelines for the placement of system components.

- Place air supply registers away from return grilles.
- Counteract heat lost or gained through glazing. Place supply registers or radiators near windows.
- Locate thermostats in thermally stable places—protected from cold drafts and direct sunlight; away from supply registers, radiators, and other heat-producing equipment.
- Keep return air grilles out of bathrooms, kitchens, and other odor-, humidity-, or air pollution-producing spaces.
- Take advantage of the tendency of cold air to sink. In cooling-dominated zones, locate supply registers near the ceiling and return grilles near the floor.
- Take advantage of the tendency of hot air to rise. In heating-dominated zones, locate supply registers near the floor and return grilles near the ceiling.
- Channel return air through lighting fixtures to minimize heat gain from lights or to redistribute the heat for use elsewhere.

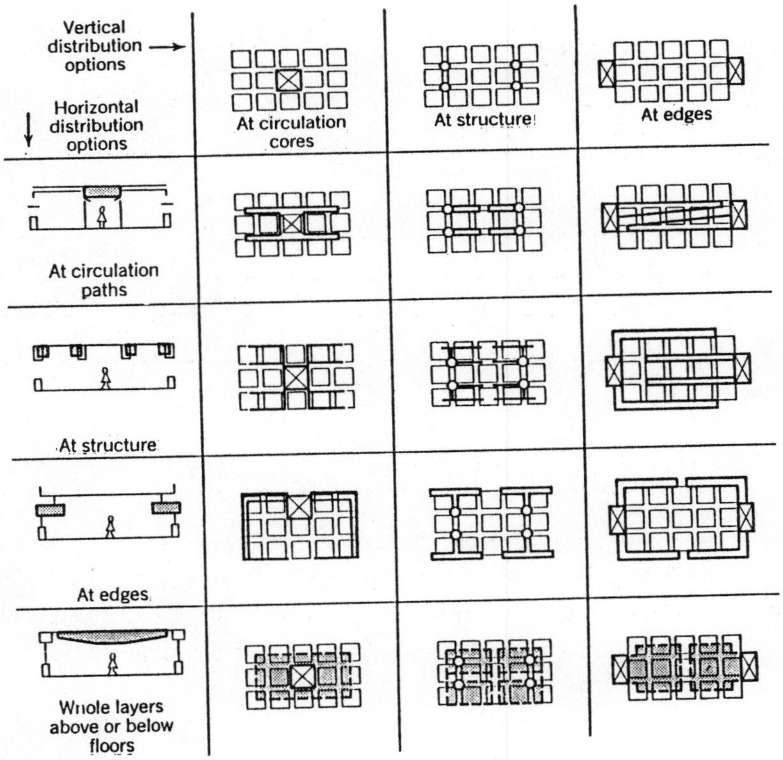

Figure F1.5.1 **Distribution Tree Placement Options. Vertical (impact on plan) and horizontal (impact on section). Reprinted, by permission, from Stein and Reynolds, *Mechanical and Electrical Equipment for Buildings*, 8th ed., 295, copyright © 1991, John Wiley & Sons, Inc.**

HVAC SYSTEM LAYOUT (continued)

PROCEDURE

Evaluate how well your design attains Thermal Integration Goal C—**Backup thermal systems are efficiently integrated with the structural and spatial schemes of the building.**

1. Identify your heating and cooling equipment, their capacities, and the source of your information.
2. Illustrate, in building plans and sections, the placement of all HVAC system components—boilers or furnaces, heat exchangers, cooling units, ductwork or piping, supply registers or radiators, return grilles, plenum spaces, thermostats, fan coils, or reheats. Color-code supply and return systems. Clearly identify all task areas.
3. Annotate your drawings. Describe how the system operates and any energy-conserving measures you have incorporated. Discuss the advantages and disadvantages of the system design and how well it meets Goal C.

G1 THERMAL

BUILDING EVALUATION

INTRODUCTION

G1

G1.0

GOAL

Use the appropriate evaluation methods to assess the building's final overall passive thermal performance.

THERMAL BUILDING EVALUATION GOALS

A. The building uses less fuel for heating in the winter than comparable, "energy-conserving" buildings in its climate.

B. When heating is required and the building is occupied, the inside temperatures are maintained between 68 and 80°F without the use of auxiliary fuel.

C. When cooling is required and the building is occupied, the inside temperatures are maintained between 68 and 80°F without the use of auxiliary energy.

D. The building will use less thermal energy than newly constructed, energy-conserving buildings.

DISCUSSION

Determining how well your design and its components perform is an important part of design development. Sometimes this evaluation is necessary to prove compliance with the building code. Sometimes it is necessary to help fine-tune the building design. Hand calculation methods, like those presented here, usually model the performance of specific elements of the building design. It is an extremely complex and time-consuming proposition to hand calculate the total, integrated performance of all the thermal and energy components of a building, but it is desirable to do these calculations. Now that microcomputers are a commonly used tool, the once onerous task of total energy performance evaluation has been eased by the use of computer-based analysis tools.

FOR MORE INFORMATION

MEEB, Appendix H, pp.1534–1539 (discussion of computer-based, thermal analysis tools).

IN THIS SECTION YOU WILL:

1. Complete the design development of your building to accomplish the thermal building evaluation goals.

2. Evaluate the passive solar performance of your building.

3. Evaluate the role of thermal mass in controlling temperature swings in the building during both the cooling and heating cycles.

4. Calculate the building's total thermal performance using a computer-based analysis tool.

G1 THERMAL

BUILDING EVALUATION

THERMAL DESIGN NARRATIVE

G1

G1.1

DISCUSSION

In the course of completing A1–F1, you have researched building precedents, analyzed the site and the program, designed a building, and evaluated its performance. Just as completing ALL those steps is important, it is equally valuable to assess this design process, particularly if this building was the first one you designed in response to thermal determinants in addition to aesthetic and programmatic considerations.

PROCEDURE

Source: D. A. Brentrup, University of North Carolina—Charlotte

Report on what you have accomplished so far and what you have learned. Be concise, coherent, and clear. Integrate illustrations with your discussions in a well-organized format. Include:

1. An introduction that defines your perceptions and understandings of thermal design before you began the exercises in this workbook.
2. A description of the evolution of (1) your design and (2) your awareness of thermal considerations. Cite and discuss examples that illustrate:
   a. your understanding of thermal contexts
   b. the tradeoffs between aesthetic and thermal design considerations
   c. the integration of heating and cooling systems
   d. detail design development that made the schematic design successful.
3. A conclusion that summarizes or highlights what you have learned. Comment on the effectiveness of this learning process.

HEAT LOSS CRITERIA

Discussion

The maximum heat loss criteria (MHLC) are standards for the rate of heat loss in energy-conserving buildings. MHLC are concerned only with the thermal performance of the whole building. They are based on a skin-dominated load building with a balance point temperature of about 65°F. MLHC must be adjusted upward as the balance point temperature decreases. This reduction in the standard heat loss rate is compensated by the higher internal heat gains.

Procedure

Evaluate how well your design attains Thermal Building Evaluation Goal A—**The building uses less fuel for heating in the winter than comparable, "energy-conserving" buildings in its climate.**

For each zone:

1. Adjust the maximum heat loss criteria if your building has a balance point temperature of less than 65°F.

   a. Using the average balance point temperature for January, calculate the adjusted maximum heat loss criteria.

   $$MHLC_{adj} = (MHLC)\,[(65°F - T_{Jan}) / (T_{BP} - T_{Jan})]$$

   where:

   $MHLC_{adj}$ = adjusted maximum heat loss criteria (Btu/HDDft²)
   $MHLC$ = maximum heat loss criteria (Btu/HDDft²) [heat loss criteria table]
   T_{Jan} = average January temperature (°F)
   T_{BP} = average January balance point temperature (°F)

NOTE: If the result of this calculation is negative, your building does not require heating.

Heat Loss Criteria

Heating Degree Days (HDD)	Maximum Heat Loss Criteria* (Btu/HDDft²)
<1,000	7.6
1,000–3,000	6.6
3,000–5,000	5.6
5,000–7,000	4.6
>7,000	3.6

* Standard for passive, solar-heated, residential buildings, exclusive of losses through solar apertures.

Reprinted from Balcomb et al., *Passive Solar Design Handbook*, 2:24.

HEAT LOSS CRITERIA (continued)

2. Calculate the total heat loss density per degree day.

   a. **MULTIPLY** the heat loss rate [E1.7, step 3] by 24 hours (per day) to obtain the daily heat loss rate (Btu/HDD).

   b. **DIVIDE** the daily heat loss rate by the zone's floor area to get the daily heat loss density (Btu/HDDft²).

Adjustment of Heat Loss Criteria [suggested format]

Zone	MHLC (Btu/HDDft²)	T_{BP} (°F)	T_{Jan} (°F)	$MHLC_{adj}$ (Btu/HDDft²)	Heat Loss Rate (Btuh)	Heat Loss Rate (Btu/HDD)	Heat Loss Density (Btu/HDDft²)
A	*5.6*	*43*	*27*	*13.3*	*205.8*	*4,939*	*6.18*

3. Compare the overall heat loss [step 2] to the adjusted maximum heat loss criteria [step 1].

4. Discuss how well your building envelope insulation strategy meets the maximum heat loss criteria, and indicate any design changes that you view as necessary to meet Goal A.

SOLAR HEATING

Discussion

Now that you have calculated the heat loss efficiency [G1.2], you will be able to determine whether passive solar energy would be an appropriate heating source. The solar savings fraction (SSF) compares the heating performance of your passively solar heated building with a non-solar, yet energy-conserving, building.

For example, if a solar building requires 25 units of energy and the energy-conserving building needs 70 units of energy, the difference is 45 energy units, which is 64% of the energy-conserving building's total requirement. The solar building has an SSF of 64%. However, due to its substantially greater glazing area (which causes more heat loss), the solar building needs more total heat (100 units) than the energy-conserving building.

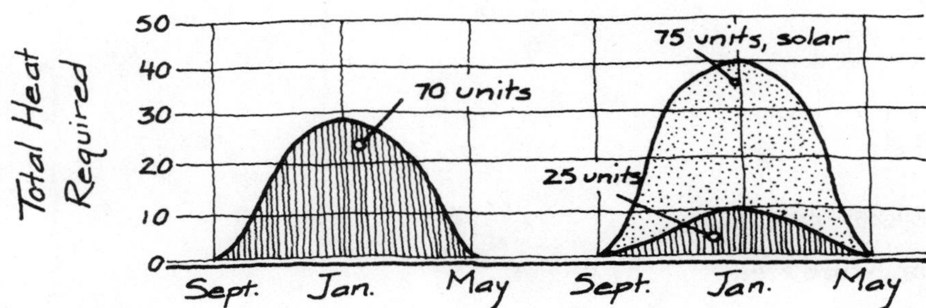

Figure G1.3.1 Comparison of Heating Requirements. *Left,* "conserving" building; *right,* nearly identical, solar-heated building.

A comparison between the building's need for heat to maintain comfort and the area of its heat-admitting, south-facing aperture is called the load collector ratio (LCR). The smaller the LCR (large solar window area in comparison to the heating load), the higher the SSF. If your SSF is greater than zero, your building performs better than an energy-conserving building.

For more information

Balcomb et al., *Passive Solar Heating Analysis* (complete SSF tables for the passive solar heating systems listed in *MEEB*, Table C.1).

SOLAR HEATING (continued)

PROCEDURE

Evaluate how well your design attains Thermal Building Evaluation Goal A—**The building uses less fuel for heating in the winter than comparable, "energy-conserving" buildings in its climate.**

For each zone using passive solar heating:

1. Determine the building load coefficient (BLC).

   a. Find the total daily heat loss rate, HL_T (Btu/HDD) [G1.2, step 2a].

   b. Calculate the south-facing solar aperture's daily heat loss rate (Btu/HDD).

   $\mathbf{HL_S = (U)\ (A)\ (24)}$

   c. Calculate the BLC.

   $\mathbf{BLC = [HL_T] - [HL_S]}$

2. Determine the load collector ratio (LCR).

   LCR = BLC / solar aperture area

3. Identify your passive heating system [*MEEB*, Table C.1, pp.1465–1467].

   For example, if you have a semi-enclosed sunspace with a masonry common wall, vertical glazing, and night insulation, the system designation is SS–C2.

4. Determine the SSF.

   Find the appropriate value [*MEEB*, Table C.2, p.1468]. If necessary, interpolate values.

   For the example described in step 3, located in Chicago, Illinois: if the LCR is 27, the SSF is 35%.

SSF Determination [suggested format]

Zone	Heat Loss Rate (Btu/HDD)	South Aperture Area (ft^2)	South Window Loss (Btu/HDD)	BLC (Btu/HDD)	LCR ($Btu/HDDft^2$)	Passive Heating System	SSF (%)
A	*4,939*	*137*	*1,235*	*3,704*	*27*	*SS–C2*	*35*

5. Discuss the success of your passive heating strategy. Does your building meet Balcomb's suggested SSF [Figure D1.7.1]? Indicate any design changes that you view as necessary to meet Goal A.

INTERIOR TEMPERATURE SWING

G1.4.1: HEATING

DISCUSSION

To maintain conditions in the comfort zone, walls or floors that have stored solar energy should provide radiant temperatures higher than 60°F. The average interior temperature is the sum of three temperature effects:

- average outside temperature (TA)
- change in air temperature due to the incidence of solar radiation (ΔT_{solar})
- change in air temperature heated by internal heat gains ($\Delta T_{internal}$).

PROCEDURE

Evaluate how well your design attains Thermal Building Evaluation Goal B—**When heating is required and the building is occupied, the inside temperatures are maintained between 68 and 80°F without the use of auxiliary fuel.**

For each thermal zone:

1. Find the January ambient outdoor temperature, TA [Appendix C, Climatological Data].
2. Find ΔT_{solar} [*MEEB*, Figure 5.17, p.246]. ΔT_{solar} depends on:
   a. latitude
   b. LCR [G1.3]
   c. type of passive solar strategy (if you are using a roof pond strategy, use water-wall values).
3. Calculate $\Delta T_{interior}$.

   $$\Delta T_{interior} = HG_{interior} / [(HL)(24)]$$

   where:

   $HG_{interior}$ = total daily internal heat gain (Btu/day) [E1.4]
   HL = total heat loss (Btuh/ft²) [E1.7, step 1a]

   NOTE: $\Delta T_{interior}$ averages 5–7°F for most residences.

4. Calculate the average January interior temperature.

   $$\text{avg Jan } T_{interior} = TA + \Delta T_{solar} + \Delta T_{interior}$$

   where:

   TA = January ambient outdoor temperature (°F) [step 1]
   ΔT_{solar} = temperature difference from solar energy (°F) [step 2]
   $\Delta T_{interior}$ = temperature difference from internal gains (°F) [step 3]

INTERIOR TEMPERATURE SWING (continued)

5. Calculate the interior temperatures.

   a. Calculate ΔT_{swing} [*MEEB*, Table 5.18, p.247].

   b. The average indoor temperature [step 4] is the midpoint of the temperature swing. Calculate the high and low interior temperatures.

   $T_{high} = T_{interior} + (\Delta T_{swing} / 2)$

   $T_{low} = T_{interior} - (\Delta T_{swing} / 2)$

Heating Interior Temperature Swing [suggested format]

Zone	TA	ΔT_{solar}	$\Delta T_{interior}$	Avg Jan $T_{interior}$	ΔT_{swing}	T_{high}	T_{low}
	(°F)	(°F)	(°F)	(°F)	(°F)	(°F)	(°F)
A	*37*	*27*	*5*	*59*	*8*	*63*	*55*

6. Does your building meet the criteria of staying within the 68–80°F temperature range on a clear, January day without mechanical heating? Discuss any mitigating circumstances or design changes that you deem necessary to meet Goal B.

BUILDING EVALUATION

INTERIOR TEMPERATURE SWING (continued)

G1.4.2: COOLING

DISCUSSION

Buildings which have ventilation apertures that must be closed during daytime hours, can be provided with a means of controlling daytime temperatures through night ventilation of the thermal mass. The heat stored in the thermal mass is removed by passing cool, outdoor air over it. Thus flushed of heat, the mass is cooled and ready by morning to absorb the day's heat gain. If winds are calm, an electric fan may be necessary to ventilate the mass. Energy used for such a fan is substantially less than energy used for mechanical cooling.

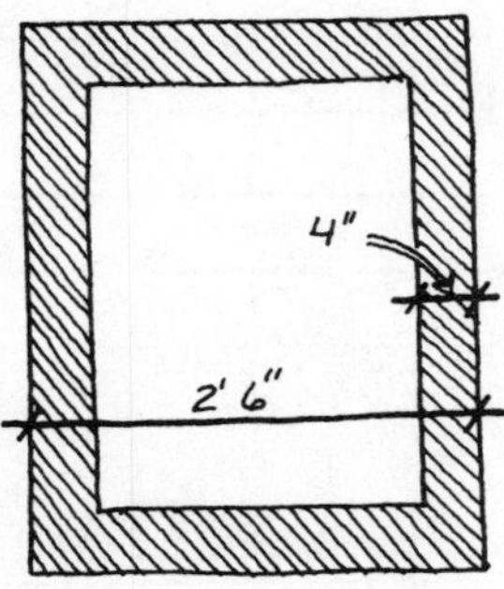

Figure G1.4.1 Effective Thermal Mass of Chimney. Plan view.

A building using night ventilation will be "closed" when the temperature outside is warmer than the building's mass temperature. It will be "open" after the outside temperature has peaked and dropped to a comfortable level (e.g., 80°F when the temperature crosses into the comfort zone or is cooler than the building's mass temperature).

FOR MORE INFORMATION:

MEEB, Section 5.9(a), pp.261–269 (discussion of night ventilation of thermal mass).

PROCEDURE

Evaluate how well your design attains Thermal Building Evaluation Goal C—**When cooling is required and the building is occupied, the inside temperatures are maintained between 68–80°F without the use of auxiliary energy.**

For each thermal zone that uses a "closed" or "mixed" strategy:

1. Graph the final heat gain patterns [E1.5].
   a. Plot heat gain patterns [Figure G1.4.2].
   b. Indicate the hours when the building is: (1) "closed" and heat is being stored and (2) "open" to ventilate the excess interior heat.
2. Calculate the total cooling load stored in thermal mass during the "closed" hours. This total is the sum of the Btus stored during each "closed" hour.
3. Calculate the total amount of mass (surface area and volume). This total should include the amount you have itemized for passive heating as well as all of the mass (4" or less thick) in your building. Even if the thermal mass is thicker than 4", use a 4" thickness for each exposed surface when calculating the volume of the mass [Figure G1.4.1].

BUILDING EVALUATION
INTERIOR TEMPERATURE SWING (continued)

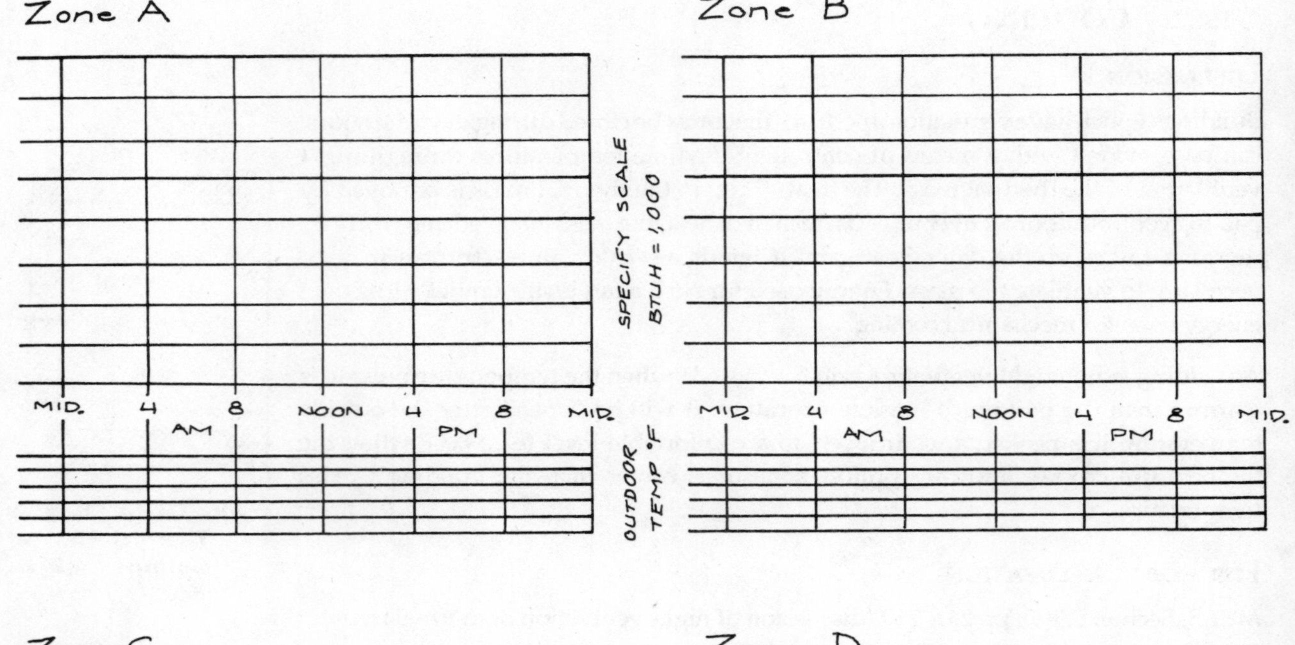

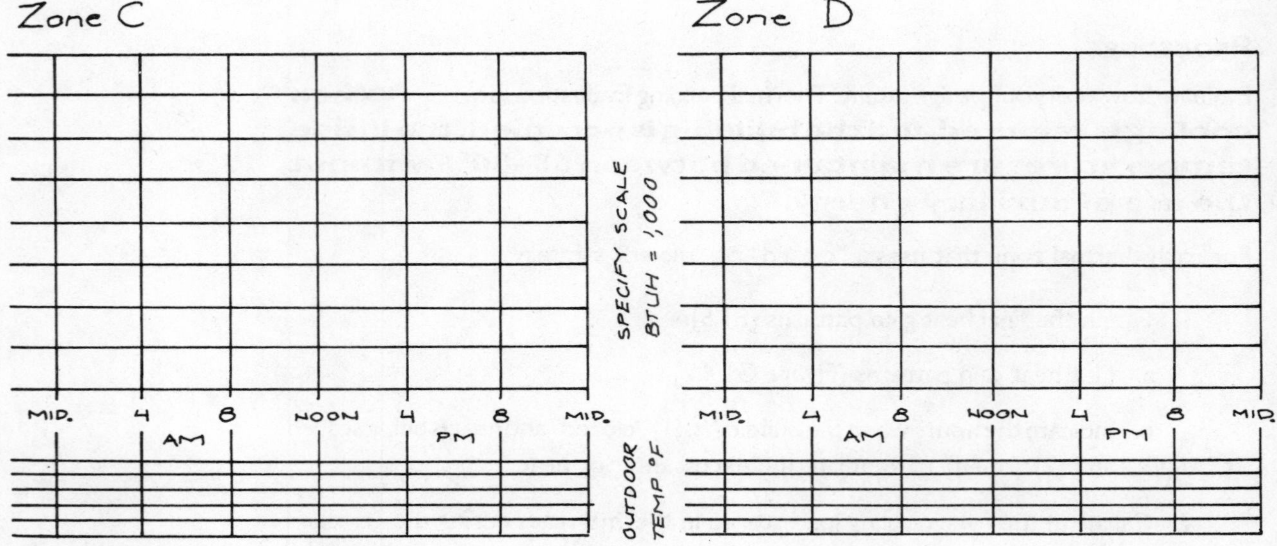

Figure G1.4.2 Peak Heat Gain Charts

INTERIOR TEMPERATURE SWING (continued)

4. Calculate the mass heat storage capacity (Btu/°F). For common building materials the product of mass density times mass specific heat is calculated in *MEEB*, Table 5.29, p.267. Density and specific heat values for other materials are given in *MEEB*, Table 4.2, pp.136–142.

   heat storage capacity = (mass volume) (mass density) (mass specific heat)

5. List the hourly outside air temperatures from the first hour in the evening when it drops to 80°F in August as plotted on the temperature curve [Figure E1.8.1].

6. Calculate the hourly cooling Btus and resultant mass temperature change beginning with the first hour the outside air temperature drops below 80°F until the mass temperature is lower than the outside air temperature.

   $$Btu_{cooling} = (T_{mass} - T)(A_{mass})(C_{mass})$$

   $$\Delta T_{mass} = Btu_{cooling} / HC_{mass}$$

   where:

   T_{mass} = mass temperature from the preceding hour (°F)
   T = outside air temperature (°F)
   A_{mass} = mass surface area (ft^2)
   C_{mass} = mass surface conductance ($Btu/ft^{2}{}^{\circ}F$) [assume C_{mass} = 1.0 or use *MEEB*, Table 4.3, p.143]
   ΔT_{mass} = hourly mass cooling (°F)
   HC_{mass} = heat storage capacity (Btu/°F)

   NOTE: For the first hour's calculation, assume a preceding hour's mass temperature of 80°F.

7. Calculate the hourly adjusted mass temperature.

   $$T_{mass\ adj} = T_{mass} - \Delta T_{mass}$$

8. Check to see whether adequate cooling has been provided by comparing the total of cooling Btus with the cooling load [step 2].

   NOTE: If cooling Btus do not exceed the cooling load [step 2], the interior and mass temperatures will exceed 80°F. You will have to add cooling capacity through the addition of thermal storage mass or rely on another cooling strategy.

INTERIOR TEMPERATURE SWING (continued)

Mass Cooling Capability [suggested format]

	Climate		Building Performance	
Time	T	$Btus_{cooling}$	ΔT_{mass}	T_{mass}
8p.m.	*80*			*80*
9p.m.	*78*	*2,400*	*.27*	*79.73*
10p.m.	*76*			
11p.m.	*73*			
midnight	*69*			
1a.m.	*65*			
2a.m.	*62*			
3a.m.	*59*			
4a.m.	*57*			
5a.m.	*55*			
6a.m.	*57*			
7a.m.	*62*			
8a.m.	*68*			
9a.m.	*75*			
10a.m.	*80*			
Total Cooling Btus	=		Btu/day	
Required Cooling	=	*195,400*	Btu/day	
Mass Surface Area	=	*1,200*	ft²	
Mass Heat Capacity	=	*9,000*	Btu/°F	

9. Discuss the success of your strategy and indicate any design changes that you view as necessary to meet Goal C.

G1 THERMAL

BUILDING EVALUATION

COMPUTER SIMULATION OF THERMAL PERFORMANCE

G1

G1.5

Discussion

For a clear understanding of the proposed heating and cooling systems' interaction, you must assess their combined thermal performance through an evaluation of the interior thermal environment under typical climatic conditions, which are constantly changing. The most feasible way to test the systems is to use a computer-based analysis tool. Tools such as *SOLAR–5*, *Energy Scheming*, *CALPAS–3*, *F–Chart*, or *DOE–2* are generally recommended for this purpose. These tools vary widely in their complexity and analysis techniques.

For more information

MEEB, Appendix H, pp.1534–1539 (descriptions and sources of these and other computer-based tools).

Procedure

Evaluate how well your design attains Thermal Building Evaluation Goal D—**The building will use less thermal energy than newly constructed, energy-conserving buildings.**

1. Use an appropriate, computer-based analysis tool to model your building's overall thermal performance.
2. Annotate and highlight the computer simulation output, pointing out significant aspects of your building's performance.
3. Discuss the success of your thermal design strategies as tested by the computer. Indicate any design changes that you deem necessary to meet Goal D.

A2 LIGHTING

PRECEDENTS OF LIGHTING RESPONSE

INTRODUCTION

A2

A2.0

Goal

Assessing technologies in well understood ethnological and climatic contexts can yield principles of design technology that can be generalized and applied to similar contexts. Thus, through the study of less energy-intensive building technologies in either traditional or industrialized cultures, you will gain insights for solving energy-related architectural problems.

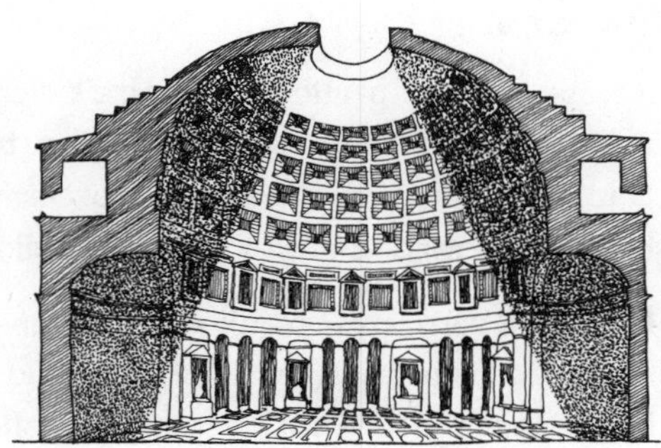

Figure A2.0.1 Toplighting Precedents. ***Above***, the Pantheon under cloudy sky; ***below***, the Hagia Sophia under clear sky.

Discussion

Prior to the twentieth century, buildings depended on daylight as their primary lighting source. Haphazard window placements, though providing adequate lighting, may have allowed excessive heat in hot climates or the loss of valuable heat in cold climates. Of necessity, vernacular architecture was sensitive to both the luminous and thermal environments.

Ancient buildings that are paradigms of the advanced technology of their time reveal the connections their builders made among daylighting, building technology, and aesthetics. In the hot, mediterranean climate of Rome the Pantheon's vast domed space is adequately lighted with a single, unglazed oculus high overhead whose area is only 4% of the rotunda's floor area. The Hagia Sophia in Istanbul, which likely was modeled after the Pantheon, features a ceiling that "floats" above the rotunda. This effect is achieved by a band of small, deeply recessed clerestory windows at the spring of the dome.

This practice of blending daylighting, technology, and aesthetics has been sustained by masters of the craft. The works of Alvar Aalto, Le Corbusier, Albert Kahn, Louis Kahn, James Stirling, and Frank Lloyd Wright warrant further investigation. Excerpting principles that are appropriate to your assigned building type and climate from both vernacular and modern works can enrich your understanding and development of lighting design that is technically, culturally, and aesthetically rich.

. . . the placement of these windows in a region of the Hagia Sophia's dome where tensile hoop forces were expected to be critical seemed audacious. But now it is clear that if the base of the dome had been left solid, it would have been prone to the same meridianal cracking experienced by the Pantheon. The window openings of the Hagia Sophia are thus a prudent expedient to ward off cracking, as well as a dramatic source of light.

—Robert Mark

In this section you will:

1. Investigate precedents of daylighting response in buildings.
2. Generalize principles of daylighting response in climatic and cultural settings.

ARCHITECTURE, CLIMATE, AND PEOPLE—VERNACULAR PRECEDENTS

A2.1

PROCEDURE

Assess how a particular culture's vernacular architecture, the indigenous and traditional architecture of the region, balances culture and climate. Observe how cultural, technical, and economic forces were integrated with those of climate and the luminous environment to influence building form.

Choose a vernacular building or cluster of buildings in a climate zone similar to your site's. Consider the climate of the locale and the living patterns of its residents. Look for evidence of how such patterns are affected by climate and visual comfort. Analyze the lighting strategies used in terms of their cultural and physical contexts.

DOCUMENT YOUR CHOICE AS FOLLOWS:

1. Identify the location of your vernacular building.
2. Describe the climatic characteristics of its locale.
3. Draw a building response diagram to illustrate how seasonal or diurnal changes in the luminous environment affect the living patterns.
4. Document your findings using vignettes, photocopies, photographs, etc. with concise and clearly written annotations. Include any uncertainties, questions, or ambiguities that would require additional research. Your analysis should be no longer than two 8½" x 11" pages.

Figure A2.1.1 Dutch Row Houses, ca. 16th Century

Lighting Design Strategies

Site-Scale Strategy

- Provide an unobstructed view of the sky dome.

Cluster-Scale Strategies

- Reflect light into the building from adjacent building and ground surfaces.
- Control glare and sunlight with external shading devices, such as trees or arbors.

Building-Scale Strategies

- Place spaces that require daylight near the skin of the building.
- Position smaller spaces so they "borrow" light from larger spaces. Surround large, high-ceilinged, daylighted spaces with small, low-ceilinged spaces.

Figure A2.2.1 Cloudy Sky Strategy. Glasgow School of Art, Glasgow, Scotland, C. R. Mackintosh (architect), 1899–1909.

Component-Scale Strategies

- Bounce light deeper into the building with shading devices used as lightshelves.
- Make sidelighted rooms no more than twice as deep as the window head is high.
- Place small windows high in the room or out of the field-of-view in clear-sky climates.
- Ensure unobstructed access to the sky dome and shading appropriate to the building's balance point temperature for large windows in cloudy climates.
- Exclude direct sun penetration through skylights into critical task areas.
- Provide light from north- and south-facing apertures to avoid glare and excessive heat gain.
- Provide balanced illumination with lighting from two sides or directions to minimize glare problems.

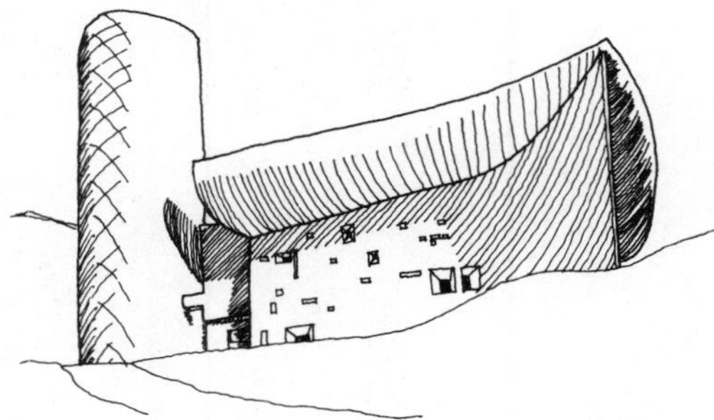

Figure A2.2.2 Clear Sky Strategy. Notre Dame du Haut, Ronchamps, France, Le Corbusier (architect), 1950–1955.

LIGHTING PRECEDENTS (continued)

Procedure

Choose an existing building or site that has: (1) a building program, luminous environment, or both that is similar to your assigned building program or climate and (2) a clear, conceptual approach that incorporates some of the design strategies above.

Document your choice as follows:

1. Identify the location, program, architect (if known), and source of your information.
2. Include photocopies or drawings (whichever is quick and easy for you) to illustrate the design.
3. Evaluate the building or site design with a building response diagram and short annotations that explain how this design is organized for lighting.

B2 LIGHTING

CLIMATE AND SITE ANALYSIS

B2

INTRODUCTION

B2.0

Goal

Determine how much light is seasonally available on your site and what potential benefits or problems may arise with the changing sky conditions. Generate a glazing strategy responsive to your climate and site.

Discussion

It is never too early in the design process to address the effect of the sun's light and heat. Is your site's climate predominantly sunny or cloudy or does it change seasonally? Look at vernacular buildings in your climate for clues about well-tested responses to the luminous environment.

The amount of exterior illumination available for daylighting depends on latitude and prevailing sky condition (clear or cloudy). The lower the latitude, the brighter the light. Also, the lower the latitude, the more intense and less thermally desirable the direct sunlight. Once you determine illumination levels, perform lighting program analyses to determine which spaces you can successfully daylight and to approximate how much glazing will be required.

In this section you will:

1. Investigate your site's seasonal sky conditions.
2. Determine how much exterior illumination is available seasonally.
3. Propose a conceptual daylighting design that takes advantage of your site's sky conditions to provide the maximum use of daylighting.

B2 LIGHTING

CLIMATE AND SITE ANALYSIS

SKY CONDITIONS

B2

B2.1

Discussion

The prevailing sky condition is an important factor when developing a daylighting scheme. Clear days have higher available illumination levels and a directional light source (the sun) which, in most cases, must be prevented from shining directly into the building. Aside from the sun, the brightest part of a clear sky is near the horizon. Cloudy days have less available illumination, and light distribution does not vary with window orientation. Generally, on cloudy days the zenith is three times as bright as the horizon.

Procedure

p. 275 Mean # of days

1. Record the number of clear, cloudy, and partly cloudy days in each month for your site's climate according to NOAA climatological data [Appendix C].

   For each month, graph and color code the number of clear, partly cloudy, and cloudy days. To enhance the readability of your chart, plot the number of cloudy days from the right, the number of clear days from the left, and the number of partly cloudy days in the middle.

Sky Conditions Chart

	Number of Days						
	0	5	10	15	20	25	30
Jan							
Feb							
Mar							
Apr							
May							
Jun							
Jul							
Aug							
Sep							
Oct							
Nov							
Dec							

KEY: ◯ Clear ◯ Partly Cloudy ◯ Cloudy

2. For each season, determine whether cloudy or clear skies predominate. (If neither is dominant, you may analyze either your worst case or both sky conditions using the technique discussed in B2.2.)

B2 LIGHTING

CLIMATE AND SITE ANALYSIS

B2

AVAILABLE ILLUMINATION

B2.2

Discussion

The available exterior illumination for any location varies both seasonally and with its sky conditions. On clear days window orientation is important when determining interior light levels. North-facing windows give less intense, more even light over the course of a day and minimize solar gain (a good cooling strategy). South-facing windows may maximize solar gain (a good heating strategy) and yield a broad range of illumination levels over the course of a day. Since it is difficult to block direct sunlight in east- and west-facing windows, use them with great care in daylighting schemes. On cloudy days light is distributed more evenly across the sky and window orientation makes little difference.

Procedure

For each season:

1. Determine your exterior light levels (footcandles) throughout the day.

   a. If you've determined the season is predominantly cloudy, use Table B2.2.1 below or *MEEB*, Figure 19.6, p.975.

   b. If you've determined the season is predominantly clear, use Table B2.2.2 or *MEEB*, Figure 19.9(b)(c) or Table B.11–B.13, pp.978, 1450–1455. Note the illumination levels for all exposures you intend to use for daylighting.

   c. If your skies are neither predominantly clear nor cloudy, use the data that give you a worst-case situation or use both sky conditions' data.

2. Plot the illumination data [Figure B2.2.1]. On each graph indicate whether you are using clear or cloudy sky data.

Table B2.2.1 Cloudy Sky Illuminance Data

Equivalent sky luminance in footlamberts--average overcast day

Latitude (°N)	December 21					March 21 or September 21					June 21				
	8 a.m. 4 p.m.	9 a.m. 3 p.m.	10 a.m. 2 p.m.	11 a.m. 1 p.m.	Noon	8 a.m. 4 p.m.	9 a.m. 3 p.m.	10 a.m. 2 p.m.	11 a.m. 1 p.m.	Noon	8 a.m. 4 p.m.	9 a.m. 3 p.m.	10 a.m. 2 p.m.	11 a.m. 1 p.m.	Noon
30	420	740	1020	1210	1270	910	1320	1710	2010	2140	1270	1730	2250	2250	2250
32	350	700	960	1150	1200	880	1290	1650	1940	2070	1280	1730	2240	2240	2240
34	320	650	910	1100	1140	860	1250	1600	1870	1980	1290	1730	2220	2220	2220
36	260	600	840	1020	1070	840	1220	1560	1800	1900	1290	1730	2200	2960	2960
38	230	550	790	940	1000	800	1200	1500	1740	1840	1290	1720	2160	2840	2840
40	190	500	740	900	930	790	1140	1460	1670	1760	1290	1700	2120	2650	3060
42	150	450	660	820	860	760	1120	1410	1600	1690	1300	1690	2080	2540	2860
44	100	380	600	760	790	740	1080	1340	1540	1620	1190	1670	2050	2430	2660
46	60	340	550	680	730	710	1030	1290	1470	1550	1290	1640	2010	2330	2520
48	40	290	470	630	650	690	990	1240	1410	1480	1290	1620	1960	2250	2400
50	0	240	420	560	580	650	940	1180	1330	1400	1260	1590	1900	2160	2280

Reprinted, by permission, from Kaufman (editor), IES Lighting Handbook, 5th edition, 7-6.

AVAILABLE ILLUMINATION (continued)

B2.2

Table B2.2.2 Clear Sky Illuminance Data

Equivalent sky luminance in footlamberts--clear days (average values, direct sun excluded)

Latitude (°N)	December 21					March 21 and September 21					June 21				
	8 a.m.	10 a.m.	Noon	2 p.m.	4 p.m.	8 a.m.	10 a.m.	Noon	2 p.m.	4 p.m.	8 a.m.	10 a.m.	Noon	2 p.m.	4 p.m.
								North							
30	450	600	600	600	450	700	1000	1050	1000	700	1550	1400	1000	1400	1550
34	350	550	550	550	350	800	800	900	800	800	1350	1400	950	1400	1350
38	300	550	550	550	300	750	800	900	800	750	1350	1300	950	1300	1350
42	250	500	500	500	250	700	750	800	750	700	1300	1300	950	1300	1300
46	150	450	500	450	150	700	750	750	750	700	1300	1250	950	1250	1300
								South							
30	1100	1950	2250	1950	1100	1700	2300	2800	2300	1700	1200	1600	2400	1600	1200
34	1100	1900	2200	1900	1100	1700	2650	2900	2650	1700	1350	1650	2300	1650	1350
38	900	2300	2200	2300	900	1700	2700	2950	2700	1700	1350	1650	2300	1650	1350
42	600	2100	2150	2100	600	1700	2700	2450	2700	1700	1350	2000	2500	2000	1350
46	400	1900	2100	1900	400	1700	2700	2900	2700	1700	1350	2100	2700	2100	1350
								East							
30	1550	1500	1000	700	400	2000	2500	1500	900	700	2800	2650	1400	1000	700
34	1350	1400	950	700	400	2400	2600	1600	950	650	2800	2700	1450	1000	700
38	1200	1300	900	650	350	2500	2600	1500	900	600	2800	2700	1400	1050	700
42	750	1200	850	600	250	2400	2400	1450	800	600	2900	2600	1400	1000	700
46	500	1100	800	500	150	2300	2100	1400	700	600	2850	2600	1400	1000	700
								West							
30	400	700	1000	1500	1550	700	900	1500	2500	2000	700	1000	1400	2650	2800
34	400	700	950	1400	1350	650	900	1600	2600	2400	700	1000	1400	2700	2800
38	350	650	900	1300	1200	600	900	1500	2600	2500	700	1050	1400	2700	2800
42	250	600	850	1200	750	600	800	1450	2400	2400	700	1000	1400	2600	2900
46	150	500	800	1100	500	600	700	1400	2100	2300	700	1000	1400	2600	2850

Reprinted, by permission, from Kaufman (editor), IES Lighting Handbook, 5th edition, 7-7.

CLIMATE AND SITE ANALYSIS
AVAILABLE ILLUMINATION (continued)

NOTE: For seasons where clear sky conditions predominate, graph both the north and south windows.

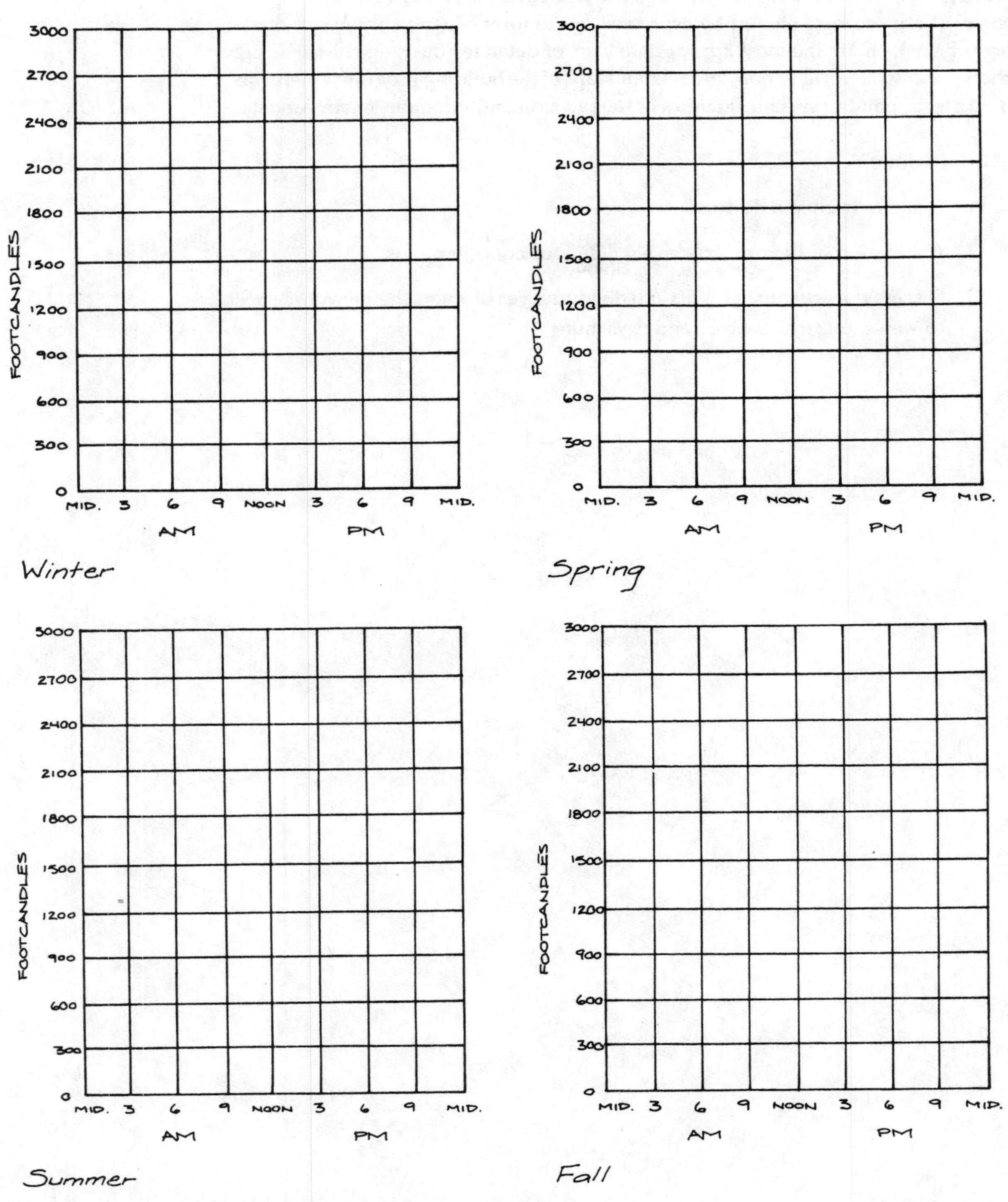

Figure B2.2.1 Available Illumination Graphs

SITE CONCEPTS

Discussion

The sky conditions and available illumination graphs help you gain insights about the luminous environment on your site. You can now explore alternate placements of your building on the site and organizations to provide access to daylight for appropriate spaces. These concepts should be expressed in the form of very simple, annotated diagrams, which are the most appropriate level of detail for this stage of the design process. Show size, orientation, and relationships of the building program's elements. Be sure to respond to specific qualities of the physical and luminous environments.

Procedure

1. Diagram each site design.
2. Annotate the diagrams to explain luminous considerations.
3. Discuss the potential benefits and disadvantages of your scheme with respect to views, thermal control, and daylighting.

C2 LIGHTING

PROGRAM ANALYSIS

INTRODUCTION

GOAL

Determine how much light your building's occupants need and whether the available external illumination is adequate to daylight your building.

DISCUSSION

Different tasks have different lighting requirements—the greater the visual difficulty, the higher the illumination level required. Areas closer to the windows and beneath the skylights receive more daylight than areas farther from the windows. The amount of internal illumination you achieve varies with the amounts of glazing and available exterior illumination. To maximize the effectiveness of daylighting design, group tasks with similar lighting needs and place those task groups that require more light closest to apertures. These decisions and climate and site analyses determine whether a particular fenestration scheme will provide adequate light.

IN THIS SECTION YOU WILL:

1. Determine what tasks are performed in your building and set a recommended lighting level and lighting energy budget for each task.
2. Determine how much exterior illumination is required to successfully daylight each task according to your glazing strategy.
3. Propose a schematic daylighting design for your building.

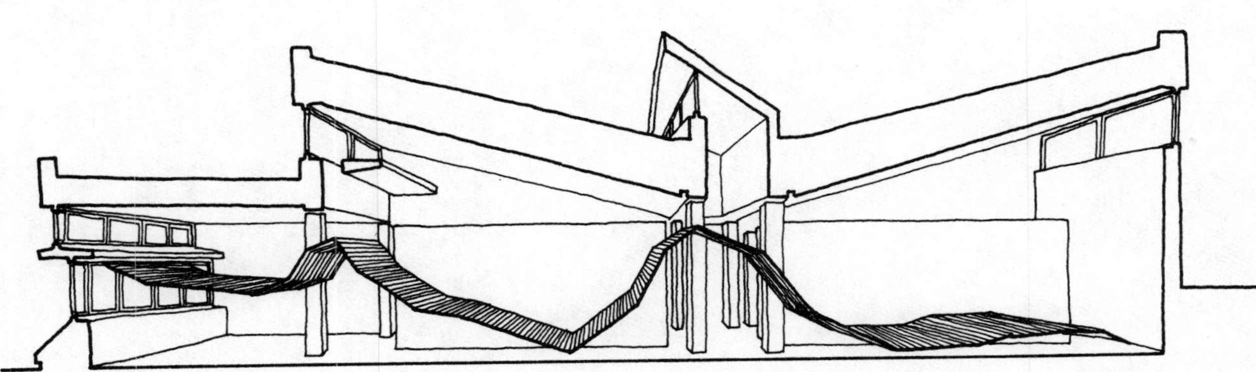

Figure C2.0.1 Lighting Section. Mount Airy Public Library, Mount Airy, North Carolina, Ed Mazria (architect), 1981.

C2 LIGHTING

PROGRAM ANALYSIS

RECOMMENDED FOOTCANDLE LEVELS

C2.1

Discussion

Space use determines lighting levels (footcandles). Spaces may be illuminated by uniform (ambient), localized (task), or a combination of both ambient and task lighting. For example, if a task requires a high light level, it may be best to use lower-level, ambient lighting and provide small areas of higher-level, separately operable, task lighting.

Procedure

1. Identify all the major tasks performed in your building. Use energy-conscious lighting recommendations [*MEEB*, Table 18.5 or 18.7, p.941 or 943] to assign a footcandle level for each task.
2. Note the hours during which each task is performed. This information will be useful later when you determine whether daylighting can be used to provide the necessary light levels.

Lighting Schedule [suggested format]

Task/Space	Light Level (fc)	Hours of Operation
lobby	*30*	*6a.m.–midnight*

RECOMMENDED LIGHTING ENERGY BUDGET

Discussion

Most building codes include an energy code that prescribes an electric lighting energy budget based on building type. The basic energy budget for a building may be adjusted upward if electric lighting control schemes based on daylighting, occupancy, or fixture performance are used.

Specific control schemes eligible for energy credits are:

1. **Continuous Dimming + Daylighting:** A photocell senses exactly how much daylight is available and informs the controller which then compensates by continuously adjusting the lights to provide required lighting.
2. **Stepped Switching + Daylighting:** A photocell senses exactly how much daylight is available and informs the controller which then compensates by adjusting the lights only when preset thresholds are reached.
3. **On–Off Switching + Daylighting:** A photocell senses exactly how much daylight is available and informs the controller which then compensates by switching on electric lights when necessary.
4. **Occupancy Sensor Switching:** A body heat or motion detector switches off the light shortly after it senses the room is unoccupied.
5. **Lumen-Maintenance Control:** The controller reduces the wattage delivered to new electric lamps and gradually increases the wattage to full power as the lamps becomes older and less efficient.
6. **Programmable Timing Control:** A microprocessor turns the lights on and off according to a programmed building occupancy schedule.

The energy code may also limit the types of lamps and fixtures you may use in your building.

Procedure

1. Select the basic lighting energy budget for the spaces in your building or for the building as a whole [Table C2.2.1].

2. Calculate your adjusted lighting energy budget based on control schemes appropriate to your design [Table C2.2.2]. Specify at least two alternatives. Call these schemes "Y" and "Z."

Lighting Energy Budget [suggested format]

Applicable Energy Code *MCS*

Space	Basic Energy Budget	Scheme Y	Scheme Z
office	*1.5 watts/ft²*	*1.95 watts/ft²* *(continuous dimming + daylighting)*	*2.16 watts/ft²* *(occupancy sensors + daylighting)*

RECOMMENDED LIGHTING ENERGY BUDGET (continued)

Table C2.2.1 Model Codes for Interior Lighting Power Budgets

Comparison of codes for interior lighting allowances (watts/ft²)

	MCS	OR.1986	WA.1986	ASHRAE 90.1P		DOE STANDARD			CEC
				ULPA	LPB	ULPA	LPB		
							1987	1992	
Office	1.5	1.7	1.7	1.9	2.3	1.9	2.3	1.6	1.5
Retail									2.0
<1000 ft²	4.5	4.0	4.0						
1000-6000 ft²	3.5	4.0	4.0						
6000-20,000 ft²	2.5	2.3	3.0	3.1	3.1	3.1	3.1	2.6	
>20,000 ft²	1.5	2.3	2.0						
Warehouse	0.7	0.7	0.7	0.4	0.3	0.4	0.3	0.3	1.0
Restaurant	1.85	1.2	1.85					1.0	
Fast Food			1.4	1.3	1.4	1.3	0.8		
Leisure Dining			2.1	2.5	2.1	2.5	1.4		
School	2.0	2.0/1.8	1.7	1.8	NC	1.8	NC	NC	NC

NC = not covered

Reprinted from Northwest Power Planning Council, "Staff Briefing Paper on Commercial Lighting," 15.

Table C2.2.2 Lighting Code Credits for Model Lighting Codes

Control credits (comparison of codes for lighting requirements)

	MCS	OR.1986	WA.1986	ASHRAE 90.1	DOE BEPS	CEC
Control Credits		none				
Occupant Sensor	30%		30%	30%	30%	30%
Continuous + Daylighting	30%		30%	30%	30%	30%
Stepped + Daylighting	20%		20%	20%	20%	20%
On/Off + Daylighting				10%	10%	
Lumen Maintenance	10%		10%	10%	10%	10%
Occupancy + Daylighting	44%		44%	40%		
Occupancy + Lumen	37%		37%	35%		
Programmable Timing				15%		

Reprinted from Northwest Power Planning Council, "Staff Briefing Paper on Commercial Lighting," 19.

REQUIRED EXTERIOR DAYLIGHT LEVELS

Discussion

There is a range of possible daylighting strategies to choose from. By using "lots of big windows," about 50% of the floor area in glazing, you can produce interior illumination levels of about 10% of exterior levels (a daylight factor of 10%); call this one "Strategy A." At the other extreme, by using "a few, small windows," about 5% of the floor area in glazing, you can produce interior illumination levels of about 1% of exterior levels (daylight factor of 1%); call this one "Strategy C." The recommended glazing level for your location will probably fall between these extremes. You can experiment with different ratios of glazing to floor area for each space; call these "Strategy B."

Procedure

For each task:

1. State the recommended interior illumination level [C2.1, step 1].
2. Investigate the range of possible glazing configurations.
   a. Calculate the minimum required exterior illumination for daylighting with "lots of big windows" (Strategy A) and "a few, small windows" (Strategy C).
   b. Determine the minimum exterior illumination required to supply the recommended interior footcandle level. For "big windows" DF is 10% or .10, for "small windows" DF is 1% or .01.

   $$fc_{ext} = (fc_{int}) / (DF)$$

   where:

   fc_{ext} = minimum required exterior illumination (fc)
   fc_{int} = recommended interior illumination (fc)
   DF = daylight factor
3. Estimate the percentage of exterior illumination that will be available as interior illumination (DF).
   a. Find your location's recommended glazing level or choose a glazing level you wish to investigate (Strategy B) [*MEEB*, Table 5.6, pp.206–210]. You may change the glazing percentage for each task.
   b. Calculate the minimum required exterior illumination levels. Use the nomograph on the next page [Figure C2.3.1] to help convert percentage of floor area in glazing to DF.

PROGRAM ANALYSIS

REQUIRED EXTERIOR DAYLIGHT LEVELS (continued)

Required Exterior Daylight Level Chart [suggested format]

Task	Recommended Interior Illumination (fc_{int})	Strategy A 50% of Floor in Glazing DF = 10% (fc_{ext})	Strategy B Recommended Glazing Level % Glass	DF	(fc_{ext})	Strategy C 5% of Floor in Glazing DF = 1% (fc_{ext})
lobby	*3*	*30*	*15*	*3%*	*100*	*300*

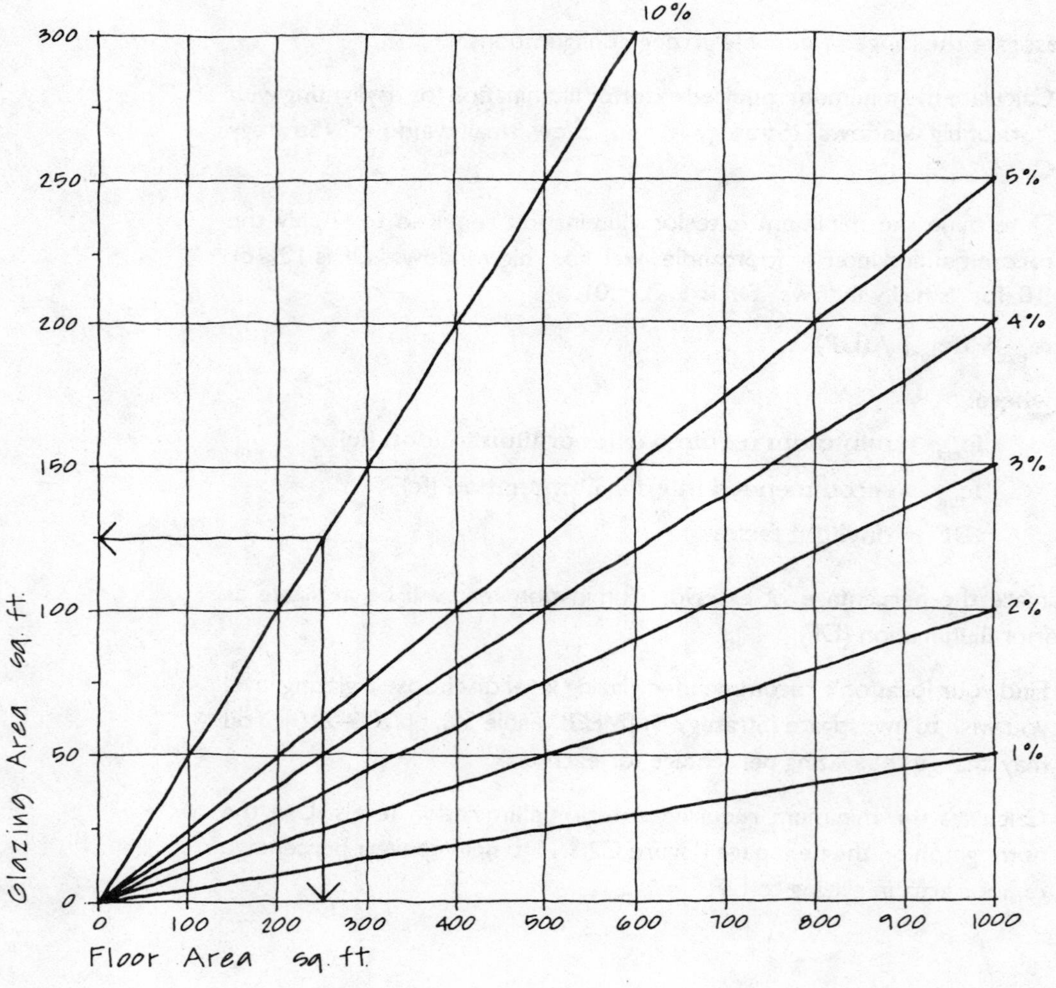

Figure C2.3.1 Daylight Factor Nomograph. Sidelighting and vertical monitors. Reprinted, by permission, from Brown, *Sun, Wind, and Light*, 127, copyright © 1985, John Wiley & Sons, Inc.

PROGRAM ANALYSIS

ILLUMINATION POTENTIAL

Discussion

Now that you know the required exterior illumination for several daylighting strategies and your actual seasonal exterior illumination levels [B2.2], you can determine which daylighting strategy is appropriate for each task in your building.

Procedure

For each task or group of tasks:

1. Photocopy the graphs you completed in B2.2.
2. Plot and label the seasonal minimum required exterior illumination at the required footcandle level for each daylighting strategy (A, B, and C) [C2.3, step 3b]. These plots, which represent each task's hours of operation [C2.1, step 2], should be horizontal lines. Your graph should be similar to that shown in Figure C2.4.1.
3. Identify the times of effective daylighting as plotted on the graph for each strategy for each season.
4. Select the daylighting strategy or glazing percentage that would adequately meet your lighting needs. Circle or highlight your choice in the hours of effective daylighting table.

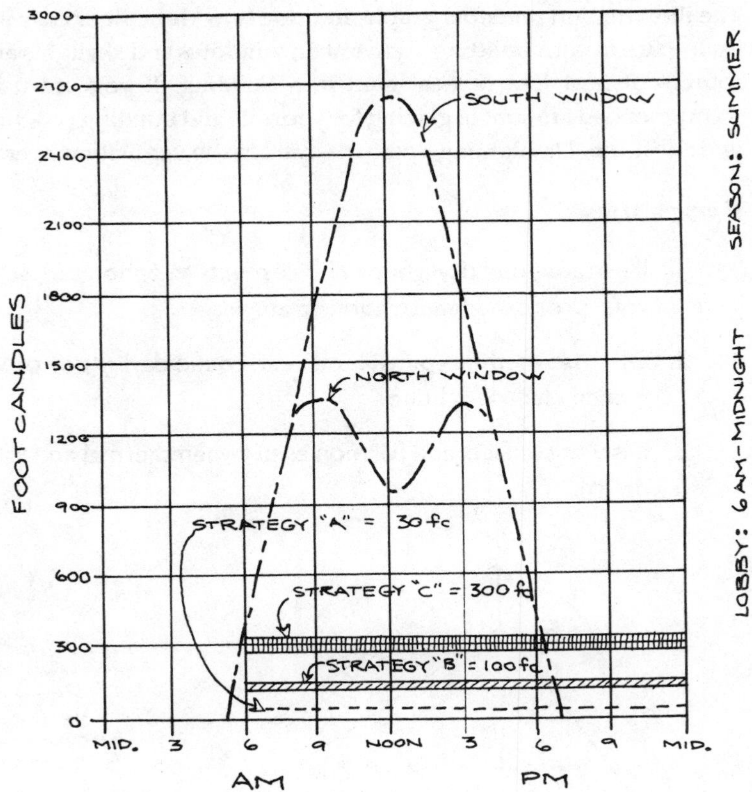

Figure C2.4.1 Sample Available Illumination Graph. Required summer illumination levels support strategies A–C in the lobby from 6 a.m. to midnight. All three strategies provide adequate illumination from 6 a.m. to 6 p.m.. Therefore, the design can be greatly influenced by other factors.

Hours of Effective Daylighting (for each task) [suggested format]

Task *office work*

Season	Hours of Operation	Hours of Effective Daylighting: Strategy A (big windows)	Strategy B (medium windows)	Strategy C (little windows)
Spring	*8a.m.–5p.m.*	*7a.m.–5p.m.*	*8a.m.–4p.m.*	*9a.m.–3p.m.*

PROGRAM ANALYSIS

DAYLIGHTING STRATEGIES

Discussion

The illumination potential graphs indicate how difficult or easy it will be to daylight each space in your building. Remember, windows and skylights are potentially major sources of heat loss or heat gain in a building. If you are using more than the recommended amount of glazing for your site and building program, you must justify your decision. Daylighting apertures can also be ventilation apertures.

Procedure

1. Illustrate your daylighting choices with an annotated, schematic diagram of your proposed fenestration strategies.
2. Discuss whether you will use year-round daylighting or will require supplemental electric lighting.
3. Discuss conflicts and harmonies between thermal and lighting needs in your design.

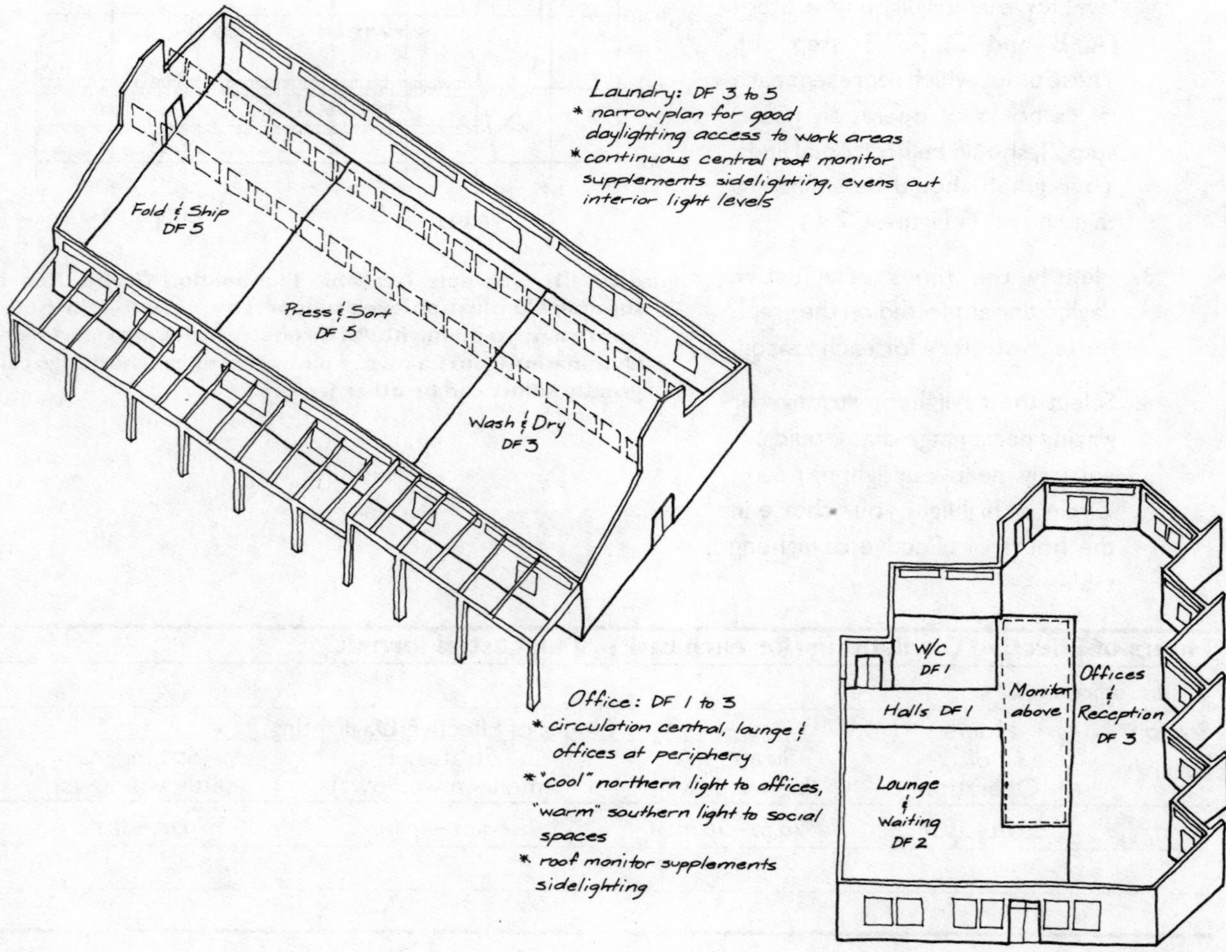

Figure C2.5.1 Daylighting Strategies. Linen supply, Charleston, South Carolina.

D2 LIGHTING

SCHEMATIC DESIGN

INTRODUCTION

Goal

Design a building that synthesizes the information you have acquired through the study of precedents; the analyses of your site, climate, and program requirements; and the application of lighting design strategies.

Lighting Schematic Design Goals

A. Daylighting is distributed so there is adequate light where people need it.
B. Gradual changes in lighting levels inside, outside, and within rooms prevent glare problems.
C. Window apertures are sized to attain the minimum levels of illumination required for tasks and spaces, the recommended daylight factor (DF).
D. Solar collecting and ventilating apertures are maintained as required for heating and cooling.

Discussion

Until 100 years ago, daylighting was the prevalent means of lighting buildings. Artificial lighting sources were installed, but provided far less light than daylight and tended to be used only at night. Once mechanical cooling systems and inexpensive fluorescent lighting were developed it became possible to exclude daylight from buildings. The heat from electric lighting had to be expelled with the heat from the occupants. The outdoors became hotter while the indoors stayed cool.

Why Would Anyone Exclude Daylight?

Daylight is highly variable; there is much less of it at 8 a.m. than at noon and less in winter than in summer. An overcast sky distributes daylight very differently than a clear sky, and sky conditions can shift several times a day. If absolutely uniform, steady light is desirable, electric lighting is a more reliable source.

Why Would Anyone Use Daylight?

Electric lighting is energy intensive. Electricity production by nuclear and fossil thermal processes captures, at best, about 40% of the fuel's value; the remainder is dumped in the environment as waste heat. Hydroelectric, geothermal, and wind-generated power, though more efficient in producing electricity, lose energy as waste heat

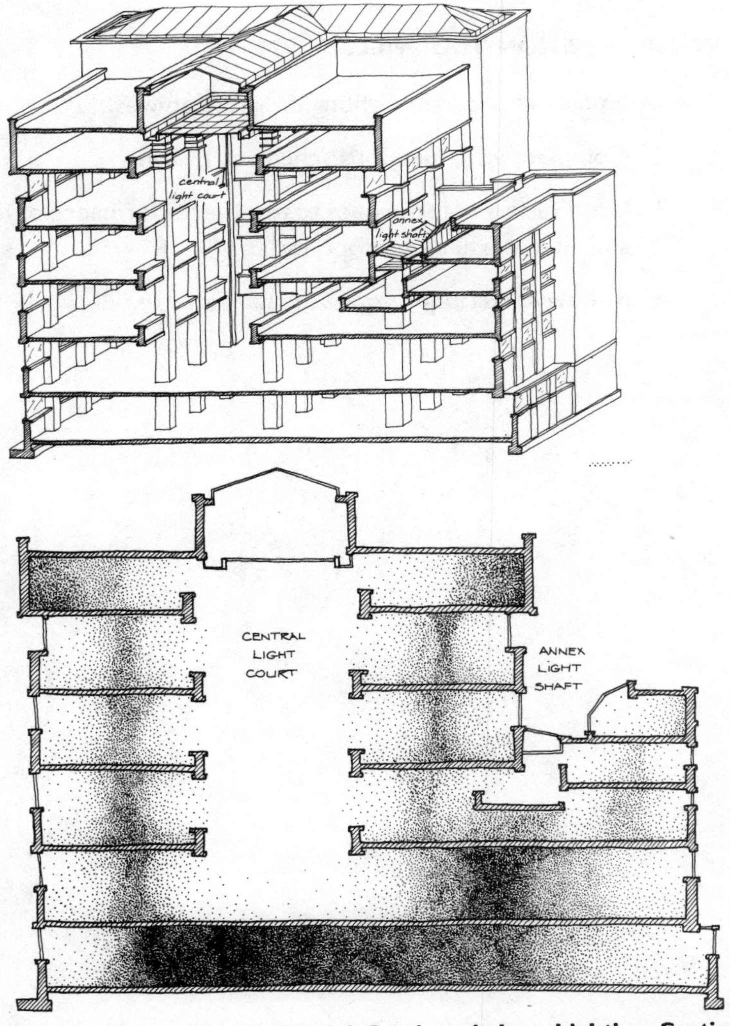

Figure D2.0.1 ***Above,*** **Spatial Section;** ***below,*** **Lighting Section. Larkin Building, Buffalo, New York, Frank Lloyd Wright (architect), 1905.**

when transmitted from the point of production to the point of use. The farther the energy must travel, the more energy is lost. Finally, as electric light is created, it brings about twice as much heat per unit of light to the space as does daylighting. Because energy conservation is important, daylighting is a more appropriate, economical choice.

What Are the Tradeoffs Between Using Electric Light *vs.* Daylight?

More electric lights mean more internal heat gain year-round. This increased load becomes the winter friend of the skin dominated load building but the year-round enemy of the internally dominated load building. More windows to provide more daylight mean more heat loss in winter. In summer daylight windows can be rendered relatively harmless to heat gain if properly shaded. In skin dominated load buildings neither lighting method is more advantageous than the other from the standpoint of energy use. However, daylighted, internally dominated load buildings usually require less energy to operate successfully.

In this section you will:

1. Employ appropriate lighting design strategies.
2. Complete your luminous schematic design.
3. Use a variety of techniques to test the performance of your schematic design and alter it where appropriate.
4. Review and critique your schematic design.

Lighting Schematic Design Strategies

Site-Scale Strategies

- Locate buildings and spaces to either use or avoid sky obstructions.
- Open your site to the sky in proportion to the sky's brightness and the amount of light you need.

Cluster-Scale Strategies

- Arrange buildings and ground surfaces to reflect or absorb light.
- Place light-colored walls north of north-facing spaces to increase reflected light.
- Control glare and heat gain by using thermal buffer zones (e.g., arbors, vestibules, courtyards, atria, greenhouses).
- Use snow as a highly effective reflector of light, but avoid its capacity for glare.

Building-Scale Strategies

- Zone tasks according to their visual difficulty. Put spaces or activities that need light near openings in the skin.
- Employ the strategy for combining daylighting and artificial lighting that is most appropriate to your building's climate and functions:
  - ❖ Supply adequate interior daylight under average sky conditions. When sky illumination conditions are below average, use supplemental lighting.
  - ❖ Provide minimum required daylighting under worst-case sky conditions (such as a cloudy, December day at 9 a.m.). Use supplemental lighting when daylighting is not adequate.
  - ❖ Maintain adequate daylight for typical working hours. Provide supplemental lighting for other hours of use.
  - ❖ Achieve the minimum lighting everywhere. Use supplemental artificial lighting in areas farthest from apertures rather than oversupplying daylight to areas nearest apertures.
- Locate tasks and proportion rooms to take full advantage of the 15' perimeter where maximum light from side windows is available.
- Avoid over-daylighting your spaces in hot climates, where darker places are associated with cooler temperatures.
- In hot climates use light reflected through windows placed near the ceiling to avoid glare at eye level and reduce heat gain from direct light.

DESIGN (continued)

Component-Scale Strategies

- **Orient openings toward the brightest part of the sky (the zenith for overcast skies, the horizon for clear skies).**
- **Prevent discomfort from excessive brightness or contrast.**
- **Provide ambient illumination from skylights to areas directly below.**
- **Use side lighting to provide task illumination to nearby areas.**
- **Use specular surfaces out of the field of view to reflect light yet avoid glare problems.**
- **Zone tasks to best use available light rather than trying to manipulate light to reach task areas.**
- **Provide glazing at a ratio of approximately 5–10% of the floor area for standard ambient lighting (office, lobby, circulation).**
- **Provide glazing at a ratio of approximately 25% of the floor area for intensive, task lighting (display, drafting, typing, factory work).**
- **Proportion window head height to equal half the depth of the room for optimal daylight penetration and distribution.**
- **Arrange shading devices so their reflective surfaces do not cause glare at eye level in the shaded spaces beyond.**

PROCEDURE

Propose a schematic design for your building based on analyses of precedent [A2], site and climate [B2], and program [C2]. Use the appropriate lighting design strategies for your design.

DOCUMENT YOUR DESIGN AS FOLLOWS:

1. Site plan, including parking and access drives (scale: 1" = 100').
2. Cluster plan, including outdoor spaces (scale: 1" = 40').
3. Floor plans (scale: 1" = 20').
4. Roof plan and elevations, or axonometrics, illustrating all building sides and roof (scale: 1" = 20').
5. Sections (scale: 1/8" = 1'-0").
6. Design diagram, annotated to identify design strategies and luminous zones.

LIGHT DISTRIBUTION RENDERING D2.2

Procedure

Evaluate how well your building attains Lighting Schematic Design Goal A—**Daylighting is distributed so there is adequate light where people need it.**

1. Render your building's shaded and lighted areas based on the predominant sky condition in your climate.
   a. Draw a plan and section on a single page. Use media that allow you to quickly approximate the relative patterns of light and dark. Pencil or charcoal on tracing paper or colored pencils on brown paper are common methods.
   b. Identify task areas and draw the furnishings to clearly show the light–task relationships.
2. Discuss whether or not your building and site improvements meet Goal A, and describe how your design must be altered if the goal has not been met.

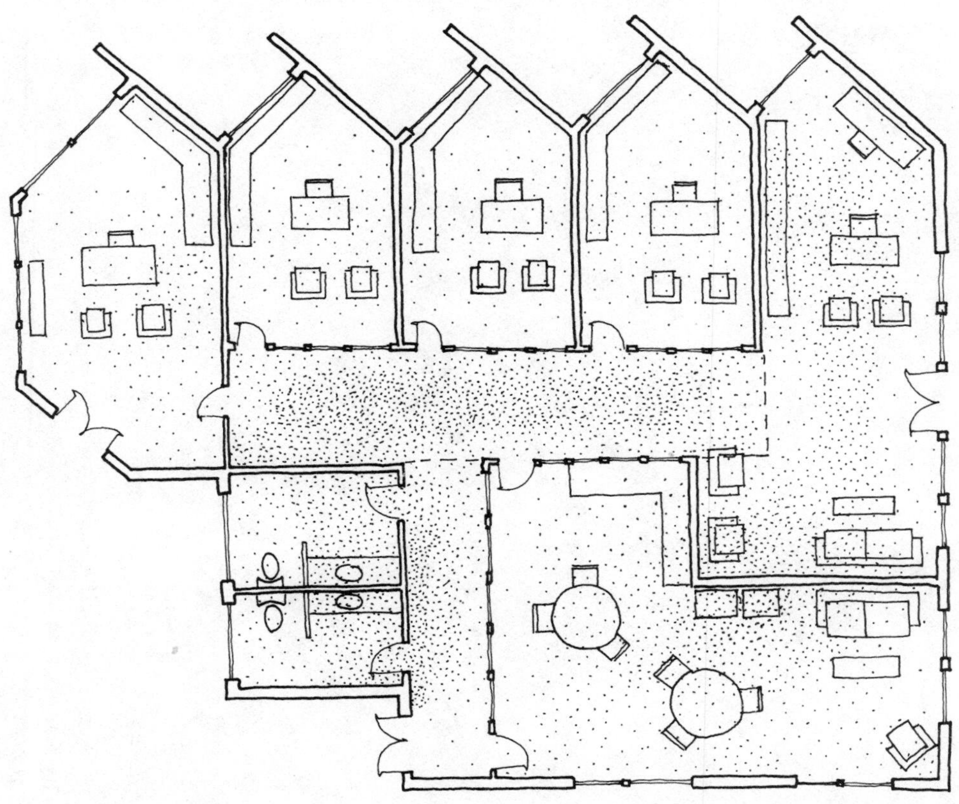

Figure D2.2.1 Light Distribution. Linen supply office, Charleston, South Carolina.

LUMINANCE EVALUATION

D2.3

PROCEDURE

Evaluate how well your building attains Lighting Schematic Design Goal B—**Gradual changes in lighting levels inside, outside, and within rooms prevent glare problems.**

1. Develop detailed renderings based on your drawings [D2.2] to show how you have designed for gradual changes in brightness. Include studies of one entry and of one room.

2. Discuss whether or not your building and site improvements meet Goal B, and describe how your design must be altered if the goal has not been met.

RECOMMENDED DAYLIGHT FACTORS

Discussion

Table D2.4.1 lists recommended minimum daylight factors under overcast sky conditions. For clear sky conditions daylight factors must be calculated by dividing the required illumination level by the "average" exterior illumination level. This "average" level is dependent on the window orientation you have chosen for daylighting. You must make an educated guess at this "average" for each space and task.

When designing for daylighting the minimum DF recommendations should be met or exceeded to provide adequate daylighting. However, if the recommendations are exceeded by twice or more, thermal problems (summer overheating or excessive winter heat loss) may result. Therefore, daylighting levels should meet or barely exceed DF recommendations.

Procedure

Evaluate how well your building attains Lighting Schematic Design Goal C—**Window apertures are sized to attain the minimum levels of illumination required for tasks and spaces, the recommended daylight factor (DF).**

For each task:

1. List the recommended DF [Table D2.4.1] if your climate has mostly overcast skies.

2. If your climate has mostly clear skies, obtain the recommended footcandle level [C2.1], and divide it by the estimated daily "average" available outdoor illumination [B2.2]. To determine the "average" available outdoor illumination, make an educated guess. Remember, it changes daily, seasonally, and with orientation. Express the result as a decimal.

   $$\textbf{DF} = \frac{\textbf{recommended interior illumination}}{\textbf{``average'' outdoor horizontal illumination}}$$

3. Compare the recommended DF [step 1 or 2] with the DF determined by the daylighting strategy selected in C2.3. If there is a large discrepancy between the two, choose the recommendation that seems more reasonable. In clear-sky climates very small daylight apertures can provide enough light from the bright sky yet prevent excessive solar heat gain. In cloudy-sky climates large, well-shaded windows are necessary to provide adequate daylighting.

D2 LIGHTING

SCHEMATIC DESIGN

RECOMMENDED DAYLIGHT FACTORS (continued)

D2.4

Recommended Daylight Factors [suggested format]		
Task Activity Area	Recommended DF*	DF [C2.3 or C2.4]
office (general)	*.04 or 4%*	*.03 or 3%*
circulation	*.01 or 1%*	*.01 or 1%*

* DFs are often expressed as percentages rather than ratios.

Table D2.4.1 Required Daylight Factors for Various Functions

Building Type	Activity	DF (%)	Comments
All Public Buildings	Circulation	0.5-3.0	Minimum is sufficient; higher levels required in transition areas from bright spaces.
	Lobbies, Foyers, Lounges	1	Individual task lighting may be necessary for reading.
	Reception Desks	1.0-2.0	Dependent on difficulty of task involved.
	Restrooms	0.0-1.0	Need not be daylighted.
Assembly, Concert Halls	General	0.0-1.0	Do not daylight if incapable of blacking out light for media presentations.
Banks	General	2.0	
Churches	Congregation	1.0	
	Pulpit	1.5-4.0	
Hospitals	Wards, Public	1.0	Provide individual task lighting for reading.
	Laboratories	3.0	Provide task lighting.
	Operating Room, Examining Room	3.0-5.0	Privacy and thermal considerations may override use of daylighting.
Industrial	High Resolution Work	5.0	Toplighting is recommended.
	Other Work	2.0-3.0	Provide task lighting where appropriate.
Libraries	Stacks	1.0	Additional artificial lighting is required.
	Reading, Study Areas	1.0	Provide individually controlled task lighting.
Musea, Galleries	General	1.0	Additional artificial lighting may be required for emphasis; daylighting may be inadvisable where light-sensitive materials are displayed.
Offices	General	2.0	Provide task lights or lower DF with photocontrolled electric lighting.
	Typing	4.0	May require task lighting; be careful of reflected glare on video displays.
	Drafting	5.0	May be difficult to provide uniformly high level with only sidelighting.
Residential	Kitchens	2.0	Provide task light at work stations; recommended levels should penetrate one-half to three-quarters of the room; if elderly people are often present, higher levels are suggested.
	Living Rooms	1.0	
	Bedrooms	0.5	
Schools	Assemblies, Classrooms	2.0	
	Art Rooms	4.0	
	Laboratories	3.0	
	Staff Areas	1.0	
Sports	Playing	2.0	
Swimming	Water	2.0	
	Deck	1.0	

Reprinted from Schiler (ed.), Simulating Daylight with Architectural Models, 211-212.

DAYLIGHTING APERTURE SIZING D2.5

Discussion

Table D2.5.1 lists recommended aperture to floor area ratios for achieving average DFs with various kinds of daylight openings. This table is included as a rule-of-thumb to help further organize your basic building form. While imprecise, it is more specific than the floor to glazing ratio calculated in C2.4.

Table D2.5.1 Aperture Sizing for Daylighting—Estimated Average Daylight Factors

SIDELIGHTING	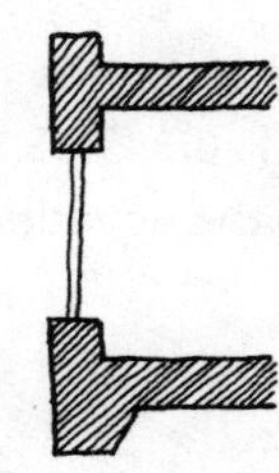	average DF = .2 (A_g / A_f) (near the window—less than $2H_{win}$ away) minimum DF = .1 (A_g / A_f) (away from the window—more than $2H_{win}$ away)
TOPLIGHTING Vertical Monitors	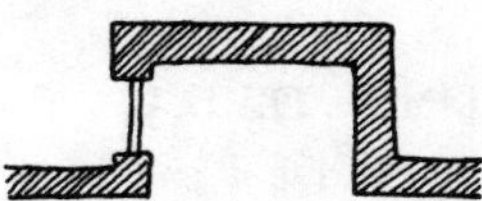	average DF = .2 (A_g / A_f)
North-Facing Sawtooth Openings	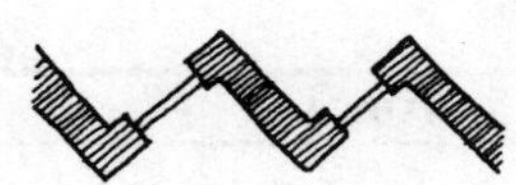	average DF = .33 (A_g / A_f)
Horizontal Skylights	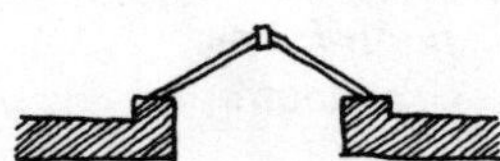	average DF = .5 (A_g / A_f)

where:

DF = daylight factor as percentage of outdoor horizontal illumination

A_g = area of exterior glazing (ft^2)

A_f = area of floor to be lighted (ft^2)

H_{win} = window-head height (ft)

Reprinted, by permission, from Millet and Bedrick, "Manual: Graphic Daylight Design Method," 5–6.

DAYLIGHTING APERTURE SIZING (continued)

PROCEDURE

Evaluate how well your building attains Lighting Schematic Design Goal C— **Window apertures are sized to attain the minimum levels of illumination required for tasks and spaces, the recommended daylight factor (DF).**

For each task area:

1. Calculate the glazing area required for adequate daylighting by solving the appropriate aperture sizing equation [Table D2.5.1] for A_g. If two or more glazing strategies are used, you can add the equations (DFs are additive), and solve for each A_g.

   If $\mathbf{DF = y\,(A_g / A_f)}$,

   then $\mathbf{A_g = [(DF)\,(A_f)] / y}$.

   where:

   y = efficiency ratio [Table D2.5.1]

2. Compare the required glazing area, A_g, with the glazing area you have designed and its resultant daylight factor.

Daylight Aperture Size [suggested format]

Task Area	Floor Area (ft^2)	Target DF [D2.5]	Glazing Strategy	Glazing Area Required (ft^2)	Glazing Area Provided (ft^2)	Resultant DF
office	*600*	*.04 (4%)*	*north-facing sawtooth*	72	*100*	*.06*

3. Discuss whether or not your building and site improvements meet Goal C, and describe how your design must be altered if the goal has not been met.

D2

SOLAR AND VENTILATION APERTURE

D2.6

PROCEDURE

Evaluate how well your building attains Lighting Schematic Design Goal D—**Solar collecting and ventilating apertures are maintained as required for heating and cooling.**

1. Calculate the total solar aperture area.
2. Calculate the total ventilation aperture area.
3. Compare your solar and ventilation total aperture area calculations with the requirements specified in D1.7 and D1.5.
4. Draw a schematic plan and section of your building. Annotate your drawings to describe your heating, shading, ventilation, and lighting strategies. Discuss whether or not your building and site improvements meet Goal D, and describe how your design must be altered if the goal has not been met.

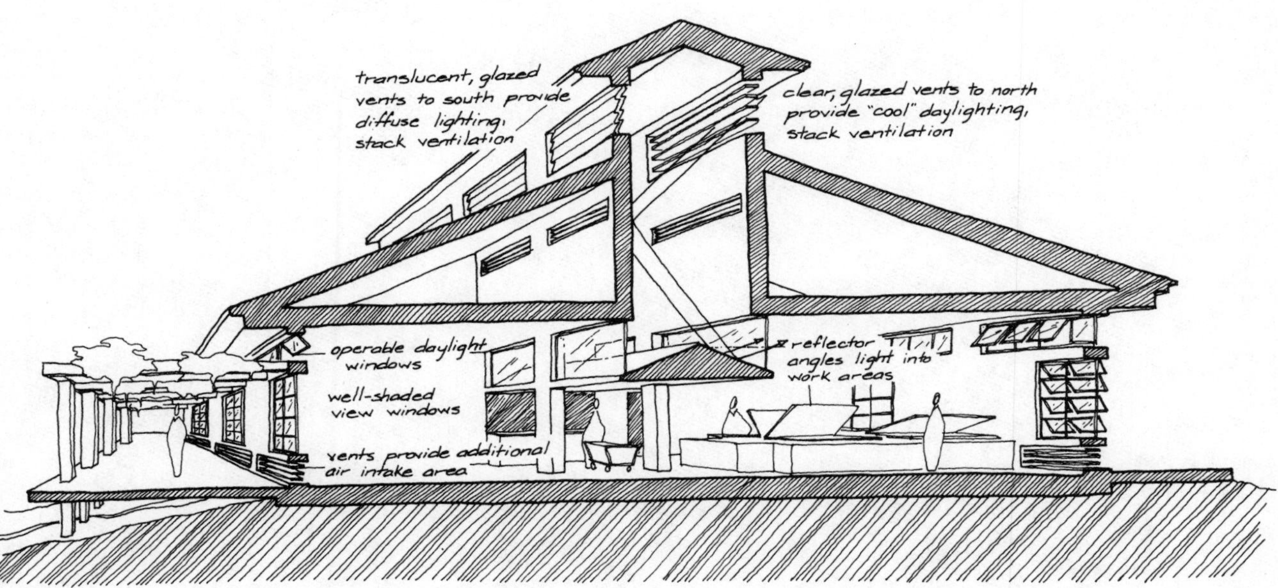

Figure D2.6.1 Integrated Lighting and Thermal Strategies. Linen supply, Charleston, South Carolina.

D2 LIGHTING

SCHEMATIC DESIGN

DESIGN REVIEW

D2.7

Discussion

You have formulated a schematic design for your building and site, chosen appropriate lighting strategies, and ensured access to appropriate environmental forces. Each step has been accomplished independently and conflicting decisions may have been made. This design review affords you the opportunity to synthesize your cumulative design decisions.

Procedure

Review how well your design accomplishes Lighting Schematic Design Goals A–D.

1. Discuss the tradeoffs between the lighting and thermal strategies that your design required.
2. Discuss situations where your lighting and thermal strategies worked effectively together.
3. Make a schematic design sketch that combines all your lighting design strategies. Indicate how and why your design has evolved in response to meeting D2.0 goals.

INTRODUCTION

Goals

Refine and further develop your building's lighting design without sacrificing desirable views or thermal performance. Use your schematic design as the foundation for your design development.

Lighting Design Development Goals

A. Recommended daylight factors are achieved in each task area.
B. The range of DFs in each task area does not exceed a ratio of three to one.

Discussion

There are four major daylight components to consider when developing a daylight design—sunlight, light from the sky, externally reflected light, and internally reflected light. The details of the fenestration and room surface finishes must interact appropriately with these four components.

In most cases, apertures should be designed to reject the sun component, which can cause thermal and glare problems. Exterior shading devices, lightshelves, interior light baffles, and deep lightwells are examples of components that effectively reject the sun, yet admit daylight.

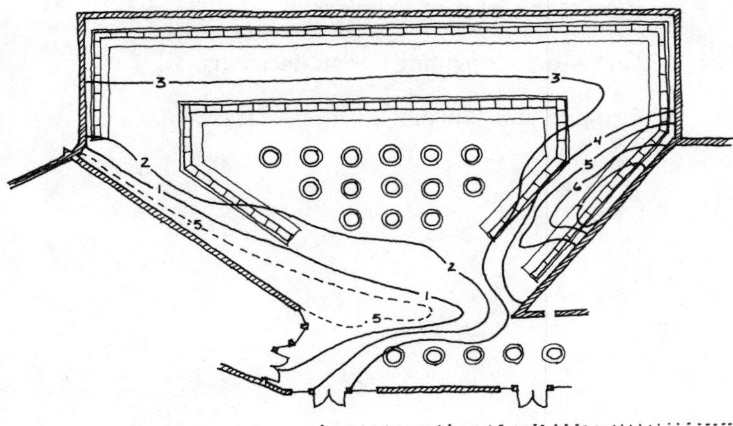

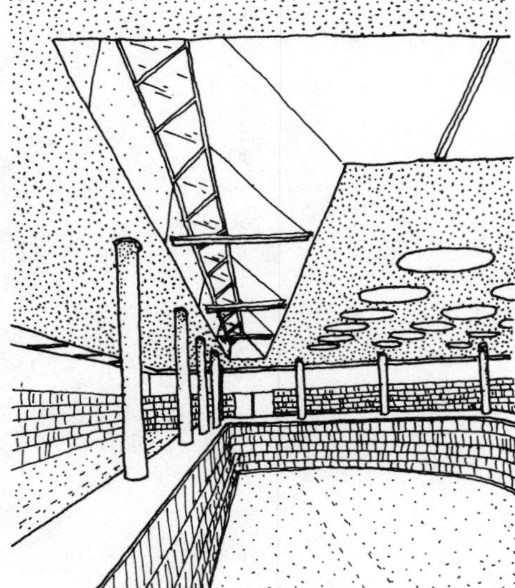

Figure E2.0.1 ***Above***, Floor Plan with DFs; ***below***, Interior View. Wolfsburg Library, Wolfsburg, Germany, Alvar Aalto (architect), 1959.

The sky component is the primary source of daylight. Placement of apertures determines how light from the sky enters the space. Apertures placed in more than one plane tend to provide more balanced light. Apertures placed close together tend to provide larger areas of more even light, while widely spaced apertures tend to create isolated pools of light.

Light reflected from surfaces outside the building can be used to alter the lighting conditions inside. The externally reflected component may originate from neighboring buildings, vegetation, walls, or external shading devices. A light-colored, garden wall can increase light through a north-facing window in a sunny climate. Or light reflected from a red, brick wall can "warm up" the light in a space.

The internally reflected component plays the major role in illuminating a space. Room surfaces receive the daylight from the apertures and act as secondary sources of illumination. Light-colored surfaces (those with a high, internally reflected component) redistribute the light more evenly throughout the space and maximize the illumination potential. Dark-colored surfaces (those with a low, internally reflected component) absorb the light and reflect little, so that the apertures, in effect, only provide pools of

INTRODUCTION (continued)

light. Surfaces adjacent to apertures can be sculpted to direct reflected light to specific areas. Apertures can be placed normal and adjacent to planar surfaces to wash them with light.

How you use the apertures and the surfaces to admit and manipulate the four daylight components determines the character of the architectural space.

In this section you will:

1. Complete the luminous design development of your project.
2. Predict the interior illuminance.
3. Evaluate your lighting design decisions.
4. Critique your building's lighting design.

DESIGN

Lighting Design Development Strategies

Site-Scale Strategy

- Use a variety of landscape surfaces for their reflective characteristics. These will change seasonally, especially in climates with snow. Remember, snow can cause glare on sunny days.

Cluster-Scale Strategies

- Provide reflected light in hot, sunny climates where shading devices block direct sun and greatly reduce light from the sky.
- Avoid extremely large, exterior, daylight obstructions in overcast climates.

Building-Scale Strategies

- Proportion rooms to include the lighting effect of overhangs. The room effectively begins at the edge of the overhang, not where the glazing is located.
- Use sidelighting for task lighting to minimize veiling reflections and supply useful modeling.
- Use toplighting for uniform, ambient lighting.

Component-Scale Strategies

- Bevel the wall adjacent to a narrow, vertical window to reduce glare caused by the contrasting bright window and the darker wall.
- Reduce glare by using lighter colors and higher reflectances near openings.
- Choose internal surface reflectances to take advantage of the primary source of incoming light.
  - ❖ In temperate climates use the floor to reflect light from the sky.
  - ❖ In tropical climates use the ground then the ceiling to reflect light from the sun.
- Increase light in dimmer areas by using lighter colors and higher reflectances on surfaces farthest from openings.
- Reduce daylight near the window wall and increase daylight deep in the room with lightshelves.
- Diffuse light from skylights and prevent direct sunlight from entering the space by using baffles and deep lightwells.

DESIGN (continued)

PROCEDURE

Refine your final schematic design [D2] to include component design considerations. Use the appropriate strategies to aid your lighting design development. Modify your schematic design drawings to include the components you have developed.

DOCUMENT YOUR DESIGN AS FOLLOWS:

On all drawings, state the sky condition (clear or overcast) for which you are designing.

1. Site plan, including parking and access drives (scale: 1" = 100').
2. Cluster plan, including outdoor spaces (scale: 1" = 40').
3. Floor plans (scale: 1" = 20').
4. Roof plan and elevations, or axonometrics, illustrating all building sides and roof (scale: 1" = 20').
5. Sectional perspective, labeling finishes, materials, colors, and square footage of each surface (scale: 1/8" = 1'-0").
6. Design diagram, annotated to identify design strategies and lighting zones.
7. Details of lighting components (e.g., mullions, lightshelves, baffles, glazing, and shading devices) of all daylight apertures in elevation and section.

"RAW" ILLUMINANCE FOOTPRINTS

E2.2

DISCUSSION

The graphic daylighting design method (GDDM) is the best available hand calculation method for predicting daylight distribution for most spaces. However, there are complications that arise when applying it to non-orthogonal openings or rooms of complex shape. It also does not evaluate the localized effects of interior or exterior room reflectances, exterior obstructions, or fixed shading devices.

For fixed shading devices simply estimate the percentage of daylight that gets through them. Do not use shading coefficients used for heat gain estimations, since these refer more to direct sun than daylight from all sources. Thus, the percentage of daylight will be somewhat higher than shading coefficients would indicate. For example, louvered shading devices that block the majority of direct sun (high shading coefficient) will admit a majority of the daylight available.

The GDDM uses the principle that each aperture contributes a characteristic pattern of light based on its geometry. These patterns are described by "footprints" of contours of equal illumination (expressed in daylight factors). The footprints can be added because illumination levels are additive.

The footprints [Appendix F] can be used without modification on floor plans at one-eighth scale for windows that are completely above the workplane and 5' x 10', 10' x 10', or 10' x 5'. For other sizes and shapes they must be re-scaled and combined so each pattern's solid rectangle is the same size as the window projection. Any portion of the window below the workplane will not contribute "raw" illumination to the workplane.

The GDDM involves three basic steps:

1. Match openings with available footprints.
2. Draw isolux contours that describe the light distribution and the "raw" DF range in your space.
3. Correct the DFs applied to the isolux contours.
   a. Increase the DF for factors that increase light in the space (principally interior reflectances).
   b. Reduce the DF for factors that reduce light in the space (glass transmission, dirt, and shading devices).

FOR MORE INFORMATION

Millet and Bedrick, "Manual: Graphic Daylighting Design Method," pp.23–30 (distribution patterns, refinement steps, and procedures for dealing with obstructions).

DESIGN DEVELOPMENT
"RAW" ILLUMINANCE FOOTPRINTS (continued)

PROCEDURE

Evaluate how well your design attains Lighting Design Development Goals A and B—**Recommended daylight factors are achieved in each task area (A), and the range of DFs in each task area does not exceed a ratio of three to one (B).**

For the largest and most important daylighted space in your building:

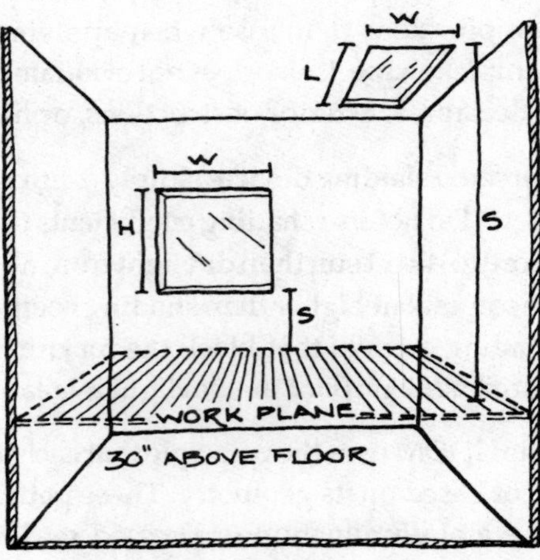

Figure E2.2.1 Dimensions for GDDM Ratios

1. Match the openings to footprints. For sloped openings, see Figure E2.2.2.

   a. List your windows and skylights and their dimensions.

      H = height of window (above the workplane only)

      W = width of window or skylight

      L = length of skylight

      S = distance from workplane to sill (if the sill is at or below the workplane, S = 0)

   b. Find the appropriate footprint for each aperture [Table E2.2.1].

      (1) For apertures with an S/H or S/W ratio between 1:3 and 3:1, choose the footprint with the ratios that most closely match the aperture's ratios.

      (2) For apertures with an S/H or S/W ratio not between 1:3 and 3:1 or of non-rectilinear shape, subdivide the aperture into rectangles with S/H or S/W ratios between 1:2 and 2:1. Then choose the most appropriate footprint for each subdivision of the original aperture.

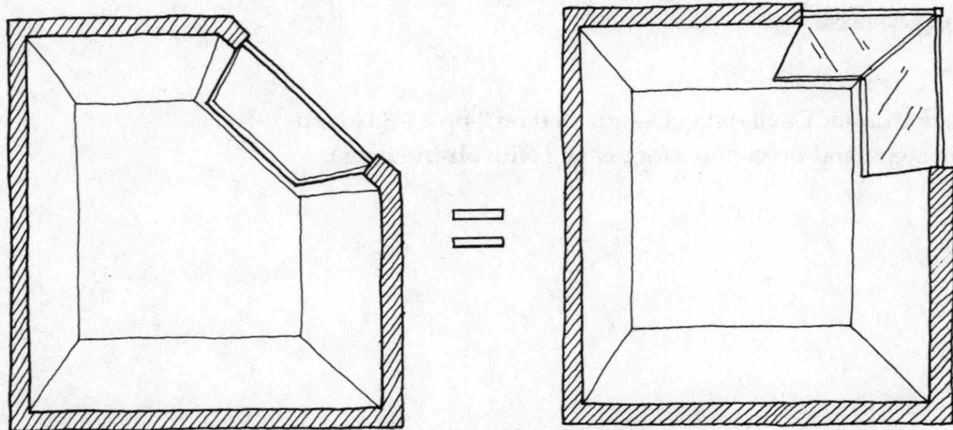

Figure E2.2.2 Sloped Skylight Simulation for Footprint Procedure. For a square skylight with an area of 141 ft² at a 45° slope, the resulting window and horizontal skylight will each be 10' x 10'. When estimating the IRC, use the actual area (141 ft²) in Table E2.3.1 for toplighted rooms.

"RAW" ILLUMINANCE FOOTPRINTS (continued)

Table E2.2.1 Index for Footprints Included in Appendix F

Windows			Skylights		
(H/W)	(S/H)	Footprint ID	(L/W)	(S/W)	Footprint ID
0.5	0.0	A-7	1.0	1.0	A-30
0.5	0.5	A-8	1.0	2.0	A-31
0.5	1.0	A-9	1.0	3.0	A-32
0.5	2.0	A-10	1.0	4.0	A-33
0.5	3.0	A-11	2.0	1.0	A-34
0.5	4.0	A-12	2.0	2.0	A-35
1.0	0.0	A-13	2.0	3.0	A-36
1.0	0.5	A-14	2.0	4.0	A-37
1.0	1.0	A-15			
1.0	2.0	A-16			
1.0	3.0	A-17			
1.0	4.0	A-18			
2.0	0.0	A-19			
2.0	0.5	A-20			
2.0	1.0	A-21			
2.0	2.0	A-22			
2.0	3.0	A-23			
2.0	4.0	A-24			

c. Specify any assumptions you made to accommodate the GDDM requirements, such as approximating sloped glazing, elongated openings, or non-rectilinear apertures with smaller rectilinear window or skylight components.

Footprint Selection Table [suggested format]

Aperture	Height or Length (H) or (L)	Width (W)	Sill Height (S)	Ratios (H/W) window (L/W) skylight		(S/H) (S/W)	Footprint Selected
skylight 1	*L = 6'*	*6'*	*18'*	*1.0*		*3.0*	*A–32*

"RAW" ILLUMINANCE FOOTPRINTS (continued)

2. Draw the footprints on an overlay of your plan at the workplane (recommended scale: $^1/_8$" = 1'-0").

   a. Sketch the apertures [Figure E2.2.3].

   For skylights project the aperture outline to the workplane.

   For windows rotate them 90° to the workplane (as though they were hinged at the intersection of the workplane and the vertical plane).

   b. Adjust the footprints. On the footprint the rectangle scribed by the solid line must be approximately the same size as its aperture on the plan. If they don't match, the footprint must be adjusted.

   (1) Mechanically: Use a photocopier to reduce or enlarge the footprint to the proper scale.

   (2) Manually: Redraw the footprint at the correct scale by using the actual dimensions of the aperture as the solid rectangle. Construct a proportional grid based on the aperture dimensions to guide redrawing the contours.

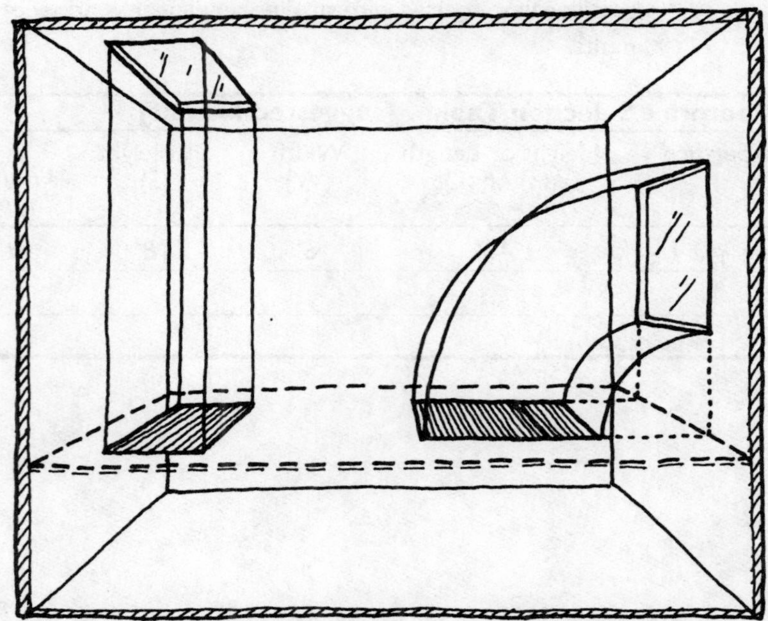

Figure E2.2.3 For GDDM, Drop a Skylight, Rotate a Window

"RAW" ILLUMINANCE FOOTPRINTS (continued)

c. Sketch the contours for each aperture.

(1) Align the footprint's solid rectangle with the outline of the aperture. For sidelight footprints, the solid line at the bottom indicates the position of the window wall. Don't position them upside-down!

(2) Note DF contour values on your plan. These values are percentages of available daylight. The dotted contour is 0.5% DF, the adjacent solid contour is 1.0% DF, and they increase by 1% each contour. The number in the center of the contours indicates the maximum DF value.

d. Combine the contours. (This overlay will be the final "raw" DF isolux contour drawing for your space.)

(1) On this overlay, mark all the points where the contours intersect, and note the sum of their DF values. If your building has many apertures that are close together, add two or three footprints at a time to derive "macro" footprints, and then add these "macro" footprints to get the final contours.

(2) Draw the resulting contour lines by connecting points of equal value. If there is no overlap, trace the original contour. Label each contour with its DF value.

3. If your DF contour distribution is not close to meeting Goals A and B, change the design now, and re-draw the contours.

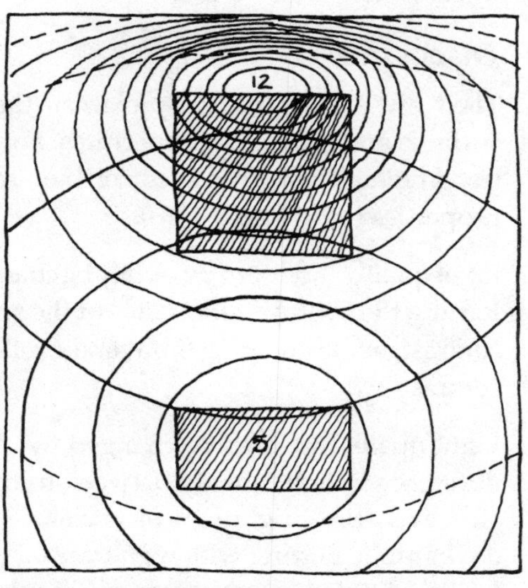

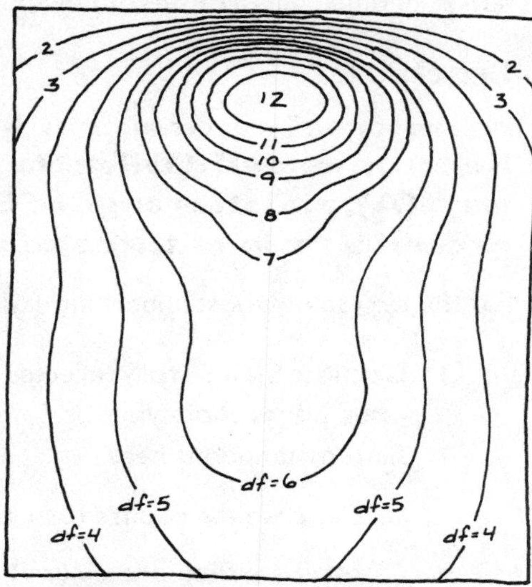

Figure E2.2.4 ***Above,*** **Contours for Window and Skylight;** ***below,*** **Combined Contours.**

DESIGN DEVELOPMENT

"CORRECTED" ILLUMINANCE FOOTPRINTS

DISCUSSION

The "raw" DF contours approximate the distribution of light in a room with all black surfaces and transparent apertures. For rooms with any other surface reflectance or transmittance, you must adjust the "raw" DFs to compensate for their luminous properties ("corrected" DFs).

Light quality in a room can easily be changed by altering the reflectances of its surfaces. Raising the average reflectance of the space will brighten the room and even out the contrast, while lowering the average reflectance will darken the room and sharpen the contrast.

Light quality can also be changed by altering the transmittance of the glazing. The difference in light quality between transparent and translucent surfaces is minimal. However, the brightness of translucent surfaces may cause glare. To maximize daylighting, glazing with high transmittance is required. Low transmittance (characteristic of tinted glass) causes dimmer interiors.

PROCEDURE

Evaluate how well your design attains Lighting Design Development Goals A and B—**Recommended daylight factors are achieved in each task area (A), and the range of DFs in each task area does not exceed a ratio of three to one (B).**

For the largest and most important daylighted space in your building:

1. Calculate the internally reflected component (IRC) for your space. If you made assumptions about your design to select the footprints in E2.2, maintain the same assumptions here.

    a. Calculate the room's total interior surface area.

    total surface area = (wall area) + (floor area) + (ceiling area)

    b. Calculate separately the net exterior glazing surface areas for sidelighting and toplighting apertures. In your calculations, use only the actual glazed area. Omit non-glazed components such as mullions or glazing bars. Alternately, you could calculate the glazing surface areas by multiplying the total opening size by a typical glazing reduction factor.

Glazing Reduction Factors

Window Type	Net Glazed Factor
all metal windows	.80
metal windows in wood frames	.75
all wood windows	.65–.70

Reprinted, by permission, from Millet and Bedrick, "Manual: Graphic Daylight Design Method," 37.

c. Calculate the ratio of glazed area to total room surface area for sidelighting and toplighting.

d. Calculate your average room reflectance. (If you are working with one typical bay or a large space such as a factory, assume 100% reflectance from the "sidewalls" that represent adjacent bays, and include their area in the total surface area for a typical bay. It is not necessary to calculate the entire factory.)

To find the values for reflectance, see *MEEB*, Tables 19.14 and 20.1, pp.1003, 1066. Assume clear glass reflectance is zero. For tinted and diffusing glass values, use colors in *MEEB*, Table 20.1.

$$R_{room} = [(SA_1)(R_1) + (SA_2)(R_2) + (SA_n)(R_n)] / (SA_{room})$$

where:

R_{room} = average room reflectance
SA_1, SA_2, SA_n = area of each surface (ft^2)
R_1, R_2, R_n = reflectance of each surface
SA_{room} = total surface area (ft^2)

e. Find the toplighted IRC values in Tables E2.3.1 and E2.3.2 below. Find the sidelighted IRC values in Table E2.3.3 below.

f. Correct the IRC values for the interior maintenance factor (IMF) which accounts for light absorption due to soiled surfaces (dusty, sooty, covered with other particulate matter).

adjusted IRC = (IRC) (IMF)

IMF Values

Work Areas	Non-Industrial	Industrial
Cleaned	.9	.8
Not Cleaned	.7	.6

Reprinted, by permission, from Millet and Bedrick, "Manual: Graphic Daylight Design Method," 35.

IRC Correction [suggested format]

Space	Total Surface Area (ft^2)	Net Sidelight Glazing (ft^2)	Net Toplight Glazing (ft^2)	Average Room Reflect-ance	Top-lighted IRC	Side-lighted IRC	IMF	Adjusted Toplight IRC	Adjusted Sidelight IRC
office	*3,800*	*114*	*38 horiz.*	*.55*	*1.09*	*1.17 avg. 0.98 min.*	*.9*	*.98*	*1.05 0.88*

"CORRECTED" ILLUMINANCE FOOTPRINTS (continued)

Table E2.3.1 Internally Reflected Component for Toplighted Rooms

Glazing from Horizontal up to 60°

Net Glazing Area to Total Room Surface Area Ratio	Average Reflectance (IRC) of Room Surfaces								
	20%	25%	30%	35%	40%	45%	50%	55%	60%
.01	.22	.32	.40	.49	.6	.72	.88	1.09	1.4
.02	.46	.62	.78	.9	1.2	1.5	1.8	2.2	2.6
.03	.6	.9	1.15	1.4	1.8	2.15	2.7	3.4	4.1
.04	.9	1.2	1.65	1.9	2.4	3.0	3.7	4.4	5.3
.05	1.15	1.5	1.9	2.4	3.0	3.7	4.3	5.4	6.6
.06	1.4	1.85	2.3	2.9	3.7	4.4	5.4	6.6	8.0
.07	1.65	2.1	2.3	3.8	4.2	5.0	6.2	7.4	8.6
.08	1.35	2.9	3.2	3.5	4.7	5.9	7.2	8.6	10.5
.09	2.0	2.8	3.6	4.4	5.4	6.6	8.0	9.6	11.7
.10	2.2	3.1	4.1	4.5	5.9	7.3	8.8	10.7	13.0

Reprinted, by permission, by Millet and Bedrick, "Manual: Graphic Daylight Design Method," 33.

Table E2.3.2 Internally Reflected Component for Toplighted Rooms

Vertical Glazing

Net Glazing Area to Total Room Surface Area Ratio	Average Reflectance (IRC) of Room Surfaces								
	20%	25%	30%	35%	40%	45%	50%	55%	60%
.01	.09	.13	.16	.20	.25	.31	.37	.45	.54
.02	.19	.25	.32	.41	.50	.62	.76	.90	1.1
.03	.27	.38	.48	.60	.75	.90	1.2	1.3	1.65
.04	.38	.50	.69	.80	1.0	1.2	1.48	1.8	2.2
.05	.47	.63	.80	.99	1.22	1.6	1.85	2.2	2.7
.06	.56	.77	.95	1.2	1.5	1.8	2.2	2.7	3.4
.07	.66	.88	.96	1.4	1.72	2.1	2.55	3.2	3.7
.08	.77	.99	1.25	1.6	1.95	2.4	3.0	3.7	4.4
.09	.86	1.15	1.42	1.8	2.2	2.7	2.45	4.1	5.0
.10	.94	1.2	1.65	2.0	2.4	3.1	3.8	4.5	5.5

Reprinted, by permission, from Millet and Bedrick, "Manual: Graphic Daylight Design Method," 34.

"CORRECTED" ILLUMINANCE FOOTPRINTS (continued)

Table E2.3.3 Internally Reflected Component for Sidelighted Rooms

Net Glazing Area to Total Room Surface Area Ratio		Average Reflectance (IRC) of Room Surfaces								
		20%	25%	30%	35%	40%	45%	50%	55%	60%
.01	avg IRC	.05	.07	.10	.13	.18	.25	.32	.40	.50
	min IRC	.02	.03	.05	.08	.12	.17	.25	.33	.43
.02	avg IRC	.09	.13	.20	.27	.36	.48	.62	.78	.96
	min IRC	.03	.06	.10	.16	.24	.34	.47	.64	.84
.03	avg IRC	.13	.20	.29	.40	.54	.74	.94	1.17	1.45
	min IRC	.03	.09	.15	.24	.36	.52	.72	.98	1.24
.04	avg IRC	.18	.27	.39	.53	.72	.96	1.25	1.60	1.95
	min IRC	.07	.12	.20	.31	.48	.70	.96	1.27	1.66
.05	avg IRC	.23	.34	.48	.66	.85	1.18	1.55	1.95	2.40
	min IRC	.08	.15	.26	.39	.60	.86	1.18	1.60	2.08
.06	avg IRC	.27	.41	.58	.80	1.07	1.43	1.80	2.35	2.92
	min IRC	.10	.18	.30	.46	.72	1.03	1.43	1.95	2.50
.07	avg IRC	.32	.47	.68	.92	1.25	1.68	2.15	2.70	3.40
	min IRC	.11	.21	.35	.56	.84	1.20	1.67	2.22	2.90
.08	avg IRC	.36	.54	.78	1.07	1.42	1.95	2.50	3.15	3.90
	min IRC	.13	.24	.41	.64	.96	1.38	1.85	2.60	3.33
.09	avg IRC	.41	.60	.87	1.18	1.61	2.15	2.80	3.52	4.39
	min IRC	.15	.27	.45	.72	1.06	1.55	2.12	2.92	3.70
.10	avg IRC	.45	.68	.96	1.30	1.74	2.40	3.10	3.90	4.80
	min IRC	.16	.29	.50	.79	1.18	1.71	2.39	3.20	4.15
.12	avg IRC	.54	.80	1.15	1.58	2.10	2.85	3.70	4.65	5.80
	min IRC	.20	.35	.60	.94	1.40	2.00	2.70	3.80	4.90
.14	avg IRC	.64	.94	1.35	1.85	2.50	3.33	4.30	5.40	6.90
	min IRC	.24	.42	.71	1.10	1.65	2.40	3.32	4.50	5.80
.16	avg IRC	.72	1.07	1.55	2.10	2.80	3.80	4.90	6.55	7.70
	min IRC	.27	.47	.82	1.25	1.90	2.72	3.80	5.10	6.60
.18	avg IRC	.81	1.20	1.70	2.35	3.20	4.30	5.50	6.90	8.30
	min IRC	.29	.53	.92	1.38	2.10	3.10	4.25	5.70	7.40
.20	avg IRC	.90	1.30	1.80	2.60	3.50	4.70	6.00	7.70	9.70
	min IRC	.33	.60	1.00	1.55	2.35	3.80	4.70	6.40	8.10

Reprinted, by permission, from Millet and Bedrick, "Manual: Graphic Daylight Design Method," 32.

2. Calculate corrections to the DFs [E2.2].

   a. Calculate the reduction in daylight resulting from glass transmission losses and shading devices.

      (1) Find the daylight reduction due to soiled surfaces.

Transmission Loss

Building Location	Inclination of Glazing	Interior Conditions	
		Cleaned	Not Cleaned
Non-Industrial Area	Vertical	.9	.8
	Sloping	.8	.7
	Horizontal	.7	.6
Industrial Area	Vertical	.8	.7
	Sloping	.8	.7
	Horizontal	.6	.5

Reprinted, by permission, from Millet and Bedrick, "Manual: Graphic Daylight Design Method," 37.

      (2) Find your glazing's transmisison reduction factor [*MEEB*, Table 19.6, p.996].

   b. Correct all the GDDM daylight factors you have found.

      (1) Correct the contours in the dimmer areas (lowest daylight factors).

corrected DF = (DF + IRC_{top} + IRC_{slm}) (TL) (GTL)

where:

IRC_{top} = toplighted IRC

IRC_{slm} = minimum sidelighted IRC

TL = transmission loss from soiled surfaces

GTL = glazing transmission loss

"CORRECTED" ILLUMINANCE FOOTPRINTS (continued)

(2) Correct the remaining contours.

corrected DF = (DF + IRC_{top} + IRC_{sla}) (TL) (GTL)

where:

IRC_{sla} = average sidelighted IRC

DF Correction [suggested format]

Raw DF	IRC_{top}	IRC_{sla}	IRC_{slm}	TL	GTL	Corrected DF
0.5	*1.09*	*NA*	*.98*	*.8*	*.85*	*1.75*
1.0	*1.09*	*NA*	*.98*	*.8*	*.85*	*2.09*
2.0	*1.09*	*1.17*	*NA*	*.8*	*.85*	*2.90*

(3) Relabel your raw DF contours with the corrected DF values that you have calculated.

3. Discuss the success of your lighting strategy, and indicate any design changes you view as necessary to meet Goals A and B.

E2 LIGHTING

DESIGN DEVELOPMENT

GDDM RENDERINGS

E2

E2.4

Discussion

Corrected GDDM contours accurately describe the daylight distribution as an abstract diagram resembling a topographical or weather map. By rendering the GDDM contours to correlate with their light levels [Figure E2.4.1] you can produce a visualization of the lighting throughout the space. Alternately, plotting the DF values in section [Figure E2.4.2] will lend context to the light distribution.

Figure E2.4.1 Value-Rendered GDDM Contours

Procedure

Evaluate how well your design attains Lighting Design Development Goals A and B—**Recommended daylight factors are achieved in each task area (A), and the range of DFs in each task area does not exceed a ratio of three to one (B).**

For the largest and most important daylighted space in your building:

1. Redraw new contours at integer intervals on an overlay by visually interpolating for the corrected values [E2.3].
2. Plot the DF contours in context. Choose a section that cuts through an aperture and vital task areas.
   a. Draw a section.
   b. Plot the DF contours on a vertical scale from zero to maximum.
3. Value- or color-render the plan and section to visualize the change and distribution of lighting.
4. Compare these renderings with your schematics in D2.2 and D2.3. Discuss any changes you deem necessary to meet Goals A and B.

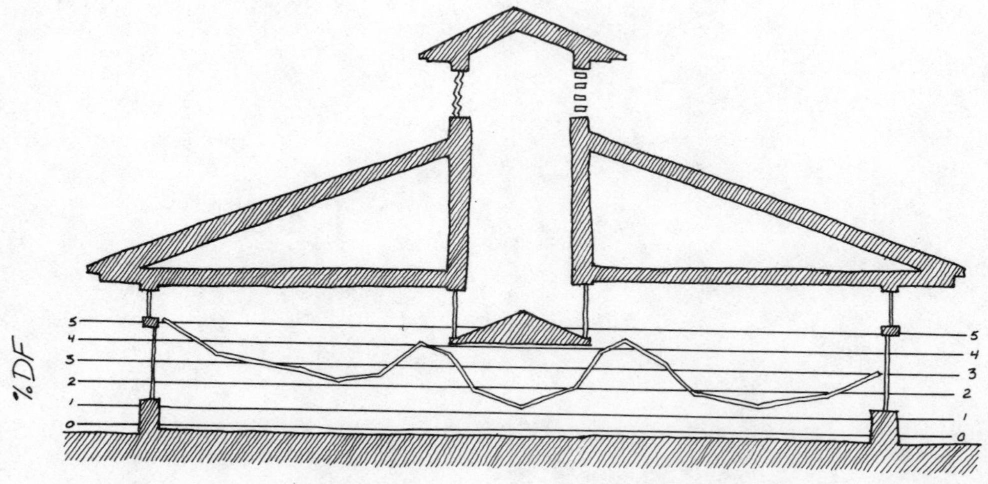

Figure E2.4.2 Daylight Section. Linen supply, Charleston, South Carolina.

DAYLIGHT MODEL

Discussion

It is advantageous to use physical models to test daylight designs. Models can be accurate representations of daylighted spaces and easily modified to test minor design changes, such as the substitution of one set of shading devices for another. They can be subjected to tests using a variety of observation techniques, such as:

- general visual assessment
- comparative photographic studies
- light meter readings.

The most critical factor when testing your model is matching as closely as possible the sky condition for which you are daylighting:

- Overcast skies should be truly overcast—the actual position of the sun should not be discernible. Avoid partly cloudy skies due to rapidly fluctuating daylight levels and unpredictable light distribution.
- Clear skies are a problem because the sun position should be at an altitude/ azimuth similar to the time of day and year in the climate for which you are designing. Do not tilt the model with a sun peg to achieve this altitude/azimuth because your building's openings will "see" a wildly different proportion of sky *versus* the ground when your model is not horizontal. Measurements must be taken to exclude the sun component yet include the sky component.

Also critical is the constancy of the outside light, against which interior measurements are compared to obtain the daylight factor. You must simultaneously measure exterior and interior levels to ensure that your DFs will be based on accurate comparative measurements. Some meters with multiple probes will automatically calculate DFs.

A less critical factor is the ground reflectance. To obviate ground reflectance, match the texture and reflectance of your site. For the first 10' height of opening above grade, a 40' width of controlled ground reflectance is necessary.

For more information

Evans, *Daylight in Architecture*, Chapter 6.

Moore, *Concepts and Practice of Architectural Daylighting*, Chapter 14.

DAYLIGHT MODEL (continued)

PROCEDURE

Evaluate how well your design attains Lighting Design Development Goals A and B—**Recommended daylight factors are achieved in each task area (A), and the range of DFs in each task area does not exceed a ratio of three to one (B).**

For the largest and most important daylighted space in your building:

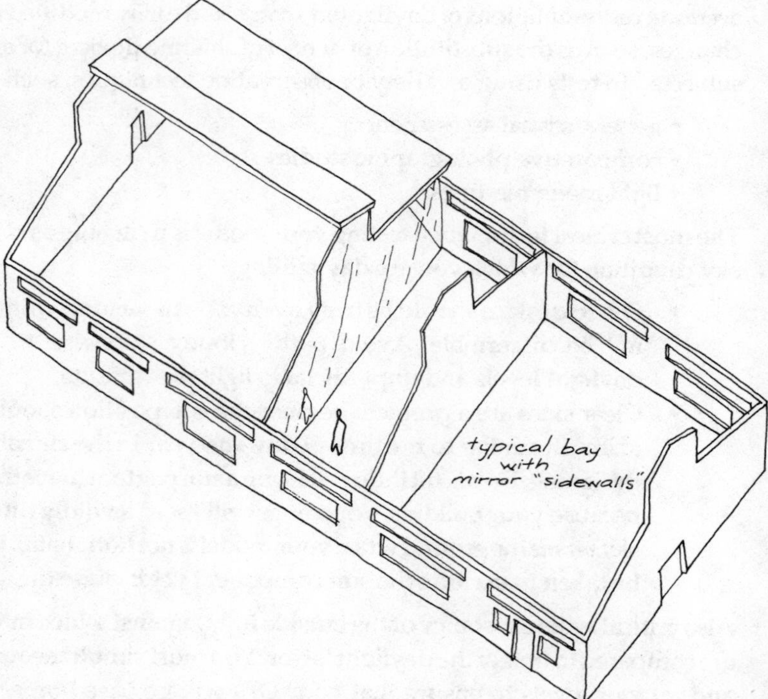

Figure E2.5.1 Modeling a Typical Bay. Linen supply, Charleston, South Carolina.

1. Build a model of the space you are investigating.

   a. Your model must be at least ½" = 1'- 0" to accurately use a standard light meter. Draw a 5' x 5' grid (minimum) on the floor of your model. Measurements will be taken at the grid intersections. Plan to take at least 15 measurements spaced equally throughout the floor area. Be accurate in your replication of details around the daylight openings—the size and depth of the mullions, the depth and reflectivity of the sill, louvers, or other shading devices, and the reflectivity of surfaces just outside the daylight openings.

   b. Choose materials that reasonably match the opacity, surface texture, and percentage reflectance of the floors, walls, and ceilings of your space. For large spaces with repetitive bays, such as a factory, build just one bay and insert mirrored "side walls" where adjacent bays would occur [Figure E2.5.1]. The reflectance of these "walls" should be at least 90%.

   c. Place glazing over model openings only if it matches the transmission characteristics of the actual glazing. Since it is difficult to obtain matching model glazing, omit it and adjust interior readings for transmission loss through the glazing [*MEEB*, Table 19.6, p.996].

   d. Tape joints between walls, floors, ceilings, etc. to avoid inaccurate measurements resulting from daylight entering through cracks.

   e. Ensure that light at each point can be measured accurately. If necessary, add a light-tight flap of some sort. To quickly and accurately position the light meter probe, determine its size and put stops for it on the model floor at each measurement point.

DAYLIGHT MODEL (continued)

2. Prepare recordkeeping and measurement tools.

    a. Draw a grid on a copy of your floor plan. This grid must be the same size as and aligned with the grid on your model. Make a copy of this gridded floor plan for each alternative you wish to test.

    b. Prepare the light meter so it will take measurements at the appropriate workplane height (usually 30" above the floor). The surface of the light meter's probe must be at this level. If necessary, add material under the probe to bring it to the proper height.

3. Test the model.

    At each grid intersection point, measure and record the daylight inside the model and, simultaneously, outside the model in a large open area. Don't affect the measurement by leaning over the probe.

4. Calculate your daylight factor for each point.

    DF = (interior fc / exterior fc) (total reduction factor)

5. Use the DF values at the grid intersections as a guide when plotting the daylight factor contours on your plan, and graph them in section as described in E2.4, step 2.

6. Discuss the success of your daylighting strategy as evaluated by model testing, and indicate any design changes you view as necessary to meet Goals A and B.

DESIGN REVIEW

Discussion

You have just completed design development of the largest, most important space in your building. The methods you have used should give you some insight for successfully lighting the rest of the building. This design review affords you the opportunity to view your lighting design holistically and to discuss the conflicts and tradeoffs with other design considerations for the entire building.

Procedure

1. Compare the developed design with the schematic design [D2]. Discuss how it has evolved and why.
2. You have explored the use of daylighting, external and internal shading devices, thermal zoning, design of the building's skin, and various cooling strategies. Comment on what tradeoffs have been made and what strategies worked well together.
3. Illustrate your discussion of points (1) and (2) with annotated, schematic diagrams of your building.

ELECTRIC LIGHTING INTEGRATION
INTRODUCTION

Goal

Provide electric lighting that complements the daylighting design for your building.

Electric Lighting Integration Goals

A. Lighting power densities comply with code, while illumination to meet the needs of users is provided throughout the building.
B. Daylighting is supplemented with electric lighting only where and when necessary.
C. Daylight provides more than half the annual required lighting.
D. Substantial savings over nondaylight-integrated electric lighting systems will be realized.

Discussion

It is highly desirable to provide electric lighting that complements and supplements the daylighting in a naturally lighted building. In a daylighted building, the entire building acts as a lighting fixture—the structure and surfaces of the building allow the light to enter and direct its distribution throughout the space. Generally, daylight is most intense near the apertures and diminishes with distance from them. Therefore, electric lighting should be switched to illuminate areas farthest from the apertures first, then areas nearer the apertures as daylight intensity fades.

In this section you will:

1. Determine how much lighting is required based on building program.
2. Design the electric lighting system to accommodate nighttime lighting.
3. Propose an electric lighting strategy to augment your daylighting.
4. Design switching controls for daytime lighting.
5. Calculate how much energy is saved because you used daylighting.

F2 LIGHTING

ELECTRIC LIGHTING INTEGRATION DESIGN

ELECTRIC LIGHTING INTEGRATION DESIGN STRATEGIES

Component-Scale Strategies

- Use architectural features to reflect and direct both daylight and electric lighting.
- Switch electric lighting according to the daylight level provided. Lights near the aperture are switched separately from those away from the aperture.
- Control electric lighting automatically with daylight or occupancy sensors. Lights are not on when daylight is adequate or when no one is using the space.
- Segregate electric light into ambient and task lighting. Ambient lighting complements daylighting, while task lighting supplements it.

PROCEDURE

Refine your daylighting design [E2] to include electric lighting systems. Generate schematic drawings illustrating your design.

DOCUMENT YOUR DESIGN AS FOLLOWS:

1. Site plan, including parking, access drives, circulation paths, exterior lighting (scale: 1" = 100').
2. Floor plans (and roof plans, if necessary), indicating electric lighting layout (scale: 1" = 20').
3. Design diagram, annotated to identify design strategies.
4. Rendered night perspectives of your building's exterior and a major interior space.

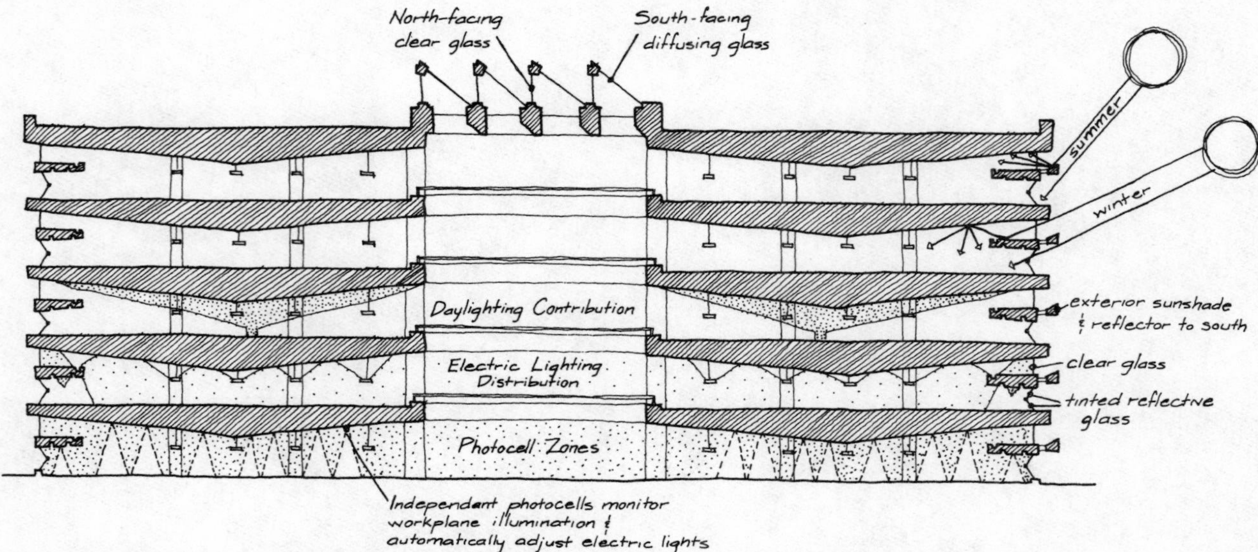

Figure F2.1.1 Lighting Section. Lockheed Building, Sunnyvale, California, Leo Daley (architect), 1983.

F2 LIGHTING

ELECTRIC LIGHTING INTEGRATION

100% ELECTRIC LIGHTING

DISCUSSION

Energy conservation guidelines call for the electric lighting system to operate at or less than a maximum power density of 2 watts/ft^2. Accomplishing this standard requires careful consideration of the task and ambient lighting systems operating under worst-case or nighttime conditions in your building.

The types of light sources you choose will have a great impact on the total lighting load. Because of poor energy efficiency, incandescent lamps are usually avoided in favor of fluorescent or high intensity discharge (HID) lamps, except where low initial cost, good color rendition, or point sources of light are required. Fluorescents are the lamps of choice for most situations, but HIDs have the advantage of longer life and higher efficacy, though their color rendition tends to be worse. In designing the layout of fixtures consider not only the task areas to be lighted but also the architectural impact of the layouts [*MEEB*, Figure 20.15, pp.1081–1084].

PROCEDURE

Evaluate how well your design attains Electric Lighting Integration Goal A—**Lighting power densities comply with code, while illumination to meet the needs of users is provided throughout the building.**

For each task area:

1. Choose appropriate lamp types for ambient and task lighting. Use the chart below and the color rendition chart [*MEEB*, Table 19.22, p.1055] to help make your selections.

Lamp Type	Application
Incandescent	Closets, confined spaces, storage, short-term use areas, performance lighting for theatres, auditoria stage spotlighting, decorative display lighting, tasks requiring a small point source of light.
Fluorescent	Multi-purpose areas, offices, classrooms and other low-ceiling applications, desk lamps and other task lighting, display cases, advertising signs, office station islands.
HID	Offices, auditoria and other high-ceiling applications, industrial areas (factories, greenhouses), outdoor area lighting, outdoor security lighting.

ELECTRIC LIGHTING INTEGRATION

100% ELECTRIC LIGHTING (continued)

2. Identify the footcandle level provided by each lighting source [C2.1 for interior spaces; *MEEB*, Table 21.5, p.1162, for exterior lighting], and determine the total wattage of electric lighting required.

   W_{total} = (watts/ft²) (area)

   where:

   watts/ft² = for each source [*MEEB*, Figure 20.3, p.1070]

   area = floor area affected by each source (ft²)

Total Lighting System [suggested format]

Task Area	Design Condition	Lamp Type	fc	watts/ft²	Area (ft²)	**TOTAL** Watts
offices	*task*	*incandescent*	*25*	*3.0*	*100*	*300*
	ambient	*metal halide*	*50*	*1.6*	*900*	*1,440*
		TOTAL		*1.93*	*900*	*1,740*

3. Calculate the total watts/ft².

$$\textbf{total watts/ft}^2 = \frac{\textbf{(total watts from all sources)}}{\textbf{(total floor area)}}$$

4. Illustrate on your building and site plans the operation of your nighttime electric system. Clearly identify all task areas, light sources, and exterior circulation paths.

5. Discuss the success of your electric lighting strategy, and indicate any design changes you view as necessary to meet Goal A.

F2 LIGHTING

ELECTRIC LIGHTING INTEGRATION

BACKUP DAYTIME LIGHTING

F2

F2.3

Discussion

Where daylighting alone is less than adequate, backup lighting is required, but a 100% electrical lighting system may be excessive. Therefore, you must determine how much backup lighting is required and where it is needed in order to design switching or dimming control systems that use all or part of the 100% electrical lighting system as needed.

For more information

MEEB, Chapter 21, Sections 21.1–21.4, pp.1140–1152 (discussion of lighting control options).

Procedure

Evaluate how well your design satisifies Electric Lighting Integration Goal B—**Daylighting is supplemented with electric lighting only where and when necessary.**

Design this backup lighting system for the major space analyzed in E2.

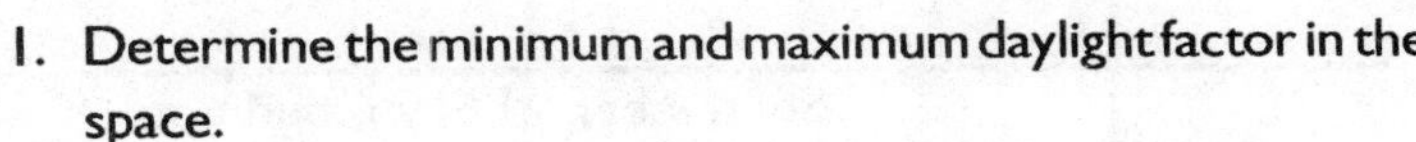

1. Determine the minimum and maximum daylight factor in the space.
   a. Review the final GDDM footprints [E2.4] for your space.
   b. Record the minimum and maximum DF in the daylight level matrix [Table F2.3.1].
2. Record the seasonal daylight availability [B2.2].
3. Calculate the seasonal internal daylight footcandles.

$$FC_{intmin} = (FC_{ext})\ (DF_{min})$$

$$FC_{intmax} = (FC_{ext})\ (DF_{max})$$

4. Compare the footcandle levels supplied by daylight (daylight level matrix) to the recommended footcandle levels [C2.1]. Determine when and how much supplemental electric lighting is required.
5. Calculate the percentage of the total electric lighting required to supply the supplemental lighting.

$$\textbf{\% electric lighting} = (fc_{reg})\ /\ (fc_{sup})$$

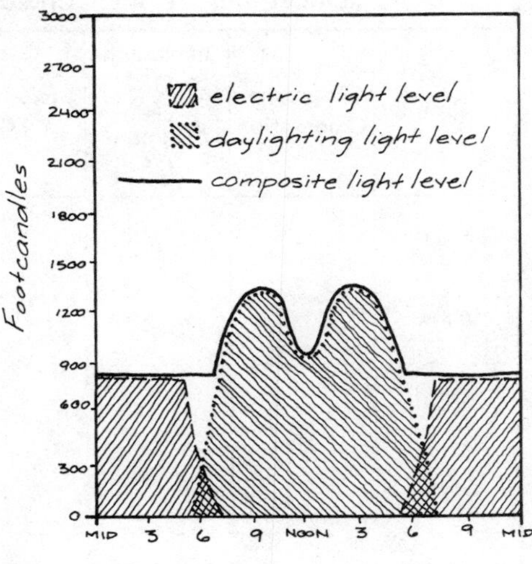

"Ideal" Dimming
(desired light level = 800 fc)

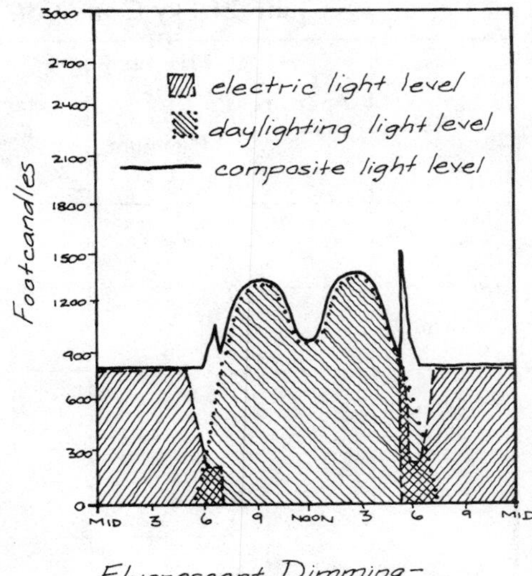

Fluorescent Dimming–
standard ballast
(desired light level = 800 fc)

Figure F2.3.1 ***Above,*** **Ideal Dimming vs.** ***below,*** **Actual Dimming for Fluorescent Lamps.**

ELECTRIC LIGHTING INTEGRATION

BACKUP DAYTIME LIGHTING (continued)

Table F2.3.1 Daylight Level Matrix

December 21 Sky Condition				
Time	Aperture Orientation	Maximum DF / Minimum DF	Exterior FC	Maximum Interior FC / Minimum Interior FC
8 a.m.				
10 a.m.				
noon				
2 p.m.				
4 p.m.				
6 p.m.				

March 21 Sky Condition				
Time	Aperture Orientation	Maximum DF / Minimum DF	Exterior FC	Maximum Interior FC / Minimum Interior FC
8 a.m.				
10 a.m.				
noon				
2 p.m.				
4 p.m.				
6 p.m.				

June 21 Sky Condition				
Time	Aperture Orientation	Maximum DF / Minimum DF	Exterior FC	Maximum Interior FC / Minimum Interior FC
6 a.m.				
8 a.m.				
10 a.m.				
noon				
2 p.m.				
4 p.m.				
6 p.m.				

September 21 Sky Condition				
Time	Aperture Orientation	Maximum DF / Minimum DF	Exterior FC	Maximum Interior FC / Minimum Interior FC
8 a.m.				
10 a.m.				
noon				
2 p.m.				
4 p.m.				
6 p.m.				

NOTES: For clear sky conditions, list the maximum and minimum footcandles for each hour for the window orientation that supplies the most light.
If the March and September sky conditions are the same, the March and September charts above are the same.

F2 LIGHTING

ELECTRIC LIGHTING INTEGRATION

BACKUP DAYTIME LIGHTING (continued)

F2

F2.3

Supplemental Illumination [suggested format]

Task Area	Season	Hours of Inadequate Daylight	Footcandles Reqd	Footcandles Daylight	Footcandles Supp	Electric Lighting Required (%)
office	*Spring*	*9a.m.–10a.m.*	*60*	*45*	*15*	*25*

6. Illustrate, with schematics, your lighting switching and control strategies. Discuss any requisite changes to the 100% electrical lighting system. On building plans indicate which lights will be used under the various supplemental conditions.
7. Discuss the success of your electric lighting control strategy, and indicate any design changes you view as necessary to meet Goal B.

ELECTRIC LIGHTING PERFORMANCE F2.4

Discussion

Since a scheme that sensitively integrates electric light and daylight will operate at less than 100% of the electric lighting capacity during the daytime, it is necessary to determine a weighted average electric load in order to compare the lighting energy used in your building with the code requirements.

Procedure

Evaluate how well your design attains Electric Lighting Integration Goal C—**Daylight provides more than half the annual required lighting.**

For each task area:

1. Identify the operating characteristics in terms of percentage of lighting in use, hours of lighting operation, and watts per square foot for each supplemental lighting condition.

2. Calculate the total watt-hours of electric lighting provided under each condition.

   watt-hours/day = (watts/ft²) (hours/day) (floor area)

   > NOTE: Dimming provides lower watts per square foot over the entire task area, while switching gives less floor area affected by the total watts per square foot.

3. Determine the total watt-hours per day for each season.

4. Calculate the weighted average lighting loads for each season in watts/ft².

   load = (WH) / [(area) (operation hours/day)]

   where:

   WH = total watt-hours per day [step 3]

   area = total floor area (ft²)

Supplemental Lighting Levels [suggested format]

Task Area	Season	Total Lighting in Use (%)	Total Lighting in Use (watts/ft²)	Hours/Day in Use	Floor Area Affected (ft²)	Watt-Hours
offices	*Winter*	*100*	*1.5*	*2*	*900*	*2,700*
		60	*0.9*	*4*	*900*	*3,240*
		20	*0.3*	*4*	*900*	*1,080*
	TOTAL					*7,020*

5. Discuss the success of your electric lighting strategy, and indicate any design changes you view as necessary to meet Goal C.

ELECTRIC LIGHTING INTEGRATION

SAVINGS FROM DAYLIGHTING

PROCEDURE

Evaluate how well your design attains Electric Lighting Integration Goal D—**Substantial savings over nondaylight-integrated electric lighting systems will be realized.**

For each task area:

1. Determine the lighting load [F2.4, step 3] in watt-hours per day for each season.

2. Compare your daylighted building design to itself as a 100% electrically lighted building (assume all the lights are on when your building is in use).

   Calculate the load in watt-hours per day.

   nondaylighted load = (night watts) (hours) (area)

   where:

   night watts = total watts/ft^2 for the 100% electric system [F2.2, step 3]

   hours = total daily operating hours

   area = total floor area (ft^2)

3. Calculate, for each season, the percentage of energy saved by employing daylighting.

$$\textbf{\% savings} = 100\% - \frac{\textbf{(daylighted watt-hours/day)}}{\textbf{(nondaylighted watt-hours/day)}}$$

Savings from Daylighting [suggested format]

Space	Season	Lighting Load		Lighting Energy
		Daylighted (watt-hours/day)	Nondaylighted	Savings (%)
office	*Winter*	*7,020*	*13,500*	*48*

4. Discuss the success of your daylighting strategy, and indicate any design changes you view as necessary to meet Goal D.

F2 LIGHTING

ELECTRIC LIGHTING INTEGRATION

DESIGN REVIEW

F2

F2.6

Discussion

Providing proper switching and control systems, the use of ambient and task lighting schemes, and specifying low-energy lamps ensure energy-conserving lighting capabilities in your design. However, the aesthetic aspects of lighting design should never be ignored. Rooms should be lighted to provide an appropriate luminous environment and rich spatial definition. It is time once again to think about and demonstrate how your lighting scheme contributes to the overall aesthetics of your design.

Procedure

1. Make schematic drawings or renderings of your lighting design that reveal the space's architectural character.
2. Annotate the drawings to explain the design.
3. Discuss changes, tradeoffs, further improvements, and how they affect the aesthetics of your design.

INTRODUCTION

GOAL

Use desirable sounds on the site, while minimizing noise (unwanted sounds).

DISCUSSION

Designing a building and site for passive heating and cooling and for natural lighting introduces design elements that have potential conflicts with acoustic design strategies. At the site-scale, openness that allows natural ventilation in the summer and adequate daylighting year-round may also allow off-site sounds, such as freeway noise, to penetrate the site. However, thick vegetation that blocks north winds and water elements for cooling can have acoustic benefits—blocking or masking unwanted sounds.

Open building plans that allow cross-ventilation, winter sun penetration, and extensive daylighting pose a problem for acoustic privacy. Adjacent areas in an open plan must not interfere with each other acoustically.

Thermal masses for passive heating and cooling introduce extensive areas of hard, acoustically-reflective surfaces that contribute to reverberant, "live" spaces that may not be appropriate. However, mass walls are excellent barriers to airborne sound transmission, as are thickly insulated, double-framed walls. Open windows for natural ventilation can allow unwanted sounds to enter the building. They also may allow indoor sounds outdoors, across an open space, and into another window—the ventilation may, therefore, by-pass the acoustic zoning and acoustic barriers.

IN THIS SECTION YOU WILL:

1. Investigate precedents of acoustic response in buildings.
2. Generalize principles of acoustic response in climatic and cultural settings.

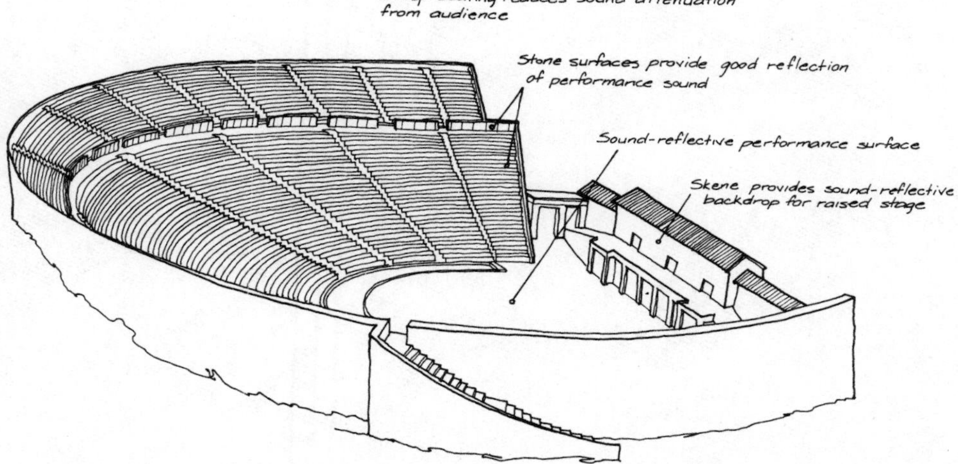

Figure A3.0.1 Typical Greek Theatre, ca. 400–350 B.C.

ACOUSTIC PRECEDENTS

Acoustic Design Strategies

Cluster-Scale Strategies

- Use buildings or other barriers to create quieter spaces, sheltered from noise sources.
- Provide quiet spaces for quiet uses.

Building-Scale Strategy

- Separate the quiet spaces from the noisy spaces.

Component-Scale Strategy

- Control noise at its source (most economical).

Procedure

Choose an existing building or site that has: (1) a building program that is similar to your assigned building program and (2) a clear, conceptual approach that incorporates some of the acoustic design strategies above.

Document your choice as follows:

1. Identify the location, program, architect (if known), and source of your information.
2. Include photocopies or drawings (whichever is quick and easy for you) to illustrate the design.
3. Evaluate the building or site design with a building response diagram and short annotations that explain how this design is organized to achieve the acoustic goal.

B3 ACOUSTICS

CLIMATE AND SITE ANALYSIS

INTRODUCTION

Goal

Determine the acoustic properties of your site and what potential benefits or problems may arise. Propose a conceptual design responsive to your climate and site.

Discussion

If you can identify the sound sources on your site and estimate their intensities, you can map the sound levels and identify acoustically favorable and unfavorable indoor and outdoor activity areas.

In this section you will:

1. Plot the sound contours for your site and building.
2. Identify acoustically favorable indoor and outdoor activity areas.
3. Develop an acoustically appropriate site design that takes advantage of features on your site.

B3 ACOUSTICS

CLIMATE AND SITE ANALYSIS

SITE SOUND CONTOURS

B3

B3.1

Discussion

There are at least three major sources of sound on your site—the freeway, the railroad, and the building itself. There also may be sources unique to your building program (e.g., running or falling water, outdoor activities). Each may be considered as either a linear or a point source of sound. For linear sources, such as the freeway and the railroad, sound is generated along a line. For point sources, such as a waterfall, sound radiates from a single point. In the case of a building, each window behaves as a point source, but a row of windows along the side of a building may be considered collectively as a line source. In any case, the sound level intensity drops 6 decibels with each doubling of distance from the point source (from 50 decibels at 1 foot to 44 decibels at 2 feet, to 38 decibels at 4 feet) and 3 decibels with each doubling of distance from the line source.

Also consider the frequency and duration of sound sources (e.g., the freeway is a constant source with peaks during rush hours, while the railroad is an intermittent source). Decide whether you need to design for the worst case or for normal conditions.

Solid obstructions on your site (e.g., earth berms, wide swathes of trees, walls, and buildings) can provide a small acoustic shadow "downwind" from the sound source.

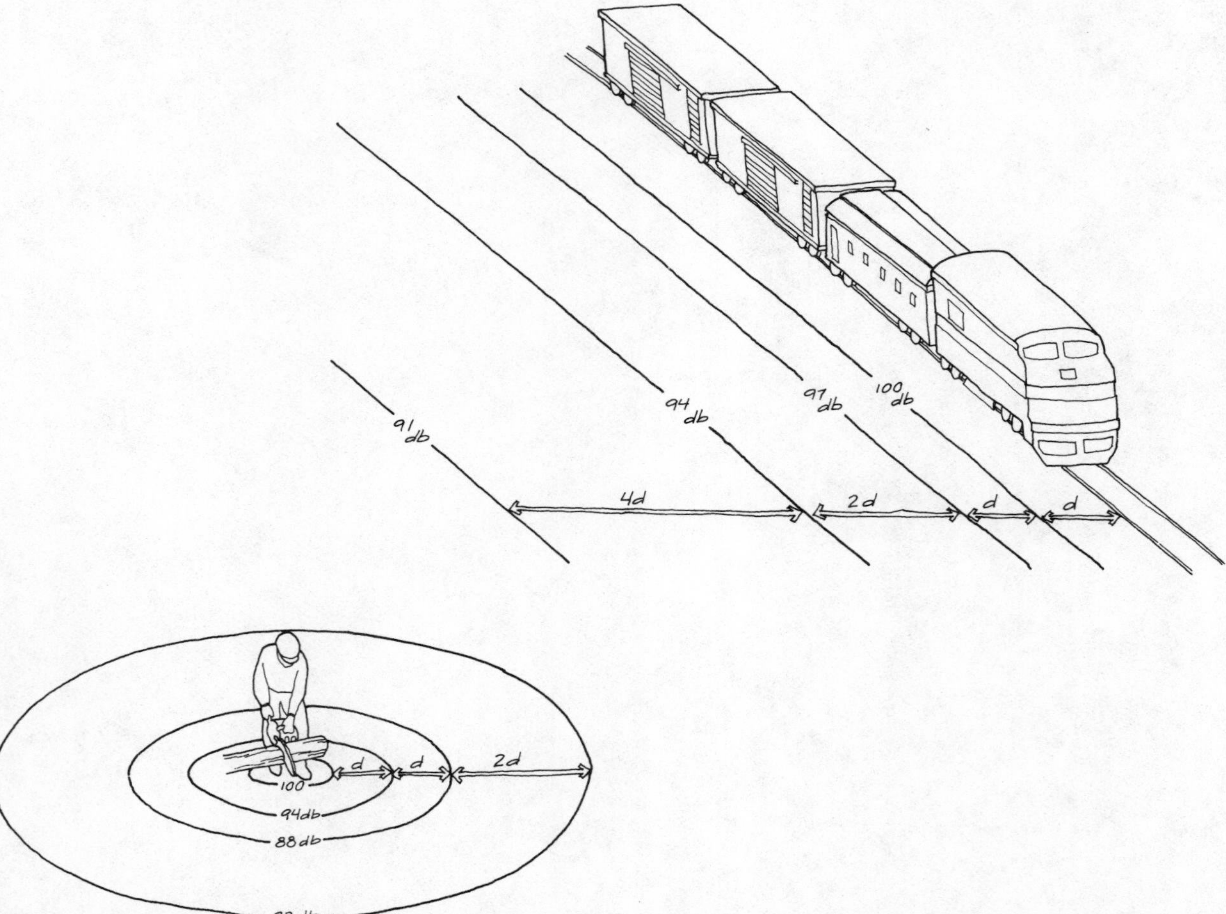

Figure B3.1.1 ***Above***, **Point Source;** ***above right***, **Line Source of Sound**

SITE SOUND CONTOURS (continued)

PROCEDURE

1. Plot the sound contours for each source (except your building) separately on your site plan.

   a. Color code or plot on a separate sheet of tracing paper each sound source.

   b. Assume a freeway sound level of 60 decibels at a distance of 50 feet; a train sound level of 70 decibels at 100 feet; and an existing solar building sound level of 50 decibels at 10 feet. Other sources may be estimated [*MEEB*, Figure 26.11, p.1335].

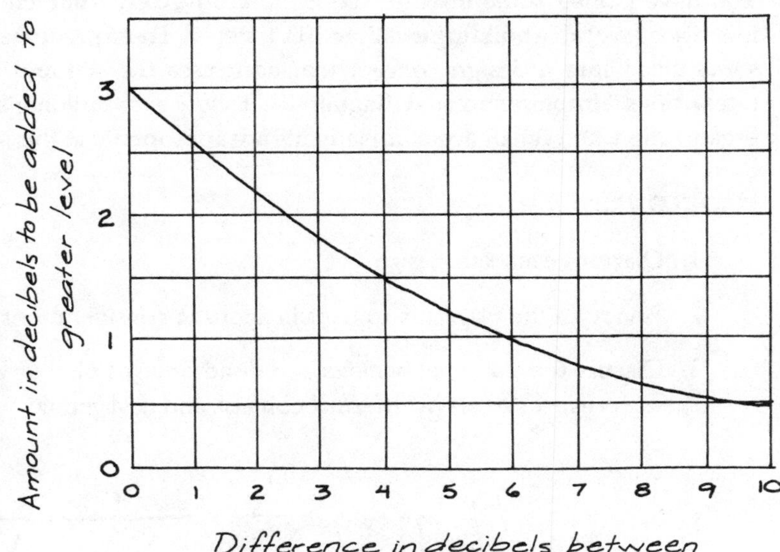

Figure B3.1.2 Nomograph for Adding Decibel Levels

2. Determine the combined sound levels, and plot them at 6-decibel intervals on your site plan.

   Add the decibel levels of overlapping contours (decibels are added logarithmically, so use the chart [Figure B3.1.2] to add the sources).

3. Plot the sound contours of your building on a separate sheet of tracing paper. You may assume a building sound level of 60 decibels at a distance of 1 foot; this sound level is a worst-case assumption for most building types.

4. Overlay all the sound contours, and identify favorable building locations. Mark these locations on your site plan.

SITE CONCEPTS

Discussion

You have gained some insights about the acoustic environment of your site and identified favorable building locations [B3.1, step 4]. These preliminary insights can be synthesized into a design concept that addresses the site and building acoustic interactions. Simple, annotated diagrams that show each building's placement, room organization, and major acoustic elements are appropriate at this stage of design.

Procedure

1. Diagram each site design.
2. Annotate the diagrams to explain acoustic considerations.
3. Discuss the potential benefits and disadvantages of your acoustic concepts with respect to views, thermal control, and daylighting.

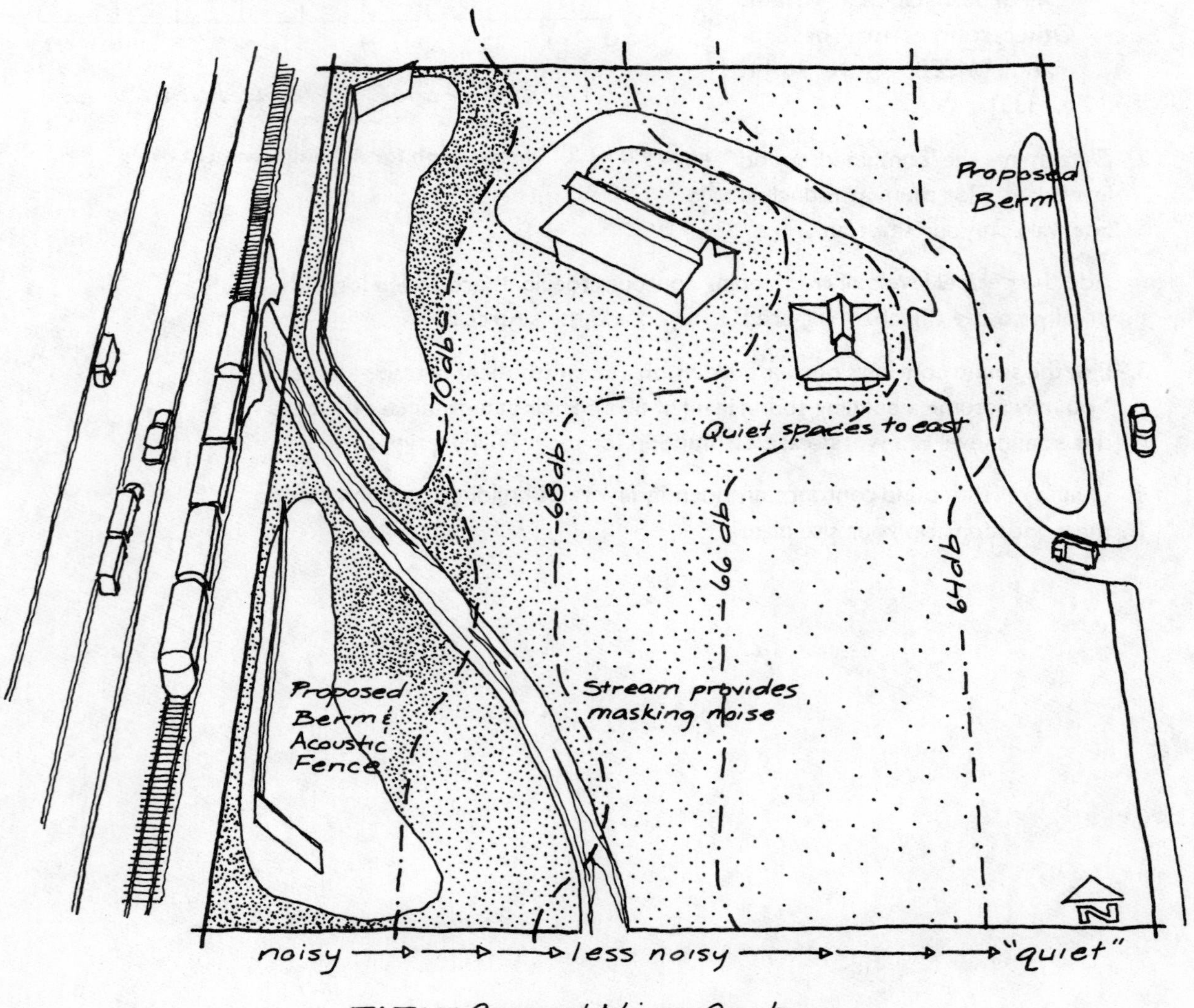

Figure B3.2.1 Acoustic Site Plan. Linen supply, Charleston, South Carolina.

C3 ACOUSTICS

PROGRAM ANALYSIS

INTRODUCTION

C3

C3.0

Goal

Determine the acoustic requirements of your building, and establish your acoustic zoning strategies.

Discussion

In order to acoustically zone your building, you must analyze the acoustic needs of each space. You can expect different sound levels in the spaces according to their functions. At the same time, you can decide how you would like each space to perform and set its acoustic goals. Using this information, you can determine which spaces are or are not acoustically compatible.

In this section you will:

1. Determine the expected sound level of each space.
2. Set a basic acoustic goal for each space.
3. Propose schematic acoustic zoning for your building.

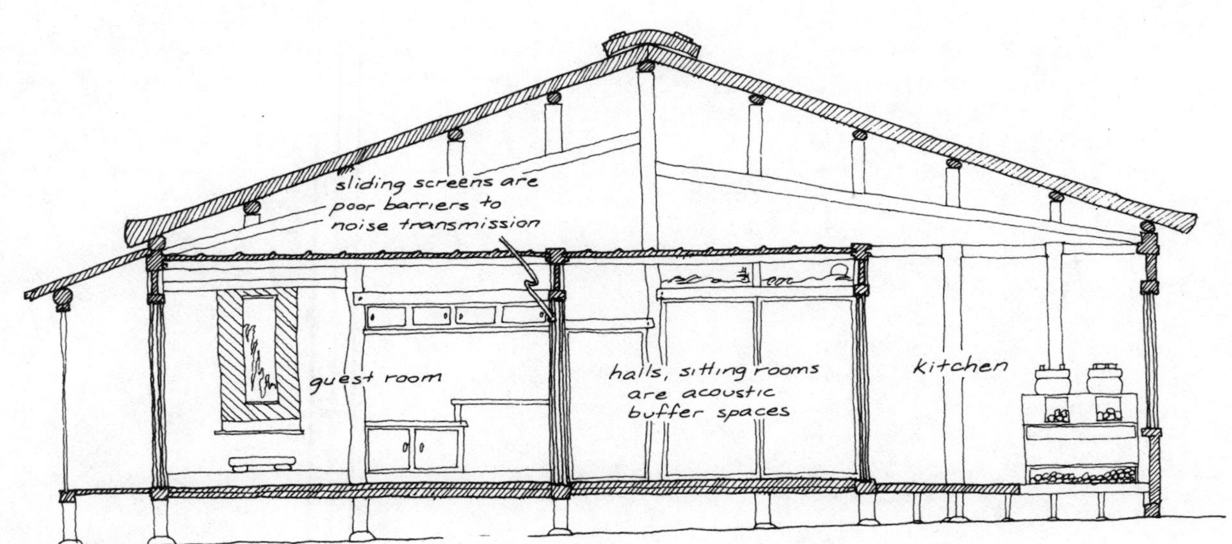

Figure C3.0.1 Acoustic Separation in a Traditional Japanese House

C3 ACOUSTICS

PROGRAM ANALYSIS

NOISE CRITERIA

C3

C3.1

Discussion

Noise criteria (NC) set the maximum background noise level for a space according to its function. Low NC (quiet) spaces should be separated from high NC (noisy) spaces, either by distance or acoustic barriers. The greater the difference between NCs, the greater the need for separation. As a guideline, sound levels for spaces whose NCs differ by more than 5 decibels are clearly different; sound levels in spaces whose NCs differ by 10 decibels seem twice or half as loud; and sound levels in spaces whose NCs differ by 20 decibels seem four times or one-fourth as loud.

Acoustic zoning also depends on the acoustic goals in your building. In silence-goal spaces, attention is focused on a single sound source, so background noise must be minimized (e.g., lecture halls and theaters). In quiet-goal spaces, background noise prevents any single sound source from predominating (e.g., offices and grocery stores).

Procedure

1. Determine an NC for each space in your building [*MEEB*, Table 27.8, p. 1403].
2. Decide on an acoustic goal—silence or quiet—for each space.

Noise Criteria [suggested format]

Space	NC (dB)	Silence/Quiet
office	*40–45*	*quiet*

C3 ACOUSTICS

PROGRAM ANALYSIS

ACOUSTIC ZONING

C3

C3.2

Discussion

In order to acoustically zone the spaces in your building you will have to identify spaces with similar or conflicting acoustic needs. Daylighting and ventilation apertures may also bring in unwanted sound from outside. Your schematic zoning proposal should address the acoustic requirements of all the spaces in your building and should mitigate all potential conflicts.

Procedure

1. Diagram your building, showing the acoustic zoning.
2. Annotate your diagram to indicate the acoustic strategies used and your rationale.
3. Discuss conflicts and harmonies among thermal, lighting, and acoustic needs in your design.

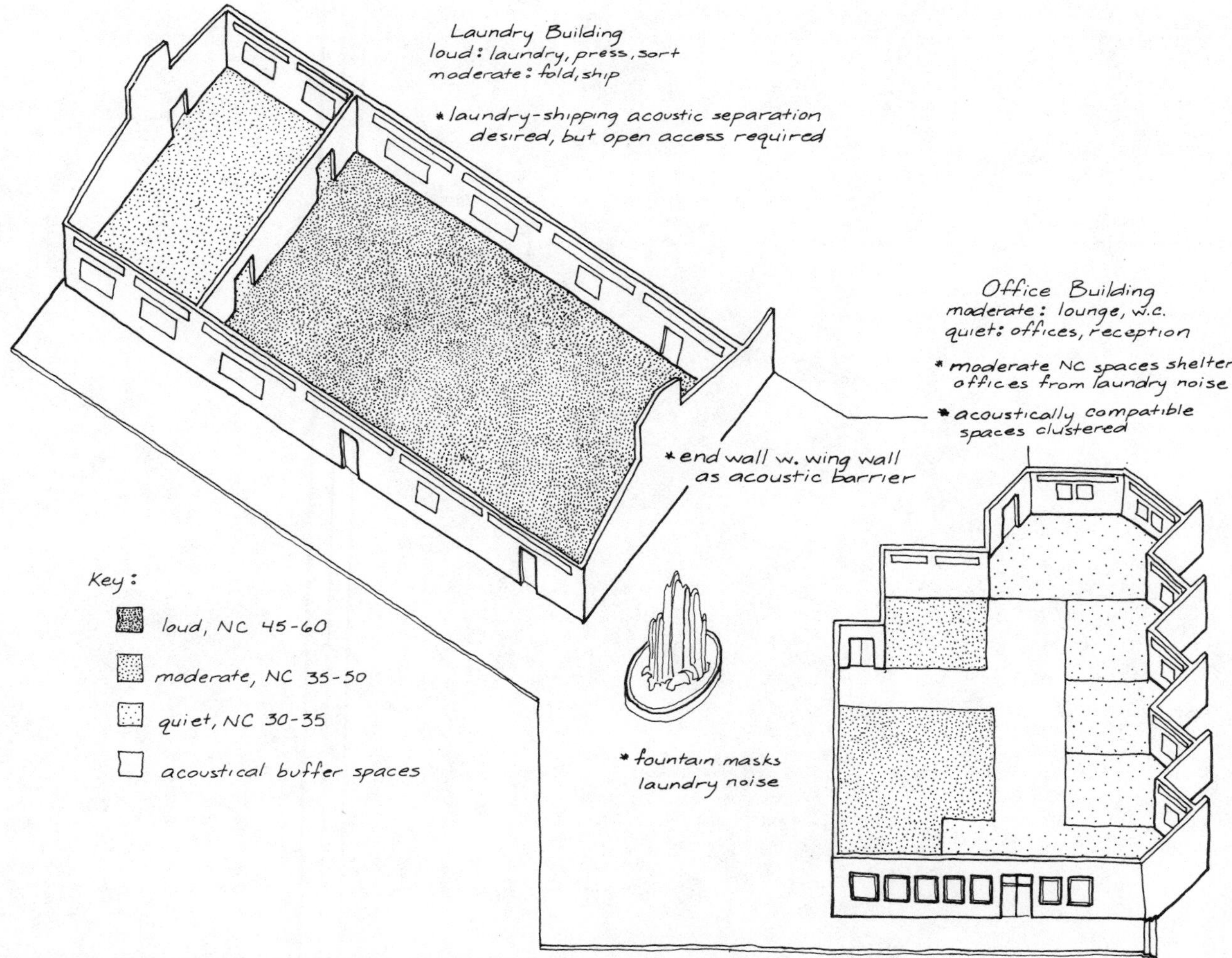

Figure C3.2.1 Acoustic Zoning. Linen supply, Charleston, South Carolina.

D3 ACOUSTICS

SCHEMATIC DESIGN

INTRODUCTION

D3

D3.0

Goal

Design a building that synthesizes the information you have acquired through the study of precedents; the analyses of your site, climate, and program requirements; and the application of acoustic design strategies.

Acoustic Schematic Design Goals

A. Potential site-scale acoustic conflicts are avoided.
B. Sound conflicts in each space are avoided.

Discussion

The acceptable background noise level in a space is indicated by its noise criteria (NC). Spaces with perceptually different, acceptable, background noise levels (>5 decibels) must be acoustically separated from each other. In most cases, acoustic walls are sufficient for this purpose. Wall and ceiling construction is rated according to its ability to reduce sound transmissions, sound transmission class (STC) [*MEEB*, Appendices F and G, pp.1514–1533]. An STC of 35 means that sound will be reduced by 35 decibels as it passes through the wall. Openings in walls (e.g., windows, doors, pass throughs, utilities) severely reduce the STC.

Rooms that contain machinery, equipment, large groups of people, or vigorous activity may have sound level peaks far above their acceptable background noise levels. These rooms may require special acoustic isolation to coexist harmoniously in your building. Sounds generated on or adjacent to the site may pose noise problems, since windows opened for natural ventilation transmit sound freely.

In this section you will:

1. Employ appropriate acoustic design strategies.
2. Complete your acoustic schematic design.
3. Illustrate the site-scale and interior acoustic features.
4. Visualize the acoustic performance of the site and building components through acoustic renderings.
5. Review and critique your acoustic schematic design.

Figure D3.0.1 Water Wall (70 dB). Paley Park, New York, New York, Robert L. Zion (landscape architect), 1966.

D3 ACOUSTICS

SCHEMATIC DESIGN

DESIGN

D3
D3.1

Acoustic Schematic Design Strategies

Site-Scale Strategies

- Use natural and manipulated landforms as sound barriers and absorbers.
- Absorb sound with dense, wide (75' minimum) swathes of vegetation.
- Mask unwanted sounds with pleasant, naturally-occurring sounds (e.g., waterfalls, rapids, rustling leaves, birds).
- Place barriers close to the sound source where they will be more effective than barriers close to the receiver.

Building-Scale Strategy

- Separate the noisiest and quietest spaces.

Component-Scale Strategies

- Use thermally massive walls as acoustic barriers.
- Control noise, especially mechanical, at the source.
- Avoid ventilation–acoustics conflicts. Windows are ears to the world. They admit sound readily when open for ventilation.

Procedure

Propose a schematic design for your building based on analyses of precedent [A3], site and climate [B3], and program [C3]. Use the appropriate acoustic design strategies for your design.

Indicate clearly on your drawings your intended acoustic design strategies.

Document your design as follows:

1. Site plan, including parking and access drives (scale: 1" = 100').
2. Floor plans (scale: 1" = 20').
3. Roof plan and elevations, or axonometrics, illustrating all building sides and roofs (scale: 1" = 20').
4. Sections (scale: 1/8" = 1'-0").
5. Design diagram, annotated to identify design strategies and acoustic zones.

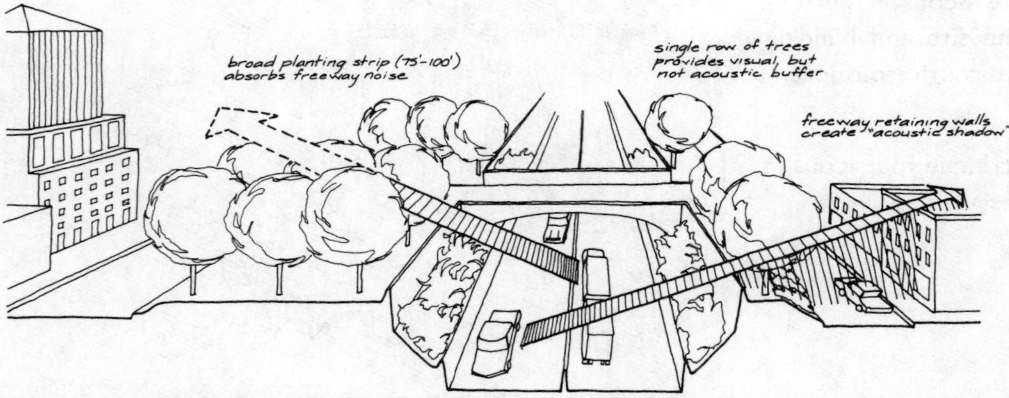

Figure D3.1.1 Acoustic Buffering Techniques in Urban Freeway Design

D3 ACOUSTICS

SCHEMATIC DESIGN

SITE SOUND RENDERING

D3

D3.2

Discussion

Sounds generated on or adjacent to the site (e.g., by trains, cars, garbage trucks) may conflict with interior noise criteria, especially when windows are open for ventilation. On the other hand, some site sounds (e.g., rustling leaves, waterfalls, fountains) may provide excellent background noise. Berms, dense vegetation, and solid walls may be used at the site scale to mitigate acoustic conflicts. Pleasant sounds may be used to mask equally loud or quieter, unwanted sounds.

Another potential site-scale acoustic conflict may occur when buildings or building wings "see" each other. When the windows are open, sound from one building or wing may cause acoustic disruption in the other.

Procedure

Evaluate how well your building attains Acoustic Schematic Design Goal A—**Potential site-scale acoustic conflicts are avoided.**

1. Draw a section through critical acoustic areas of your site and building.
2. Graphically represent the propagation of sound in the section. Show the direction of propagation, the intensity, and the quality of the sound. There is no graphic convention for representing these sound qualities, so use your imagination to devise a method of representation.
3. Discuss the site-scale strategies used and any design changes you've deemed necessary to meet Goal A. Describe acoustic problems that are not solved at the site scale.

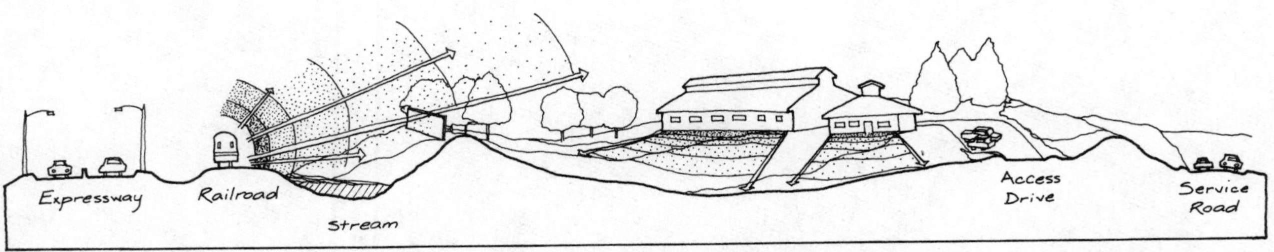

Figure D3.2.1 Acoustic Site Section. Linen supply, Charleston, South Carolina.

SCHEMATIC DESIGN

THIN-WALL BARRIER

DISCUSSION

Solid, thin-wall barriers placed near the sound source can reduce (attenuate) the sound significantly. These barriers are most effective when they are as close to the source as possible, as high as possible, and as long as possible. Thin-wall barriers are more effective in attenuating high frequency sounds (e.g., sirens, tire whine) than low frequency sounds (e.g., engine growl). Generally, a thin-wall barrier can be effectively replaced by a berm of similar height or a 75–100' wide barrier of vegetation.

FOR MORE INFORMATION

Egan, *Architectural Acoustics*, pp.253–257 (design procedure description).

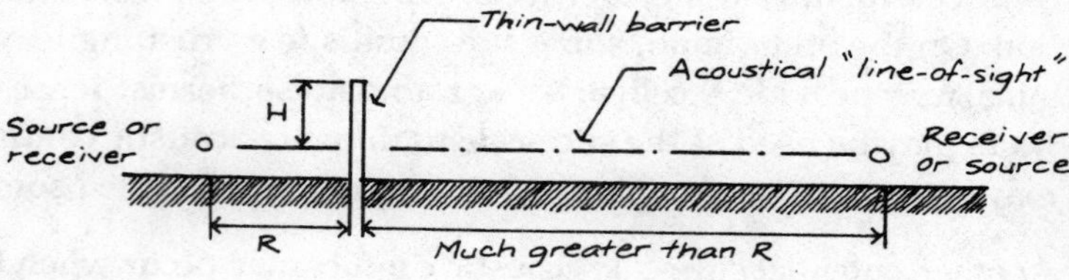

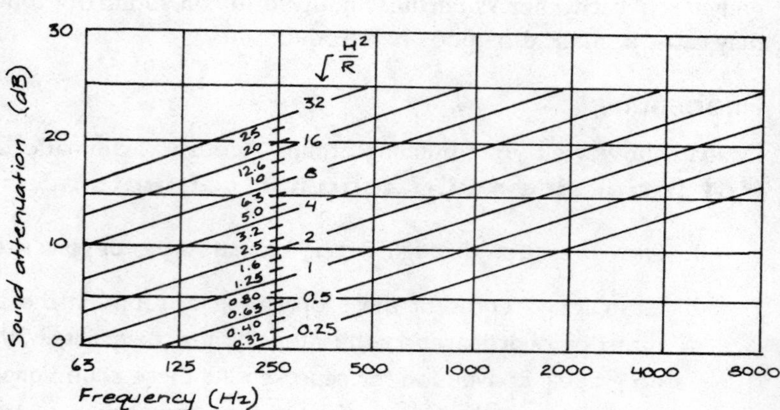

Figure D3.3.1 Thin-Wall Barrier Design Nomograph. Reprinted, by permission, from Egan, *Architectural Acoustics*, 256.

PROCEDURE

Evaluate how well your building attains Acoustic Schematic Design Goal A—**Potential site-scale acoustic conflicts are avoided.**

1. Design and draw a site section through a thin-wall barrier that mitigates one of your site-scale acoustic problems [B3.2 or D3.2].
2. Use the graph [Figure D3.3.1] to calculate the attenuation provided by your barrier.
3. Re-plot the site sound contours [B3.2] affected by your barrier.
4. Discuss the benefits and tradeoffs posed by your design in meeting Goal A. Indicate your choice of the best design resolution of this problem.

D3 ACOUSTICS

SCHEMATIC DESIGN

BUILDING SOUND RENDERING

D3

D3.4

Discussion

Activities within a building have the potential to interfere with each other acoustically. Your building should be zoned to acoustically separate activities. The zoning is successful only if the barriers between the zones are adequate. Remember, any opening, no matter how small, severely degrades the performance of an acoustic barrier. The space above a dropped ceiling is a notorious acoustic bridge between adjacent spaces.

Procedure

Evaluate how well your building attains Acoustic Schematic Design Goal B—**Sound conflicts in each space are avoided.**

1. Draw a schematic section through potentially conflicting spaces.
2. Identify the desired ambient sound level in each space [C3.1].
3. Represent the sound level in the potentially offensive room (source) and its transmission to the other room (recipient) through the acoustic barrier. Also represent the sound level in the recipient space.
4. Discuss your design strategies used for avoiding acoustic conflicts and any design changes you deem necessary to meet Goal B.

Figure D3.4.1 Acoustic Section. Linen supply office, Charleston, South Carolina.

DESIGN REVIEW

D3.5

Discussion

You have formulated a schematic design for your building and site, studied potential sound conflicts at the site- and building-scales, chosen appropriate acoustic strategies, and ensured access to appropriate environmental forces. Each step has been accomplished independently and conflicting decisions may have been made. This design review affords you the opportunity to synthesize your cumulative design decisions.

Procedure

Review how well your design accomplishes Acoustic Schematic Design Goals A–B.

1. Make a schematic design sketch that combines all your acoustic design strategies. Indicate how and why your design has evolved in response to meeting D3.0 goals.
2. Annotate your drawings to describe the systems used and the changes made.
3. Discuss the tradeoffs among the acoustic, lighting, and thermal strategies that your design required. Also discuss situations where they worked effectively together.

E3 ACOUSTICS

DESIGN DEVELOPMENT

INTRODUCTION

E3

E3.0

Goals

Refine and further develop your building's acoustic design without sacrificing desirable views, daylighting, or thermal performance. Use your schematic design as the foundation for your design development.

Acoustic Design Development Goals

A. Appropriate interior materials and finishes make spaces acoustically comfortable.

B. Wall, ceiling, and floor construction details and finish materials combine to isolate acoustic zones.

Discussion

Rooms may be classified acoustically as being "dead," "neutral," or "live." "Live" rooms have predominantly hard surfaces such as wood, masonry, plaster, glass, concrete, and metal. These hard surfaces are acoustically reflective so that sounds created in the room stay "alive" for a long time by bouncing off the surfaces. Also, reflected sounds tend to blend with new sounds so that sound in a "live" room tends to be unintelligible or noisy. Rooms that are "live" seem excessively noisy and exciting.

"Dead" rooms have predominantly soft surfaces such as upholstery, drapes, acoustic tiles, carpets, and people. The soft surfaces are acoustically absorptive so that sounds created in the room are not reflected. Sounds in "dead" rooms are distinct and clear. Rooms that are "dead" seem excessively quiet and relaxing.

"Neutral" rooms seem neither "live" nor "dead."

Effective design should provide a variety of acoustic spaces that are appropriately "live," "neutral," or "dead." For example, a concert hall or basketball court should be "live" to enhance the music or the excitement, while a lounge should be "dead" to stress the serenity it provides, and a work environment should be "neutral" to reduce stress, yet encourage productivity.

In this section you will:

1. Complete the acoustic design development.
2. Calculate the room absorbency and resultant reverberation time for three spaces in your building to determine if they are acoustically appropriate.
3. Calculate the ambient sound level of an active space in your building to determine if it is within desirable limits.
4. Design acoustic barriers between acoustically conflicting spaces and determine if they reduce noise levels sufficiently.
5. Determine if natural ventilation causes an acoustic problem.
6. Critique your building's acoustic design.

Acoustic Design Development Strategies

Component-Scale Strategies

- Reflect and direct sound with hard materials (e.g., a concrete band shell directs the music to the audience).
- Absorb sound and prevent reflective echoes with absorptive materials (e.g., an upholstered back wall in a theatre prevents echoes).
- Absorb sound with acoustic materials in the ceiling plane when other surfaces must be reflective (e.g., thermal walls and floors are highly reflective).
- Use vertical, hanging acoustic panels in the ceiling when it must be thermally massive or when it must be exposed for aesthetic or daylighting purposes.
- Ensure that acoustic walls extend from floor to ceiling to prevent sound leaking over the wall (especially through a dropped ceiling).
- Provide acoustic barriers between adjacent spaces with very different NCs (greater than 5 decibels apart) or with NCs less than 40 decibels.
- Prevent overheard conversations in adjacent, private spaces with an acoustic barrier.
- Reduce sound transmission by adding mass to a wall or by providing structural discontinuity (e.g., a double wall).
- Choose different materials for different purposes. Materials chosen for high sound absorption will generally have little value as sound barriers, due to their porous and lightweight character.

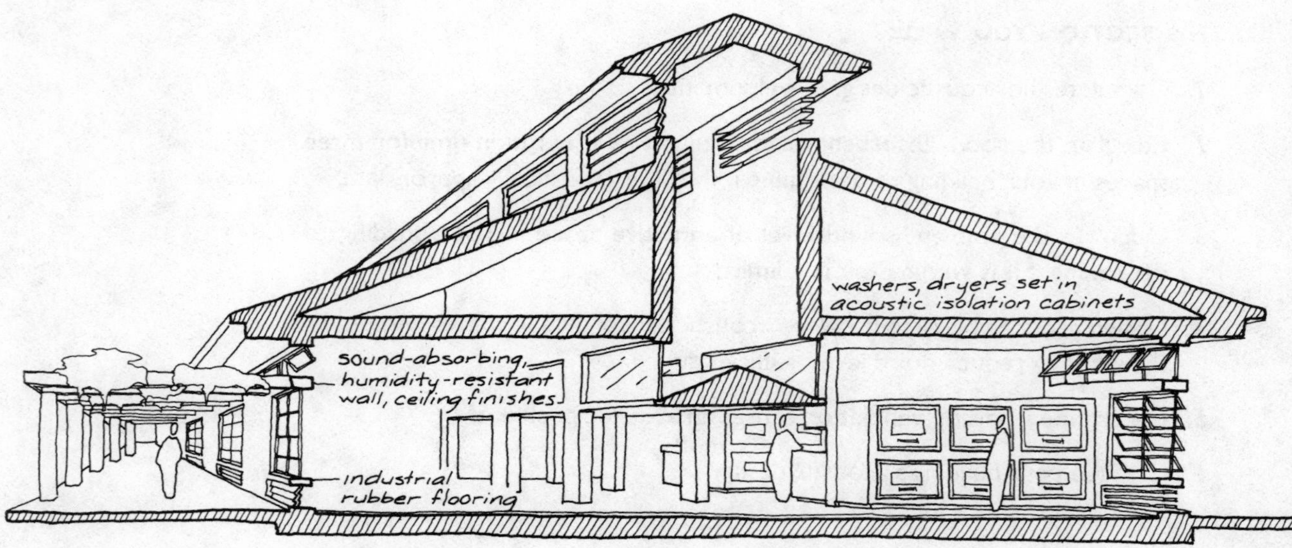

Figure E3.1.1 Acoustic Details. Linen supply, Charleston, South Carolina.

DESIGN (continued)

Procedure

Refine your schematic design [D3] to include component design considerations. Choose three spaces in your building for these acoustic studies. Select either a large room that has multiple sound sources (such as 16 cash registers) or a large room that has the acoustic goal of silence (such as a lecture hall or sanctuary). Also select two adjacent rooms that require an acoustic barrier between them.

Use the appropriate strategies to aid your acoustic design development. Develop details for your acoustic design in the three spaces you have selected. Make an educated guess of proper finish materials and surfaces to create appropriately "live," "neutral," or "dead" spaces.

Summarize your selections and design intentions.

Space	Reason for Selection	"Live," "Neutral," or "Dead"	Why?

Modify your schematic design drawings to include the components you have developed. Indicate clearly on your drawings intended acoustic design strategies.

Document your design as follows:

1. Sectional perspective of each room, indicating finish materials, surfaces, and furnishings (scale: 1/8" = 1'-0").
2. Wall or floor/ceiling section, showing the acoustic barrier between adjacent rooms (scale: 3/4" = 1'-0").

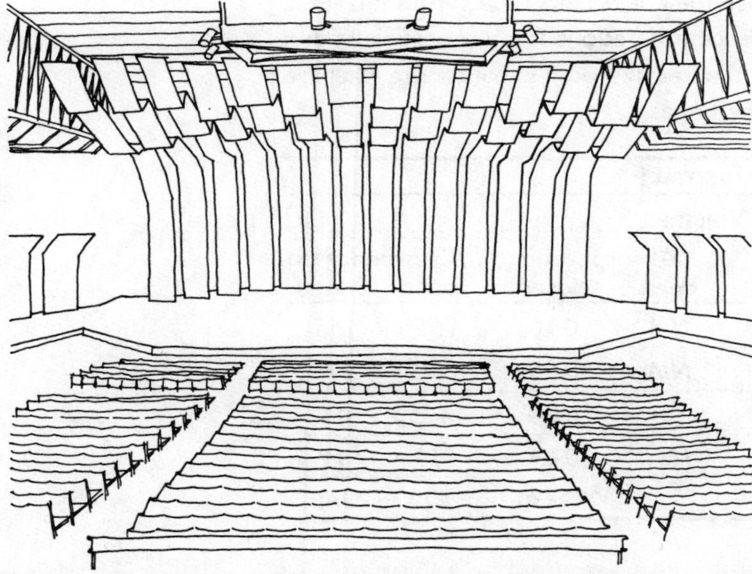

Figure E3.1.2 **Adjustable Acoustic System. Concert Theatre, C. W. Post Center, Greenvale, New York, Mitchell/Giurgola (architects), 1982.**

DESIGN DEVELOPMENT

ROOM ABSORBENCY

E3.2

Discussion

In order to determine how acoustically "live" a space is, its total absorbency must be calculated. Absorbency is measured in Sabins and can be calculated

$$A = (\alpha)\ (a),$$

where:

A = absorbency (Sabins)

α = coefficient of absorption

a = surface area (ft^2).

Total room absorbency is simply the sum of the absorbency of all the surfaces, furnishings, and people in the room. Normally, absorbency is measured at the mid-range for human speech (500 hertz). However, your space has a different characteristic sound (e.g., the whir of blenders, the cry of infants, the bleating of sheep), do your calculations using the appropriate frequency [*MEEB*, Figures 26.11–26.12, pp.1335–1336]. If the space has operable windows, calculate for both open and closed windows. An open window is totally absorptive (α = 1.0) but provides no acoustic barrier to sound. If a floor is densely covered by occupants, as in a fully occupied theatre or classroom, calculate the absorbency of the occupants, not the floor absorbency.

Procedure

Evaluate how well your design attains Acoustic Design Development Goal A—**Appropriate interior materials and finishes make spaces acoustically comfortable.**

For each of the three rooms selected:

1. Calculate the room absorbency. Use the coefficients of absorption [*MEEB*, Table 27.1, pp.1370–1371]. Include considerations for the occupants and furnishings, since they may significantly alter the room's acoustics (e.g., a bare apartment compared with a furnished one).

Absorbency Calculation [suggested format]

Room ______________________ Volume ____________________ ft^3

Surface	Material	Area (ft^2)	Absorption open	Absorption closed	Absorbency (S) open	Absorbency (S) closed
ceiling	*parquet*	*550*	*N/A*	*0.10*	*55*	*55*
occupants	*25 students*	*550*	*N/A*	*0.50*	*275*	*275*
walls	*sliding glass*	*475*	*1.0*	*0.04*	*475*	*19*
	fixed glass	*475*	*N/A*	*0.04*	*19*	*19*
				TOTAL	*824*	*368*

ROOM ABSORBENCY (continued)

2. Evaluate the liveness of your spaces using the room liveness graph [Figure E3.2.1].

3. If your room acoustics are not appropriate for the space [step 2], redesign for more or less absorption, and redo steps 1 and 2.

4. Make a schematic diagram illustrating the design changes that you deem necessary to meet Goal A, or comment on the appropriateness of your design.

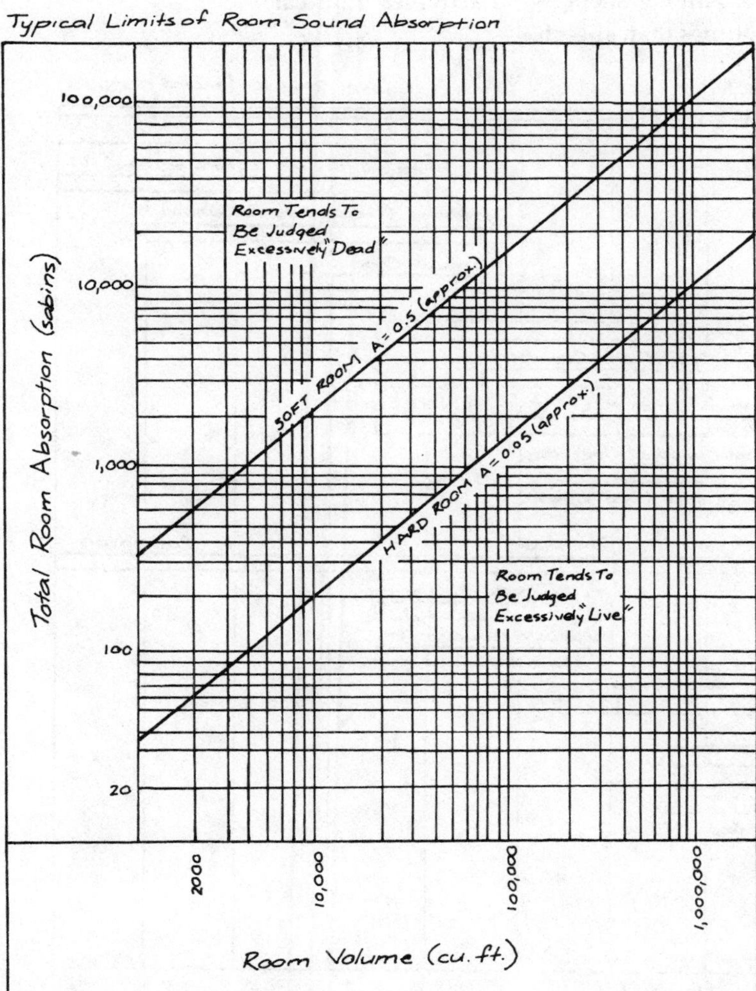

Figure E3.2.1 **Room Liveness Graph. Adapted, by permission, from Flynn and Segil, *Architectural Interior Systems*, 66.**

Discussion

Room reverberance is a factor in evaluating the suitability of a space for its function. Function can be categorized by two acoustic goals—silence and quiet.

Silence-goal rooms are designed for a monarchy of sound. One sound source dominates the room (e.g., a lecture hall, a symphony hall, a theatre, or a church). For silence-goal rooms, the reverberation time is critical in determining the fit between the room and its activities.

Quiet-goal rooms are designed for a democracy of sound. No single sound source dominates the room, so all sounds produced blend to form the background noise. Also, the room may support multiple activities (e.g., offices, supermarkets, libraries, and bowling alleys). For quiet-goal rooms, the reverberation time is useful in determining noise control characteristics.

Generally, "live" rooms have longer reverberation times than "dead" rooms and are more appropriate to silence-goal activities. Among silence-goal activities, musical performance requires longer reverberation times than speech.

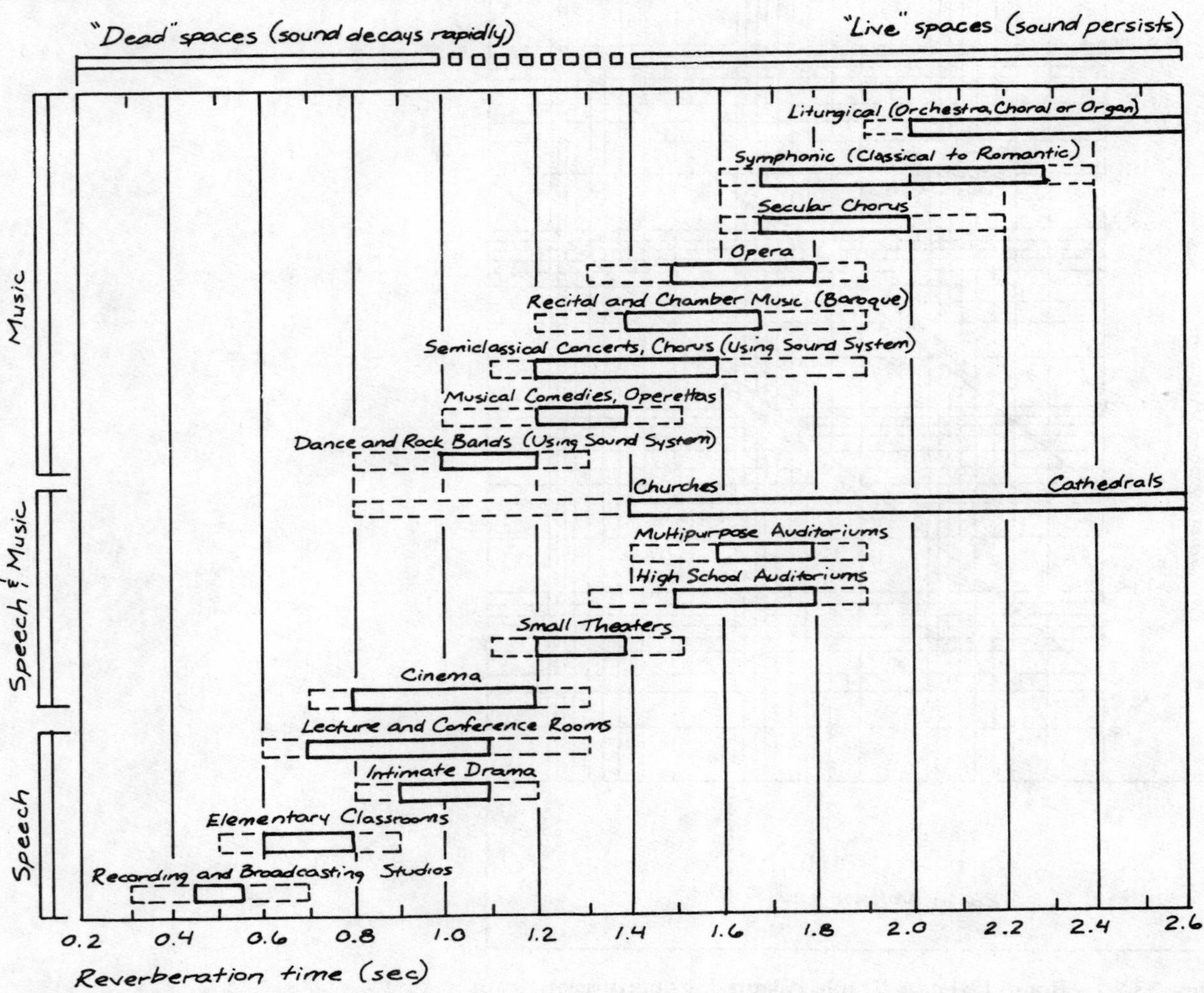

Figure E3.3.1 Recommended Reverberation Times. Adapted, by permission, from Egan, *Architectural Acoustics*, 64.

DESIGN DEVELOPMENT

REVERBERATION TIME (continued)

PROCEDURE

Evaluate how well your room attains Acoustic Design Development Goal A—**Appropriate interior materials and finishes make spaces acoustically comfortable.**

1. Determine the recommended reverberation time, T_R, for your spaces [Figure E3.3.1 or *MEEB*, Figures 26.25–26.26, pp. 1352–1353]. If you cannot determine a recommended T_R, make a note to that effect, and make an educated guess.
2. Calculate the T_R for your spaces.

   $\mathbf{T_R = [(.049)(volume)] / (absorbency)}$

Reverberation Time [suggested format]

Space	Recommended T_R (seconds)	Volume (ft^3)	Absorbency (Sabins)	Actual T_R (seconds)
lecture hall	*.7–1.1*	*7,200*	*350*	*1.0*

3. Discuss the design changes that you deem necessary to meet Goal A, or comment on the appropriateness of your design.

ROOM SOUND LEVEL

Discussion

When a room has several sound sources, it should be tuned acoustically to keep the sound level within limits as specified by its noise criteria (NC). Tuning a room is a function of the volume of the room and its reverberation time. Volume can be adjusted by changing the ceiling height, while T_R can be adjusted by changing either the volume or the absorbency of the room's materials.

The amount of sound generated is measured by the source's sound power level, L_W. This measurement is not dependent on the source being enclosed. The amount of sound generated in a specific room can be calculated, given the source's L_W and the room's T_R and volume. This measurement is the sound pressure level, L_P.

For more information

Egan, *Architectural Acoustics*, Appendix C, pp.398–399.

Procedure

Evaluate how well your room attains Acoustic Design Development Goal A—**Appropriate interior materials and finishes make spaces acoustically comfortable.**

For the room with multiple sound sources:

1. Make a schematic diagram of the space, identifying the sound sources, their L_W, and frequency band. Determine L_W [*MEEB*, Figures 26.11–26.12, Tables 26.4–26.5, pp.1335–1336, 1340, 1350].

   NOTE: The information in *MEEB* designated SPL or dBA is L_W, since it is not room-specific.

2. Determine the allowable decibel level at 500 hertz [C3.1 or *MEEB*, Table 27.8, p.1403].

3. Determine the sound pressure level, L_P, of each individual sound source. The value for $(L_P - L_W)$ is on the vertical axis of Figure E3.4.1.

   $$L_P = L_W + (L_P - L_W)$$

Room Sound Sources [suggested format]

Sound Source	L_W (dB)	Hz	T_R (seconds)	Volume (ft^3)	L_P (dB)	NC (dB)
stereo, teenager	*82*	*500*	*0.8*	*1,200*	*80*	*65*

DESIGN DEVELOPMENT

ROOM SOUND LEVEL (continued)

4. Calculate the total sound level in the space by "adding" L_P for all sound sources. Use Figure B3.1.2 as a nomograph for adding decibels.

   List the sound pressure level of the sound sources from the highest to the lowest, and add them incrementally.

Total Sound Level Calculation [suggested format]

Source	L_P (dB)	ΔL_P (dB)	Add dBs	**TOTAL** L_P (dB)
stereo #1	*80*			*80.00*
stereo #2	*80*	*0*	*3.00*	*83.00*
parent's voice	*77*	*6*	*1.28*	*84.28*

NOTE: The total sound pressure level may be calculated mathematically from the formula that defines decibels. If each L_P is converted to watts/cm^2 [*MEEB*, Table 26.2, p.1329], then

total $L_P = 10 \log [(\sum L_P) / (10^{-16})]$.

If you use this method, show your work.

5. Does your space meet the recommended noise criteria? Explain how your design must evolve to meet this criteria and Goal A.

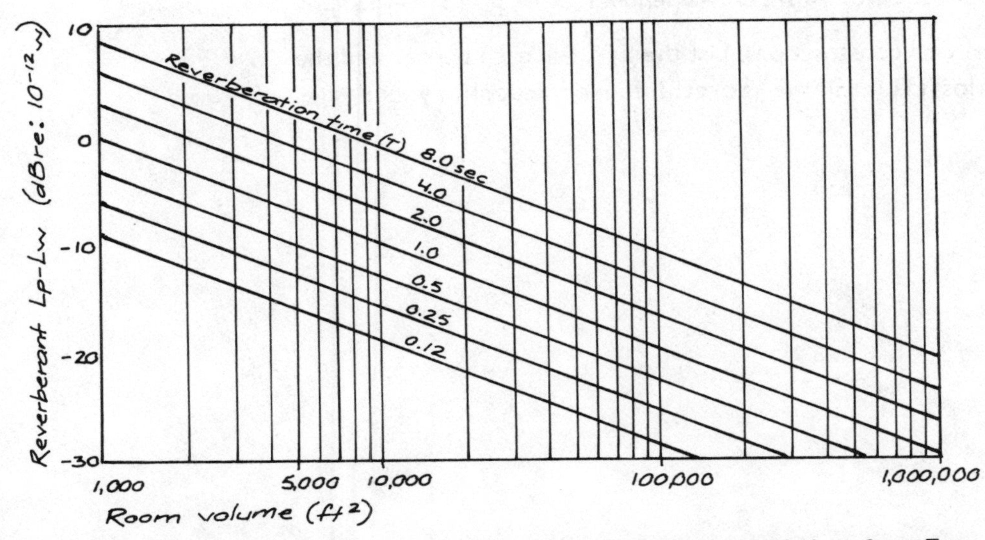

Figure E3.4.1 Reverberant $L_p - L_W$ Nomograph. Adapted, by permission, from Egan, *Architectural Acoustics*, 399.

E3 ACOUSTICS

DESIGN DEVELOPMENT

STC RATINGS

Discussion

The sound transmission class (STC) of a wall is based on its ability to reduce the decibel level of audible sound at all frequencies as the sound passes through the wall.

STC Rule-of-Thumb	
STC 35	Normal speech can be heard and understood through the barrier.
STC 40	Normal speech is heard as a murmur through the barrier.
STC 50	Very loud sounds can be heard only faintly through the barrier.

For design purposes, sound is reduced by 35 decibels as it passes through an STC 35 wall. A wall with an opening or a door usually has a significantly lower STC than a similar, solid wall.

For more information

MEEB, Section 27.16, pp.1387–1389 (STC definition).

Procedure

Evaluate how well your wall construction attains Acoustic Design Development Goal B—**Wall, ceiling, and floor construction details and finish materials combine to isolate acoustic zones.**

1. Choose at least two walls to evaluate—one a solid wall and the other a wall with an opening. One of these walls should be the barrier between the two rooms you chose in E3.1. On a sketch of your floor plan indicate where these walls are located.

2. For the solid wall:

    a. Find the recommended STC [*MEEB*, Tables 27.12–27.14, pp.1416–1418].

    b. Select appropriate barriers [*MEEB*, Appendix F or G, pp.1514–1533].

    c. Sketch the barrier constructions. List the STC for each barrier and the transmission loss (TL) at 500 hertz or at a frequency crucial for your design.

STC RATINGS (continued)

3. For the wall with openings:

   a. Select an appropriate door or window configuration [*MEEB*, Tables 27.5–27.6, pp.1393–1394].

   b. Sketch the wall and list the STC and TL at 500 hertz.

   c. Calculate the composite STC and TL for this wall [*MEEB*, Figures 27.22–27.23, p.1392].

NOTE: The highest composite TL attainable with a hole in a wall is 28.

STC of Acoustic Barriers [suggested format]

Barrier Description	Area	Recommended STC	Actual STC	TL @ 500 Hz	Composite STC	Composite TL
staggered wood stud wall	*120*	*40*	*44*	*45*		
pass-through (1' x 2')	2	*40*	*0*	*0*	*28*	*28*

4. Discuss decisions you made as a result of this evaluation and any design changes you deem necessary to meet Goal B.

OUTDOOR SOUND AND VENTILATION

Discussion

Ventilation apertures allow free movement of air for heat removal, but also allow sound to readily enter the building. Remember, a noisy room in a building may transmit sound to another room across an outdoor space when both are being naturally ventilated. The conflict between ventilation and sound control must be identified and resolved at the site scale with external noise barriers, at the building scale with appropriate acoustic zoning, or at the component scale with adequate, switchable barriers.

Procedure

Evaluate how well your room attains Acoustic Design Development Goal B—**Wall, ceiling, and floor construction details and finish materials combine to isolate acoustic zones.**

1. Select a room that uses windows to naturally ventilate on the noisiest side of the building (use the site sound contours [B3.1] to select this room).
2. Establish the sound level outside the window of the room you have selected. Consider only constant noises, like those from a freeway or an overly noisy room.
3. Record the recommended NC for your room [*MEEB*, Table 27.8, p.1403].
4. Calculate the percentage of the wall that is open window. Note the transmission loss through the solid part of the wall. Find the resulting sound level inside the space [*MEEB*, Figure 27.23, p.1392].

Sound Level through Ventilation Openings [suggested format]

Space	Sound Level Outside Window (dB)	Recom-mended NC (dB)	Wall TL or STC (dB)	Open Window (%)	Resulting TL (dB)	Resulting Inside Sound Level (dB)
office	*65*	*30–35*	*45*	*10*	*10*	*55*

5. If the recommended NC isn't met, discuss the design alternatives necessary to meet Goal B.

DESIGN REVIEW

Discussion

You have just completed acoustic design development for a few important spaces in your building. The methods you have used should give you some insight for successful acoustic design for the rest of the building. This design review affords you the opportunity to view your acoustic design holistically and to discuss the conflicts and tradeoffs with other design considerations for the entire building.

Procedure

1. Compare the developed design with the schematic design [D3]. Discuss how it has evolved and why.
2. You have explored the use of acoustic, daylighting, and thermal strategies. Comment on what tradeoffs have been made and what strategies worked well together.
3. Illustrate your discussion of points (1) and (2) with annotated, schematic diagrams of your acoustic design for the entire building.

A4 WATER AND WASTE

PRECEDENTS OF WATER/WASTE RESPONSE

INTRODUCTION

A4

A4.0

Goals

Use the aqueous resources on your site. Use rainwater when possible, conserve fresh (potable) water, and treat and recycle waste water.

Discussion

People once chose the sites for their settlements based, in part, on proximity to potable water. Until storage, plumbing, and water pumping systems were available, water was highly valued. Modern plumbing usually provides adequate quantities of high quality water with little apparent cost. The attitude of present-day water consumers in industrialized nations is much like their attitudes toward electricity and oil before the energy crisis. However, more and more people are realizing that water, too, is a limited resource. Water tables are lowering, urban land is subsiding into depleted aquifers, and regional droughts are affecting larger areas.

Even if water weren't a limited resource, the path it travels is often long, expensive, and environmentally damaging. Murray Milne (1976) describes the process. "Imagine how long it took to move through the stream, lake, river, reservoir, spillway, aqueduct, pump, main, lateral meter, pipe, valve, tube, and faucet before it finally flows out to fill your glass. Now think about the rest of the trip; down the drain, through the trap, out and down into a network of merging pipes and tunnels, and finally through a treatment plant and back into the watercourse."

A very small proportion of our direct consumption of water is actually used for drinking or cooking (approximately 5% of residential consumption). Most of the water we use transports waste. On the average during the course of a year, one person in the U.S. contaminates 13,000 gallons of fresh water to flush 165 gallons of human body waste (Stoner 1977). We also use tremendous amounts of water indirectly. For example, every time architects specify a ton of structural steel for a building they also specify the consumption of 37 tons of water (Vale and Vale 1975).

The costs of water and waste systems are more than the costs to transport the water from its source to your glass, cooking pot, or toilet. Even though most places in the U.S. have abundant water supplies, there is ample evidence to support the need for conservation. Merely reducing the amount used can have a limited effect on potable water's associated costs. More extravagant are those costs associated in support of water and waste systems.

A. The energy used for water pumping is the fastest growing line item in water system budgets. Some experts believe that within ten years, energy shortages will force the legislation of strict water conservation programs in most of the U.S.
B. It is likely that most homeowners spend more money and energy to heat water than they do to acquire the water. Because 4% of our national energy use is for domestic water heating, conserving hot water also means conserving energy.

A4 WATER AND WASTE

PRECEDENTS OF WATER/WASTE RESPONSE

A4

INTRODUCTION (continued)

A4.0

C. The monetary, material, land use, and energy costs of building additional waste treatment plants are directly related to the volume of contaminated water to be treated. Recycling water and reducing the amount used to transport waste can obviate the need for more and more waste treatment plants.

IN THIS SECTION YOU WILL:

1. Investigate the precedents of water supply and waste disposal in buildings.
2. Document your work through vignette sketches with short annotations.

WATER AND WASTE PRECEDENTS A4.1

Water and Waste Design Strategies

Site-Scale Strategy

- Minimize runoff to encourage recharging the water table.

Cluster-Scale Strategy

- Store water at an elevation higher than it is used; treat waste at lower elevations. Water moving downhill requires no pumping.

Building-Scale Strategies

- Use service spaces, which include toilet and mechanical rooms, to organize the building.
- Isolate toilet spaces acoustically; keep their ventilation patterns unshared by those of other spaces.

Component-Scale Strategy

- Group plumbing fixtures for cost savings and for convenience.

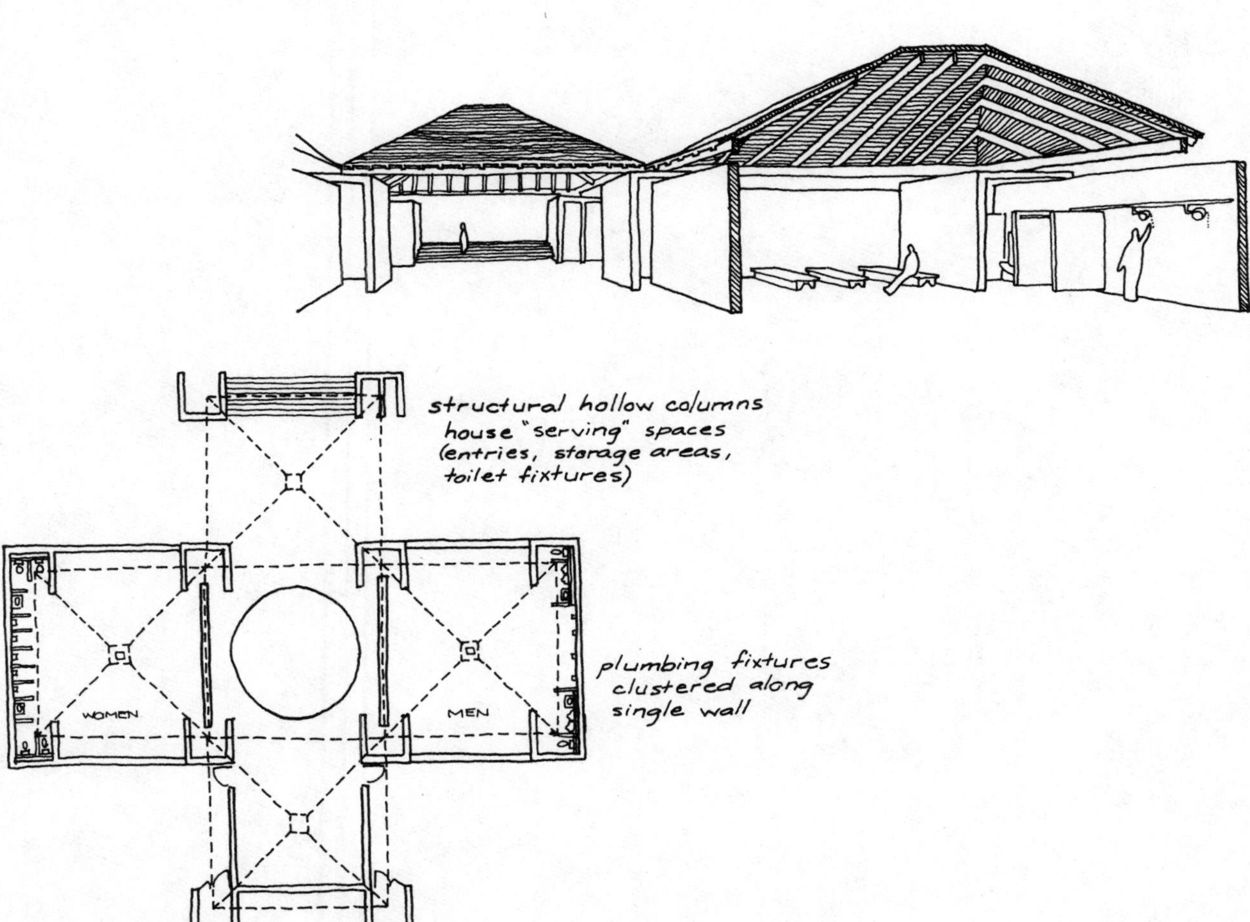

Figure A4.1.0 Trenton Bathhouse. Trenton, New Jersey, Louis I. Kahn (architect), 1956.

WATER AND WASTE PRECEDENTS (continued) A4.1

PROCEDURE

Choose an existing building or site that has: (1) a building program, aqueous environment, or both that is similar to your assigned building program or site and (2) a clear, conceptual approach that incorporates some of the design strategies above.

DOCUMENT YOUR CHOICE AS FOLLOWS:

1. Identify the location, program, architect (if known), and source of your information.
2. Include photocopies or drawings (whichever is quick and easy for you) to illustrate the design.
3. Evaluate the building or site design with a building response diagram and short annotations that explain how this design is organized for water and waste.

B4 WATER AND WASTE

CLIMATE AND SITE ANALYSIS

INTRODUCTION

Goal

Determine your site's resources for water collection and waste treatment and what potential benefits or problems may arise from the climatic conditions on your site. Develop appropriate water-conserving strategies for your building.

Discussion

By determining the availability of water on your site and the drainage characteristics of your soil you can gauge the feasibility of design strategies for collecting water and treating wastes on-site. This information will also help determine the role of water conservation in your design. Sites having dry climates, dry seasons, or poor drainage capacity especially merit water-conserving designs.

Additionally, the site you have been assigned is typical of sites beyond city limits in that it has no city-provided water supply or sewer. For such sites, on-site water supply, water conservation, and on-site water treatment are particularly critical.

In this section you will:

1. Analyze your annual and monthly rainfall patterns.
2. Identify your soil's drainage characteristics.
3. Develop appropriate water collection, waste treatment, and water-conserving strategies.

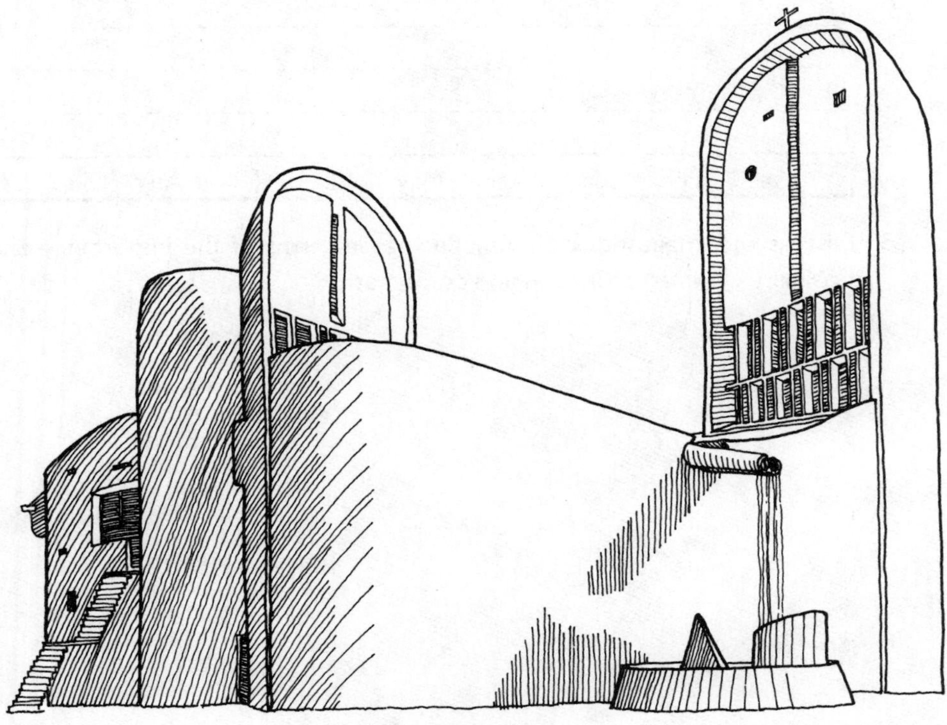

Figure B4.0.1 **Scupper and Cistern. Notre Dame du Haut, Ronchamps, France, Le Corbusier (architect), 1950–55.**

B4 WATER AND WASTE

CLIMATE AND SITE ANALYSIS

ANNUAL RAINFALL COMPARISONS

B4

B4.1

Discussion

The availability or scarcity of water on your site determines whether on-site, water collection can be effective and which water-conservation strategies are appropriate and applicable.

Procedure

1. Determine your location's annual rainfall [Appendix C, Climatological Data]. Determine whether the rainfall is unusually low, average, or unusually heavy [*MEEB*, Figure 8.5, p.494].

2. Plot the monthly rainfall data on the bar graph below. Identify any wet or dry seasons.

Monthly Rainfall Bar Graph

Average Rainfall (in inches) per Month												
10												
9												
8												
7												
6												
5												
4												
3												
2												
1												
	Jan	Feb	Mar	Apr	May	Jun	Jul	Aug	Sep	Oct	Nov	Dec

3. Discuss the implications of your findings in terms of the importance and feasibility of water collection and conservation.

B4 WATER AND WASTE

CLIMATE AND SITE ANALYSIS

SOIL CHARACTERISTICS

B4

B4.2

Discussion

Your site's soil characteristics determine the feasibility of on-site, waste-water treatment since soil type determines the requisite size of the drainage field. The more porous the soil, the smaller the drainage field. In locations with poor drainage, water-conservation strategies are imperative.

For more information

MEEB, Section 10.2(a), pp.602–610 (discussion and examples of on-site, waste treatment for small buildings).

Procedure

1. Identify your site's soil type.

Soil Characteristics

Location	Soil Type	Drainage
Madison Dodge City	sandy loam or sandy clay	medium to good drainage
Phoenix	coarse sand or gravel	excellent drainage
Charleston Spokane	clay with large amounts of sand or gravel	medium to poor drainage

2. Discuss the implications of the site's soil type and its ability to support on-site, waste-water treatment.

Discussion

You have gained some insights about the water and waste environment of your site that will help you propose strategies appropriate to your site and program for conserving and collecting water and for treating wastes on-site. These concepts should be expressed in the form of very simple, annotated diagrams, which are the most appropriate level of detail for this stage of the design process. Show the approximate size and placement of buildings and the water and waste features on the site.

Procedure

1. Diagram each site design.
2. Annotate the drawings to explain the water and waste considerations.
3. Discuss the potential benefits and disadvantages of your water and waste proposal with respect to views, thermal control, and daylighting, and acoustics.

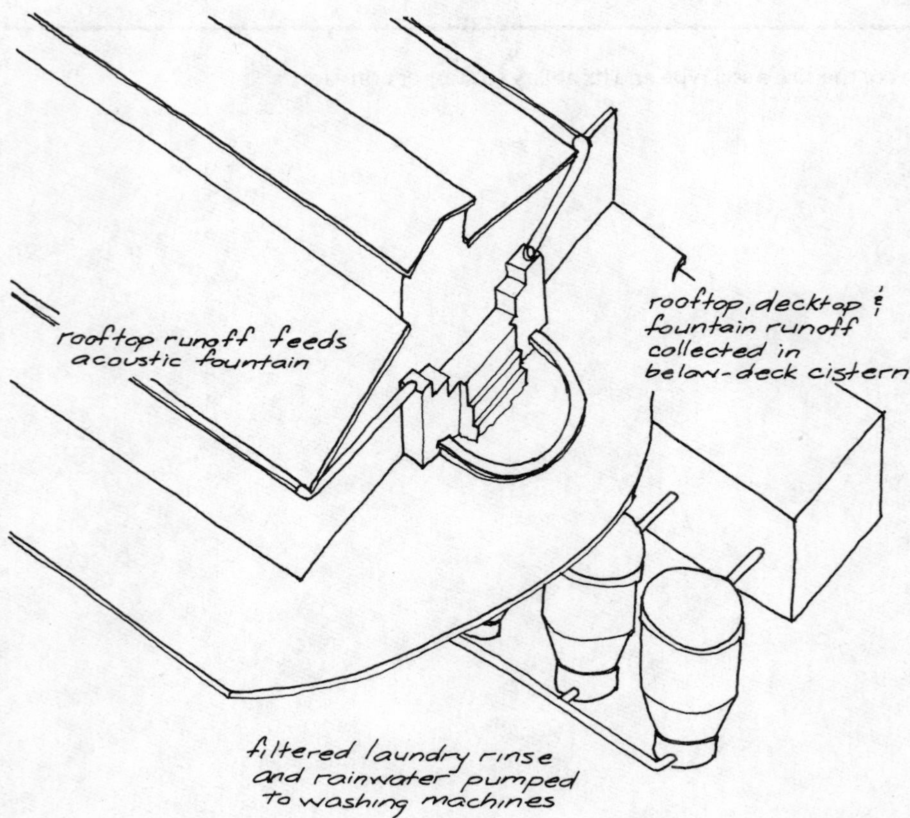

Figure B4.3.1 Water Conservation Measures. Linen supply, Charleston, South Carolina.

C4 WATER AND WASTE

PROGRAM ANALYSIS

INTRODUCTION

Goal

Determine the basic plumbing requirements for your building, estimate its conventional water use and waste production, and develop conserving design strategies.

Discussion

By determining conventional patterns of water use and waste production for your building type, you can determine how best to conserve water and recycle wastes. Since your water and waste schemes may seriously affect your building design and layout, it is best to consider them early in the design process. Additionally, the number and types of plumbing fixtures required by your building govern the size of certain building spaces, and the efficient layout of plumbing in the building affects the overall organization of building spaces.

In this section you will:

1. Determine the minimum required number and types of plumbing fixtures for your building.
2. Estimate the total water use of your building based on conventional design strategies.
3. Determine how much and what sorts of solid waste the occupants of your building will produce.
4. Generate a schematic water and waste design concept for your building and site.

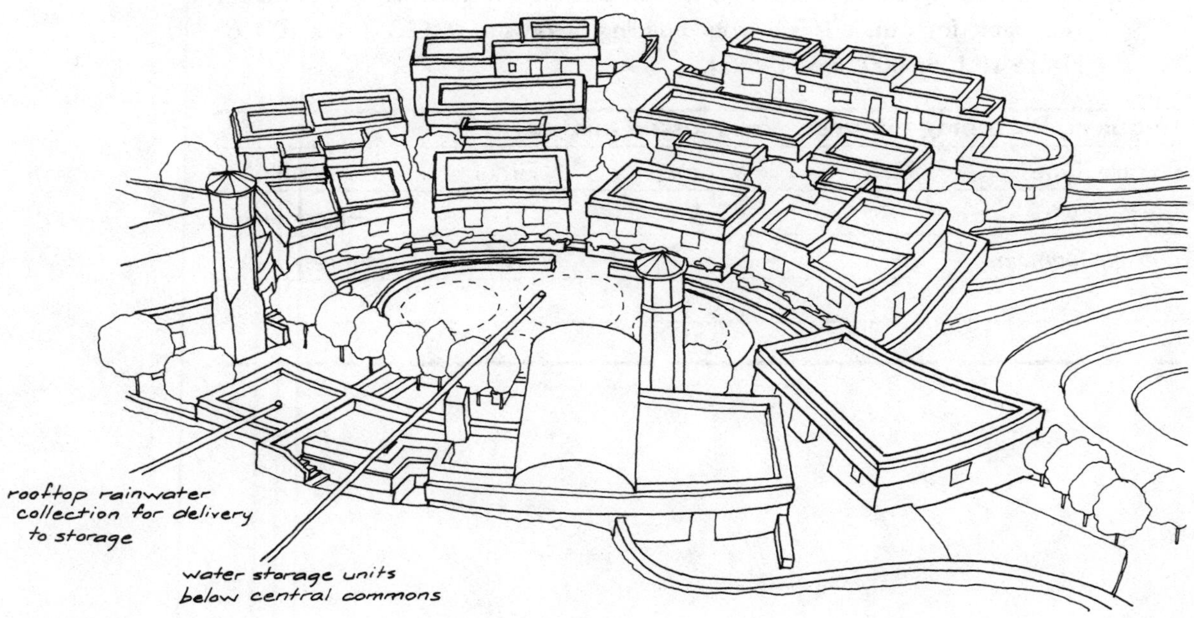

Figure C4.0.1 Foundation School (project). Ojai, California, Sim van der Ryn (architect), 1991.

C4 WATER AND WASTE

PROGRAM ANALYSIS

C4

MINIMUM REQUIRED PLUMBING FIXTURES

C4.1

Discussion

The function and population of your building determine the required number and types of plumbing fixtures you need. Plumbing codes specify required numbers of drinking fountains and toilet room facilities according to building occupancy, so your peak occupancy will be a determining factor. Since men and women are equally likely to use or occupy the building, two-thirds of the total population should be accommodated by either sex's toilet rooms. In addition, other plumbing fixtures may be needed to meet the requirements of your building program (e.g., vending machines, washing machines).

Procedure

1. Determine the minimum required number of drinking fountains and toilet room fixtures for your building type [*MEEB*, Table 8.3, pp.486–490].
   a. State the number of occupants during peak hours.
   b. If you decide to exceed the minimum number of fixtures or work with off-peak occupancy, explain your rationale.
2. Determine the number and types of any additional plumbing fixtures that your building program requires.
   a. State the number of users during peak hours.
   b. If you decide to exceed the minimum number of fixtures or work with off-peak occupancy, explain your rationale.
3. Determine whether the waste water from these fixtures is potentially recyclable for other uses in your building or on-site [*MEEB*, Table 10.1 or Figure 10.1, pp.600 or 591–592].

Required Plumbing Fixtures [suggested format]

Fixture Type	Men	Women	Either	Recyclable?
low-flush w/c	*5*	*5*	-	*no—black water*
drinking fountain	-	-	*2*	*yes—grey water*

C4 WATER AND WASTE

PROGRAM ANALYSIS

CONVENTIONAL FIXTURE AND WATER USE ESTIMATE

C4.2

Discussion

Based on your building type and occupancy, you can estimate the daily water use. The estimate is based on conventional water use practices and will serve as a basis for comparison as you develop water-conservation strategies.

Procedure

1. Determine the per capita daily water use for your building type [*MEEB*, Table 8.2, pp.483–484]. For building types not listed in the table, make an educated guess based on similar building types and your program information. To more accurately determine your water use, divide your building into a few zones.
2. Multiply the per capita daily water use by the total building population to determine the total water use.

Water Use Estimate [suggested format]

Space	Per Capita Use (gallons/day)	# People	**TOTAL** (gallons/day)
executive washroom	*35*	*1*	*35*

C4 WATER AND WASTE

PROGRAM ANALYSIS

SOLID-WASTE RESOURCE ESTIMATE

C4.3

Discussion

You can estimate, based on your building type and size, how much and what types of solid waste your occupants produce. By establishing early in the design process what may be recycled, you can better generate a design that allows for convenient waste sorting, storage, and disposal.

Procedure

1. Estimate the amount of solid waste produced in your building [*MEEB*, Table 12.3, p.687].
2. Determine the potential for recycling the solid waste. List the types of waste resources produced, and propose a recycling method for each [*MEEB*, Figures 12.4 and 12.7, pp.679 and 682].

NOTE: Certain plastics and paper products are also recyclable.

Solid-Waste Resource Recycling Potential [suggested format]

Waste Resource	Amount per Day	Recycling Method
newspapers	*2*	*recyling center*
aluminum	*50 cans*	*recyling center*
food scraps	*40 lb*	*composting*

C4 WATER AND WASTE

C4

PROGRAM ANALYSIS

WATER AND WASTE STRATEGIES

C4.4

Discussion

Your investigations of where water is used and what wastes are produced should indicate the potential for water conservation and solid-waste recycling for your building. Keep in mind that clustering plumbing fixtures can simplify water use design. Moreover, water and solid-waste streams through the building represent circulation paths that affect the building's layout and position on the site.

Procedure

1. Diagram your concept, showing optimum plumbing and solid-waste management configurations. ***MEEB***, Figures 12.5 and 12.7, pp.680 and 682, show examples of system schematic diagrams.
2. Annotate your diagram to indicate the water and waste strategies used and your rationale.
3. Discuss conflicts and harmonies in your design among water and waste, acoustic, lighting, and thermal needs.

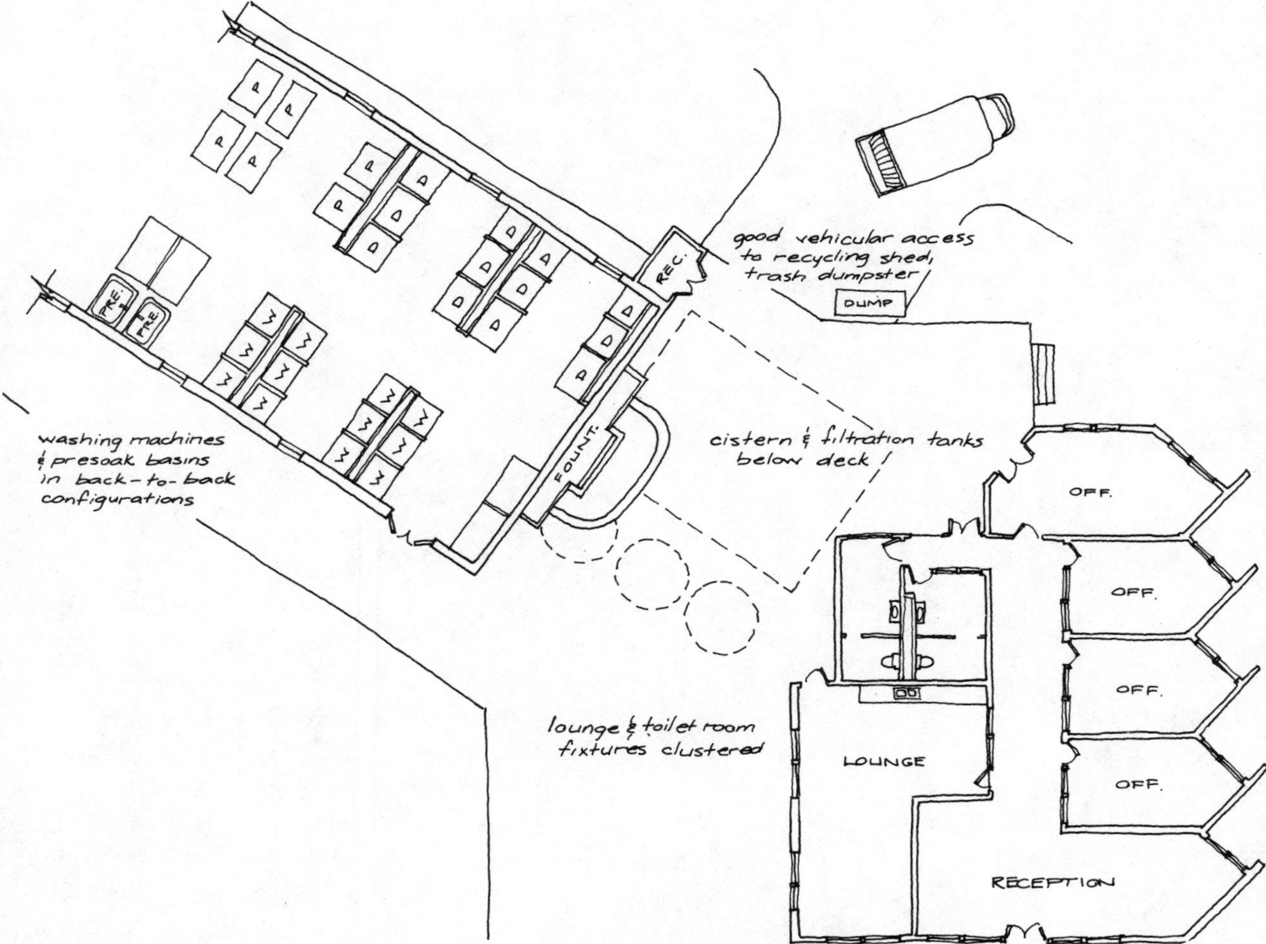

Figure C4.4.1 Schematic Water and Waste Design. Linen supply, Charleston, South Carolina.

D4 WATER AND WASTE

D4

SCHEMATIC DESIGN

INTRODUCTION

D4.0

Goals

Design a building that synthesizes the information you have acquired through the study of precedents; the analyses of your site, climate, and program requirements; and the application of water and waste design strategies. Use the site and building to the greatest degree possible to heat water, collect rainwater, and dispose of wastes.

Water and Waste Schematic Design Goals

A. Potable-water use and waste-water production are minimized through conservation and recycling.
B. At least 50% of the hot water is heated by solar collectors.
C. Adequate rainwater collection and storage are provided.
D. The septic tank and drainfield adequately treat the waste generated on-site.
E. Composting toilets meet the program requirements.
F. Facilities for solid-waste recycling are on-site.

Discussion

It is important to know if your site's climate is generally wet or dry or if it changes seasonally. Incorporate the effects of climate during the schematic design phase.

In this section you will:

1. Employ appropriate water and waste design strategies.
2. Compare a conventional water system for your building with a water-conserving system.
3. Size water and waste components (solar collectors, cistern, and septic field).
4. Evaluate whether composting toilets are appropriate for your building.
5. Layout a solid-waste recycling center.
6. Review and critique your design.

D4 WATER AND WASTE

SCHEMATIC DESIGN

DESIGN

D4

D4.1

WATER AND WASTE SCHEMATIC DESIGN STRATEGIES

Site-Scale Strategies

- Locate the septic tank and drainfield at an elevation lower than the building's.
- Use the naturally irrigated area over the drainfield as a flat, unpaved, open area for views or activities.
- Keep trees and shrubs away from drainfields. Roots may block the lines.
- Do not use the drainfield for parking.
- Provide adequate clearances from septic tank and drainfield to other site elements [*MEEB*, Table 10.5, p.608].

Cluster-Scale Strategies

- Arrange roof slopes so rainwater converges at a single location above the cistern inlet.
- Celebrate the water flow from the collection area to the storage area.
- Group the water-using fixtures adjacent to, but at a lower elevation than, the storage area.

Building-Scale Strategies

- Use solar, hot-water collectors to shade the roof if they aren't integrated with the roof.
- Provide convenient access to the collection chamber below composting toilets for easy, regular compost removal.
- Provide adequate, solid-waste storage space with convenient access to the service entry.

Component-Scale Strategies

- Use water-conserving fixtures.
- Recycle water for other uses.
- Provide unshaded south-facing exposure for solar collectors; set them at an angle above the horizontal equal to or less than your site's latitude.

PROCEDURE

Propose a schematic design for your building based on analyses of precedent [A4], site and climate [B4], and program [C4]. Use the appropriate water and waste design strategies for your design. Rely solely on the resources on-site for your water and waste systems—assume that on-site water collection and storage will meet your supply needs and that a septic tank, drainfield, and composting toilets will handle waste.

DOCUMENT YOUR DESIGN AS FOLLOWS:

1. Site plan or cluster plan, showing building, parking, drives, and, depending on your water and waste systems, cistern, rain collection areas, septic tank, drainfield, and solar collectors (scale: 1" = 100').

2. Floor plans, showing location of hot water storage, toilet rooms, plumbing lines, and solid-waste recycling space (scale: 1" = 20').

3. Building section, showing solar collectors, hot water storage, plumbing lines, cisterns, and composting toilets (scale: 1/8" = 1'-0").

D4 WATER AND WASTE

SCHEMATIC DESIGN

CONVENTIONAL WATER SYSTEM

Discussion

You have already determined your building's minimum required plumbing fixtures [C4.1] and estimated your conventional fixture water use [C4.2]. Now determine how your system would work as a conventional system, and identify ways to improve your design through conservation.

The supply fixture unit value determines the share of the total water use for each fixture. It is a unitless, weighted ratio based on the average comparative use of various plumbing fixtures. The number of gallons per day that one fixture unit represents varies from one plumbing system design to the next.

Procedure

Evaluate how well your building attains Water and Waste Schematic Design Goal A—**Potable-water use and waste-water production are minimized through conservation and recycling.**

1. List the number and types of fixtures in your building [C4.1], and assign each fixture a weight in conventional supply fixture units [*MEEB*, Table 9.17, p.587].

> NOTE: For fixtures not listed in *MEEB*, Table 9.17, the number of fixture units may be estimated based on fixtures that use water in similar quantities at similar rates. Alternately, if a gallons-per-day figure for the fixture can be estimated, the weight in fixture units may be calculated from the equation in step 3. For these buildings you must determine fixture units now, and use those values in D4.3.

2. Determine how many gallons per day per supply fixture unit your building requires.

GPFU = (WU) / (fu)

where:

GPFU = gallons/day/supply fixture unit

WU = total conventional water use (gallons/day) [C4.2]

fu = total number of conventional supply fixture units

SCHEMATIC DESIGN

CONVENTIONAL WATER SYSTEM (continued)

3. Determine how many gallons per day each supply fixture uses.

GPF = (GPFU) (FU)

where:

GPF = gallons/day/fixture

GPFU = gallons/day/supply fixture unit [step 2]

FU = weight in supply fixture units

Conventional Supply Fixture Estimates [suggested format]		
Fixture	Conventional Supply Fixture Units	gallons/day
3 tank-type, flush toilets	*5*	*75*

4. Draw a schematic diagram of how your system would operate as a conventional system based on Figure D4.2.1. You know input gallons per day, estimate output gallons per day (the two aren't necessarily equal).

5. Discuss whether your strategy is appropriate, and indicate any design changes you view as necessary to meet Goal A.

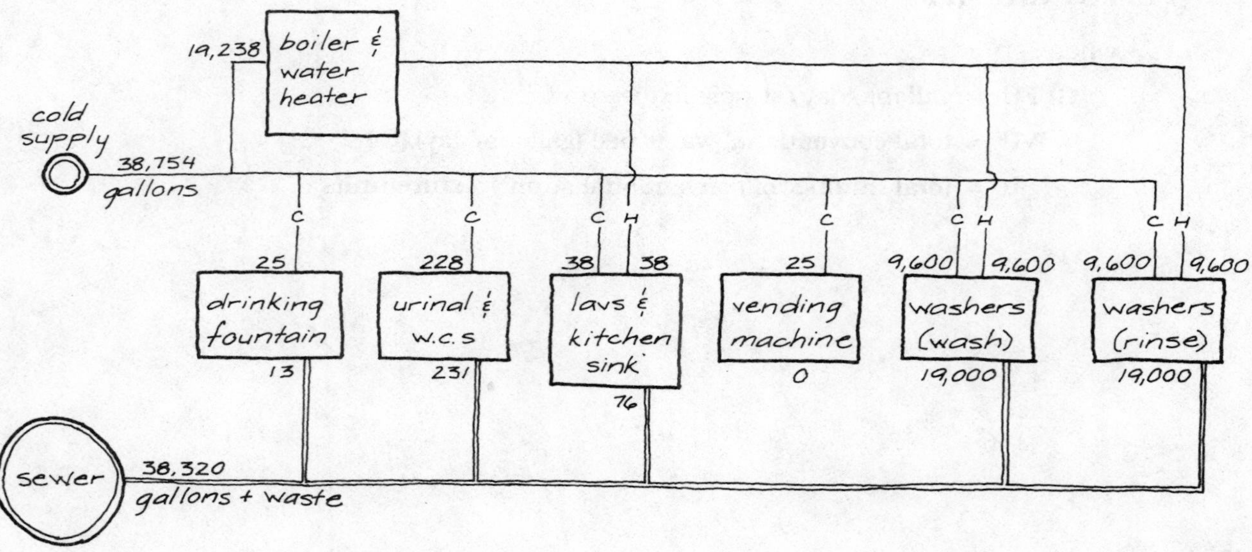

Figure D4.2.1 Conventional Water System Design Diagram

D4 WATER AND WASTE

SCHEMATIC DESIGN

WATER-CONSERVING SYSTEM

D4

D4.3

Discussion

The conventional plumbing system you designed [D4.2] reveals the expected water use for each fixture and, more important, establishes the value in gallons per day for a single supply fixture unit. This value remains the same for the conserving plumbing system. To determine the water use of the individual conserving fixtures and of the entire conserving system, the supply fixture unit value for the conserving fixtures must be adjusted to compensate for reduced use.

Your conservation strategies should include conserving fixtures and recycling water [C4.1 and *MEEB*, Section 10.1, pp.590–600].

Procedure

Evaluate how well your building attains Water and Waste Schematic Design Goal A—**Potable-water use and waste-water production are minimized through conservation and recycling.**

1. Outline a conservation strategy that includes conserving fixtures, flow restricters, and recycling.
2. Calculate conservation supply fixture units for each fixture.

   $$\mathbf{FU_{cons} = (FU_{conv})\ [(use_{cons})\ /\ (use_{conv})]}$$

   where:

   FU_{cons} = weight in conservation fixture units for the fixture

   FU_{conv} = weight in conventional fixture units for the fixture

   use_{cons} = water use by conserving fixture (gallons)

   use_{conv} = water use by conventional fixture (gallons)

3. Determine how many gallons per day each supply fixture uses.

   GPF = (GPFU) (FU_{cons})

   NOTE: Alternately, the conservation fixture units may be determined according to how many gallons per day the fixture uses.

Conservation Fixture Estimates [suggested format]

Fixture	Conservation Supply Fixture Units	gallons/day
3 low-flush toilets	*2.5*	*37.5*

4. Draw a schematic diagram of your water-conserving system. Discuss why your strategy is appropriate, and indicate any design changes you view as necessary to meet Goal A.

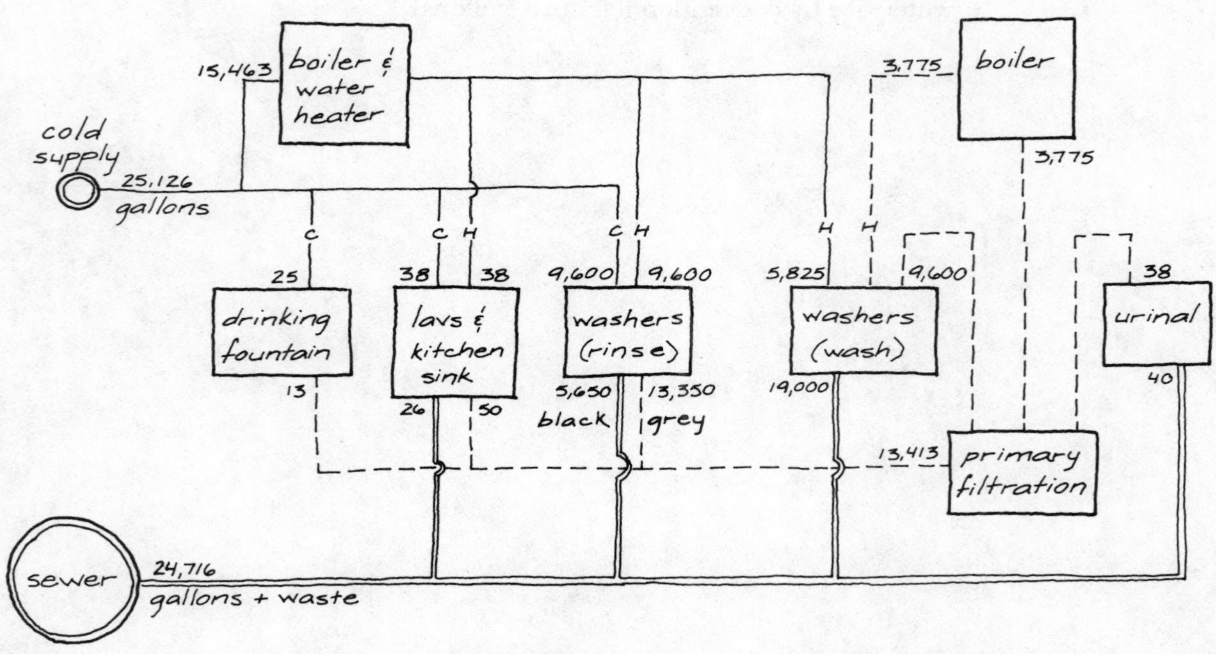

Figure D4.3.1 Water-Conserving System Design Diagram

D4 WATER AND WASTE

SCHEMATIC DESIGN

SOLAR COLLECTOR SIZING ESTIMATE

D4

D4.4

Discussion

The type of solar, hot-water system you choose affects the sizing of your solar collectors and the placement of plumbing components. Systems are classified as either "passive" (reliant on gravity for circulation) or "active" (pumps used for circulation) and either "direct" (hot water used in the building is circulated through the collector) or "indirect" (heat is transferred from the collector to the stored hot water by means of another fluid). From the simplest, direct, passive or "batch" system to the more complex active, indirect systems, there is a broad range of available systems, each with advantages and disadvantages [*MEEB*, Section 9.4(d), pp.547–565].

System efficiency is also affected by the seasonal availability of insolation on your site, which will be accounted for in E4.6 and E4.7.

Procedure

Evaluate how well your design attains Water and Waste Schematic Design Goal B—
At least 50% of the hot water is heated by solar collectors.

1. Choose an appropriate solar hot water system [*MEEB*, Table 9.12, pp.556–557].
2. Estimate hot water storage tank size based on hot water needs [D4.3]. The storage tank must be large enough to meet the peak hourly hot water use.
3. Estimate solar collector size based on your hot water system type according to the following rules-of-thumb:
   - Batch Systems: .45–.65 ft^2 collector area/gallon of water stored
   - Other Systems: .65–1.0 ft^2 collector area/gallon of water stored

   } (The lower figure applies to warmer climates.)
   - Swimming Pools: .5 ft^2 collector area/1.0 ft^2 of pool surface area
4. Discuss whether your solar water heating strategy is appropriate, and indicate any design changes you view as necessary to meet Goal B.

D4 WATER AND WASTE

SCHEMATIC DESIGN

D4

CISTERN SIZING ESTIMATE

D4.5

Discussion

Rainwater is suitable anytime potable water is not required (e.g., toilet flushing, laundry, site irrigation). The catchment area, generally all or part of the roof, will deliver 75% of the rainfall it receives. The remainder is lost to evaporation, spillage, or contamination. A comparison of your monthly water needs and the monthly rainfall catchment can help you gauge the effectiveness of rainfall storage as a design strategy for your building.

Procedure 1

Evaluate how well your design attains Water and Waste Schematic Design Goal C—**Adequate rainwater collection and storage are provided.**

1. Determine your average monthly water use based on your daily water use estimate [D4.3] for a 31-day month.
2. Determine your average annual rainfall catchment in gallons [*MEEB*, Figure 8.4a, p.485].

   NOTE: Assume the roof catchment area equals the corresponding ground floor area. Total precipitation equals two-thirds of the annual precipitation [Appendix C, Climatological Data]. These data give "dry year" or worst-case "design" precipitation.

   For very low levels of rainfall, use the formula

   $$G = [(P)(A)] / (2.15),$$

   where:

   G = rainfall collected (gallons)

   P = total precipitation (inches)

   A = roof catchment area (ft^2).
3. Determine your average monthly rainfall catchment.
4. Discuss whether your design provides adequate catchment area, and indicate what design changes you deem necessary to meet Goal C.

CISTERN SIZING ESTIMATE (continued)

PROCEDURE 2

1. Determine your cistern capacity in gallons based on the following rules-of-thumb:

   a. If the average monthly catchment exceeds the average monthly use,

   G = (U) (1.5),

   where:

   G = cistern capacity (gallons)
   U = average monthly use (gallons).

   b. If the average monthly use exceeds the average monthly catchment,

   G = 2C,

   where:

   C = average monthly catchment (gallons).

2. Determine your cistern volume.

   V = G / (7.48)

   where:

   V = cistern volume (ft^3)

3. Propose dimensions for your cistern (it should be at least 5 feet deep), and indicate schematically its placement in your design. Discuss any design changes necessary to accommodate your cistern and to meet Goal C.

D4 WATER AND WASTE

SCHEMATIC DESIGN

SEPTIC TANK AND DRAINFIELD SIZING I

Discussion

Waste-plumbing system design is based on drainage fixture units. Like supply fixture units, they are unitless ratios. The values are not equal to supply fixture unit values of comparable fixtures. In your water-conserving system, fixtures that recycle water or use no water will have drainage fixture unit values of zero because they produce no waste.

Procedure

Evaluate how well your design attains Water and Waste Schematic Design Goal D—**The septic tank and drainfield adequately treat the waste generated on-site.**

1. Determine the number of drainage fixture units in your conserving plumbing system [D4.3 or *MEEB*, Table 10.9, p.639].
2. Determine the minimum septic tank capacity [*MEEB*, Table 10.3, p.606].

> NOTE: If drainage fixture units are unknown, you may calculate septic tank capacity according to your daily sewage flow (in gallons per day):
>
> a. For flows up to 1,500 gallons per day,
>
> **C = 1.5F,**
>
> where:
>
> C = septic tank capacity (gallons)
>
> F = daily sewage flow (gallons).
>
> b. For flows over 1,500 gallons per day,
>
> **C = .75F + 1,125.**

3. Repeat the procedure to determine the required septic tank capacity for your conventional water system design [D4.2]. Discuss the impact of your water-conserving system design on septic tank sizing and what design changes you deem necessary to meet Goal D.

Drainfield Sizing [suggested format]

Fixture	Conserving Drainage (fu)	Conventional Drainage (fu)
10 toilets	*(10) (0) [Clivus Multrum] = 0*	*(10) (4) [tank-flush] = 40*
5 lavatories	*(5) (1) = 5*	*(5) (1) = 5*
TOTALS	*5*	*45*
Septic Tank Capacity	*720 gal*	*2,000 gal*

D4 WATER AND WASTE

SCHEMATIC DESIGN

SEPTIC TANK AND DRAINFIELD SIZING II

DISCUSSION

Your soil type determines how rapidly water is absorbed by your site—soils with good absorption capacity will require a smaller drainfield area than soils with poor absorption characteristics. The maximum allowable septic tank size is also determined by soil type.

PROCEDURE

Evaluate how well your design attains Water and Waste Schematic Design Goal D—
The septic tank and drainfield adequately treat the waste generated on-site.

1. State what soil type you have at your site (U.S. Soil Conservation Service data or the table below).

Soil Characteristics

Location	Soil Type	Permeability*
Madison or Dodge City	sandy loam or sandy clay	5 minutes
Phoenix	coarse sand or gravel	1 minute
Charleston or Spokane	clay with considerable sand or gravel	10 minutes

* Time for water level to drop 1" in a test hole.

2. Determine whether your septic tank is sized within the allowable limits for your soil type [*MEEB*, Table 10.4, p.606]. If not, indicate what design changes are necessary to stay within the limits.

3. Determine your required drainfield size based on your site's soil type and septic tank capacity [*MEEB*, Table 10.4, p.606]. Discuss the ability of your design to supply adequate on-site waste-water treatment. Indicate any design changes you view as necessary to meet Goal D.

4. Determine the required drainfield size for your conventional water system design. Discuss the impact of your water-conserving system design on drainfield and septic tank sizing for your building and any changes necessary to meet Goal D.

D4 WATER AND WASTE

SCHEMATIC DESIGN

COMPOSTING TOILETS

Discussion

The ultimate example of effective water conservation is the composting toilet. The Clivus Multrum system [*MEEB*, Figure 10.8, p.598] requires organic kitchen waste as an agent for composting. The large composting chamber below the unit must be located to allow easy, periodic compost removal. The system uses neither water nor energy, yet produces compost suitable for landscape vegetation.

Procedure

Evaluate how well your design attains Water and Waste Schematic Design Goal E—
Composting toilets meet the program requirements.

1. Schematically illustrate the composting toilets' place in your design, indicating how access to the composting chamber is achieved.
2. Discuss the impact of the composting system on your design, and indicate any design changes necessary to meet Goal E.

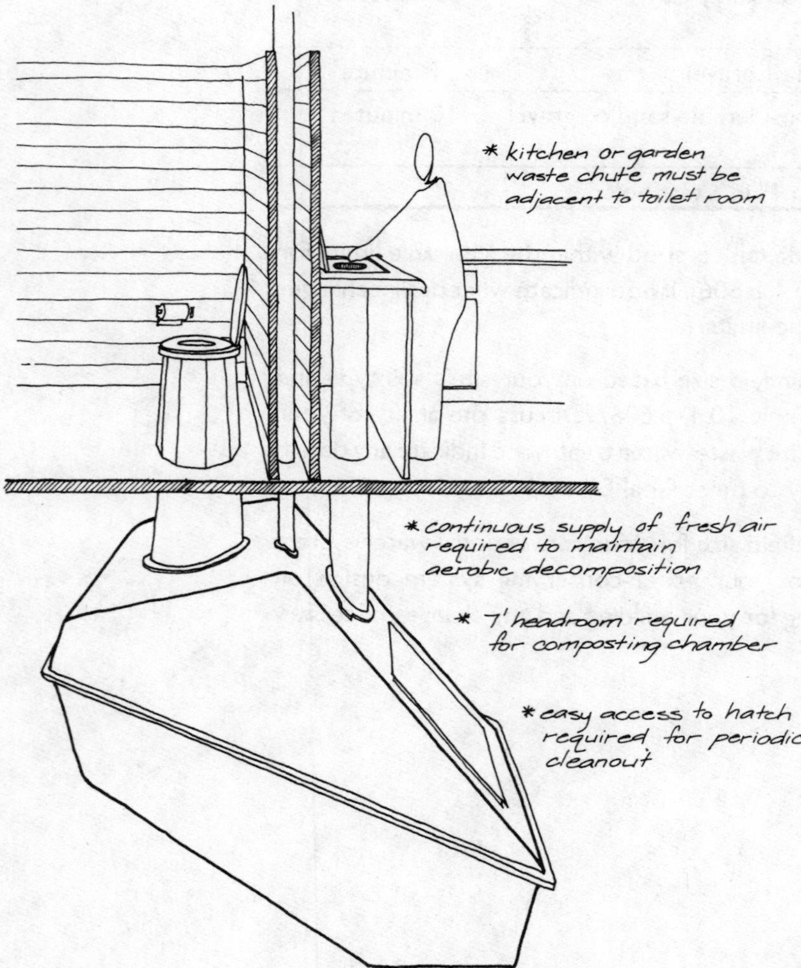

Figure D4.8.1 Composting Toilet Installation

D4 WATER AND WASTE

SCHEMATIC DESIGN

SOLID-WASTE CENTER

D4

D4.9

Discussion

Recycling solid wastes from residential and commercial buildings can provide a source of low-cost, raw materials for industry (e.g., waste paper for pulp mills, aluminum cans for aluminum plants), fuel for utilities, and a source of income for the recycler. Recycling may occur either on-site or at a central collection point. The advantages of on-site recycling are that the recycling can be done manually rather than mechanically, and the recycler is the direct beneficiary of income derived from sales of recycled goods. Moreover, separation of wastes on-site reduces the amount of garbage sent to public landfills and can provide the safeguard of removing hazardous waste from the trash.

Procedure

Evaluate how well your design attains Water and Waste Schematic Design Goal F—
Facilities for solid-waste recycling are on-site.

1. Design a solid-waste center to accommodate waste production [C4.3]. Indicate schematically its layout, position relative to waste-producing activities, and how wastes are removed. Designate storage areas for each class of recyclable materials, hazardous wastes, and garbage [*MEEB*, Section 12.4, pp.681–686].
2. Discuss the impact of the solid-waste center on your overall design, and indicate any design changes necessary to meet Goal F.

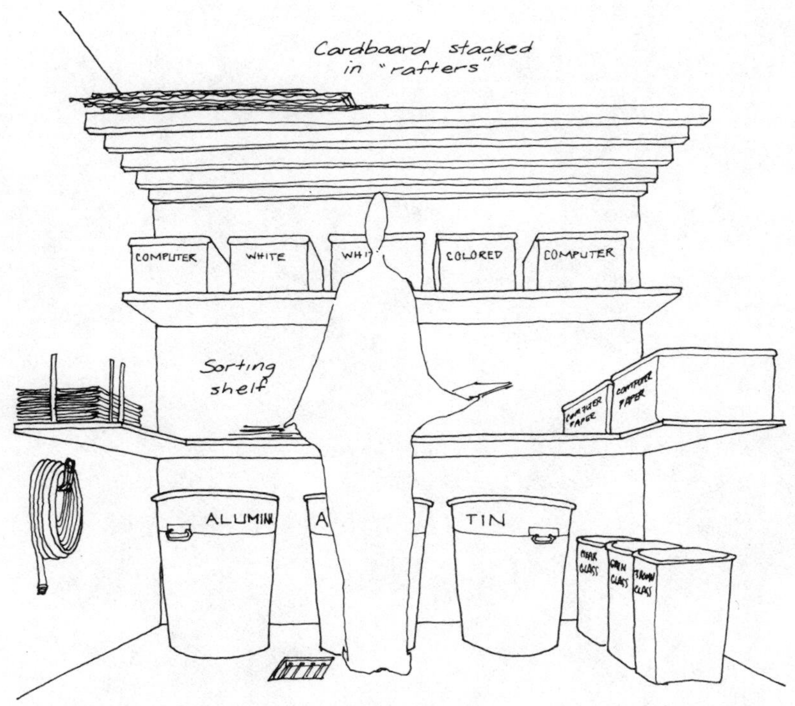

Figure D4.9.1 Recycling Center. Linen supply office, Charleston, South Carolina.

DESIGN REVIEW

Discussion

You have studied the issues of water conservation, water supply, and solid-waste management and formulated a schematic design for your building and site. Each step has been accomplished independently and conflicting decisions may have been made. This design review affords you the opportunity to synthesize your cumulative design decisions.

Procedure

Review how well your design accomplishes Water and Waste Schematic Design Goals A–F.

1. Make schematic drawings that show how your design has evolved in response to meeting D4.0 goals.
2. Annotate your drawings to describe the systems used and the changes made.
3. Discuss the tradeoffs you have made among water and waste, acoustics, lighting, and thermal design considerations.
4. Discuss situations where your water and waste, acoustics, lighting, and thermal design strategies worked effectively together.

E4 WATER AND WASTE

DESIGN DEVELOPMENT

INTRODUCTION

E4

E4.0

Goals

Refine and develop the design to treat waste in an environmentally safe manner and to conserve water without sacrificing desirable views, thermal performance, lighting, or acoustics. Use your schematic design as the foundation for design development.

Water and Waste Design Development Goals

A. Toilet rooms are configured to ensure privacy and to allow access for all users.
B. The supply pipe system is designed and sized to carry the expected load.
C. The waste pipe system is designed and sized to carry the expected load.
D. The septic tank and drainfield are properly located on the site.
E. Solar collectors are sized to provide 100% of the hot water requirements in the month with the most solar radiation.
F. The cistern capacity is adequate to meet expected water needs year-round.

In this section you will:

1. Complete the water and waste design development.
2. Complete the design of toilet rooms.
3. Design both supply and waste pipe systems.
4. Select and design the appropriate domestic hot water system.
5. Complete the cistern and catchment system design.
6. Critique your building's water and waste design.

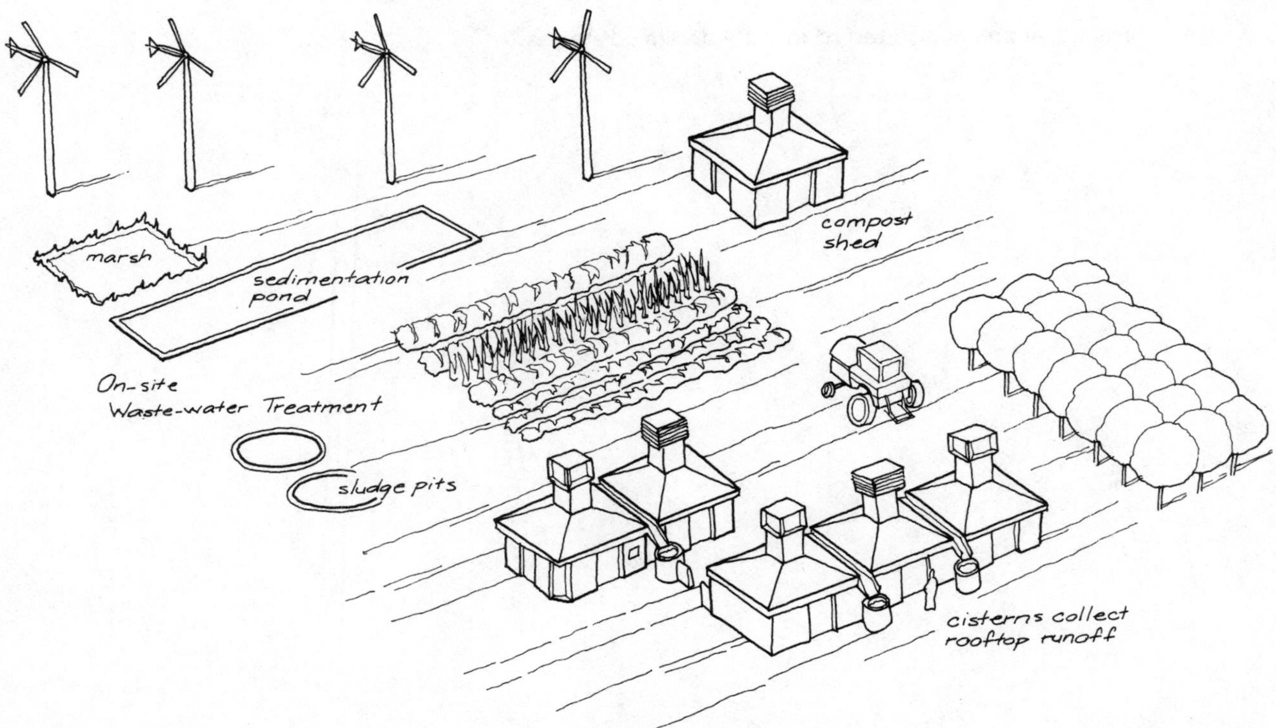

Figure E4.0.1 Blueprint Farm. Laredo, Texas, Pliny Fisk III (architect), 1991.

DESIGN DEVELOPMENT

DESIGN

WATER AND WASTE DESIGN DEVELOPMENT STRATEGIES

Component-Scale Strategies

- Ensure handicapped access to toilet rooms by designing in accordance with the *Uniform Building Code*.
- Block the line-of-sight into the toilet rooms.
- Locate fixtures on common walls to reduce piping and increase ease of installation.
- Minimize the distance from the water main or cistern to the farthest fixture.
- Set the solar collector tilt angle at approximately 10° less than your latitude to maximize performance.

PROCEDURE

Refine your final schematic design [D4] to include component design considerations. Use the appropriate strategies to aid your design development. Modify your schematic design drawings to include the components you have developed.

DOCUMENT YOUR DESIGN AS FOLLOWS:

1. Site plan, including parking, access drives, solar collectors, septic tank, drainfield, and cistern locations (scale: 1" = 100').
2. Floor plans, including plumbing fixtures, hot water storage tank (scale: 1" = 20').
3. Sections, showing solar collectors and hot water storage tank (scale: 1/8" = 1'-0").
4. Design diagram, annotated to identify design strategies.

E4 WATER AND WASTE

DESIGN DEVELOPMENT

TOILET ROOM

E4

E4.2

Discussion

The users of your toilet rooms should be able to use any fixture without being visible to anyone outside the rooms. A simple check of sight lines from beyond the entry way will ensure that vision into the rooms is properly obstructed.

To ensure handicapped access to your toilet rooms, certain design criteria must be achieved. The 1991 *Uniform Building Code*, Section 511(a), specifies:

- Doorways to toilet rooms:
  32" door width with 44" clearance beyond each side of the door.
- Clearance within toilet rooms:
  Sufficient to inscribe a circle of 5' diameter. Doors may encroach this circle by a maximum of 1' when open.
- WC stall:
  One for each sex must have 42" wide stall, with 48" clearance in front of stool within stall. (Stall door cannot encroach on this space so it usually swings outward.) The door width must be at least 32" if it is on the front of the stall or at least 34" if it is on the side of the stall.
- Lavatory:
  Under one lavatory, a space 30" wide, 29" high, and 17" deep for wheelchair access.

For more information

MEEB, Section 11.3, pp.653–671 (additional discussion and examples).

Procedure

Evaluate how well your design attains Water and Waste Design Development Goal A—**Toilet rooms are configured to ensure privacy and to allow ease of access for all users.**

1. Draw the floor plans of your toilet rooms (scale: ¼" = 1'-0").
2. Draw the sight lines from the entry [*MEEB*, Figure 11.10, p.654].
3. Indicate schematically any changes necessary to ensure that vision into the rooms is blocked.
4. Check your toilet room designs against the *UBC* criteria. Indicate schematically any changes necessary to bring your designs up to code standards and to meet Goal A.

SUPPLY PIPE SYSTEM

Discussion

To maintain proper functioning, each fixture in your building requires a certain waterflow rate, which is achieved by providing adequate flow pressure at the fixture. Water leaves the main or cistern pump at a fairly high pressure. However, water pressure decreases along the line because of the height and distance that the water must travel and because of friction as it passes through the pipes and the water meter.

As a designer, you can decrease pressure loss by minimizing the distance water must travel to reach the farthest fixture. Then you can ensure proper pressure by choosing the proper supply pipe size. You must choose a pipe large enough to handle your maximum waterflow at a velocity of no greater than 10 feet per second (at faster speeds water can be heard within the pipes), yet small enough to maintain adequate water pressure. In most residential and small commercial buildings, the supply pipe is 2" in diameter or smaller. The sizing procedure requires a preliminary pipe size estimate, a series of calculations to determine whether this size is adequate, and possibly a check to determine whether a larger or smaller pipe size is more appropriate [*MEEB*, Section 9.7, pp.582-589].

Procedure

Evaluate how well your design attains Water and Waste Design Development Goal B—**The supply pipe system is designed and sized to carry the expected load.**

1. Determine flow pressure, FP, required at the fixture farthest from the water main or cistern pump [*MEEB*, Table 9.16, p.585].

2. Determine pressure loss due to height, PL_H. Assume the water main is at the street, 6' below grade. The pump is submerged within the cistern or is 6' below grade in the well.

   $$PL_H = .433H$$

   where:

   H = height difference from the pump to the highest fixture (ft)

3. Estimate your water supply flow in gallons per minute [*MEEB*, Figure 9.50, p.586] according to the number of your supply fixture units [D4.3] and whether toilets are primarily flush tank or flush valve fixtures [*MEEB*, Table 9.17, p.587].

4. Estimate your supply pipe size [*MEEB*, Figure 9.49, p.586, may be helpful in narrowing the range of possibilities based on water flow].

5. Determine pressure loss due to water meters, PL_M [*MEEB*, Figure 9.48, p.585], based on your water supply flow and estimated supply pipe size. Even if city water isn't being used, water may be metered to keep an accurate record of water use.

SUPPLY PIPE SYSTEM (continued)

6. Calculate pressure loss due to friction in the pipes, PL_P.

   $$PL_P = M - (FP + PL_H + PL_M)$$

   where:

   M = cistern or well pump pressure (assume 40 psi)

7. Calculate desired friction loss, L, per 100' pipe length.

   $$L = PL_P / TEL$$

   where:

   TEL = total equivalent length of the piping or 1.5 times the distance from the main or pump to farthest fixture

8. Check your calculations [*MEEB*, Figure 9.48, p.585]. Does the chart value for pipe diameter for your friction loss and flow in gallons per minute comply with your estimate? Is the velocity less than 10 feet per second? If not, revise your estimate and recalculate to meet Goal B.

DESIGN DEVELOPMENT

WASTE PIPE SYSTEM

Discussion

The design and sizing of the waste pipe system is relatively straightforward. First, the runout for each fixture is sized by its required trap size. The runouts from each cluster of fixtures feed into a fixture branch that is sized by the number of drainage fixture units it serves. These fixture branches feed into the building drainline via vertical soil stacks. Similarly, a system of vent pipes that penetrates the roof serves all of the fixtures.

For more information

MEEB, Sections 10.5 and 10.6, pp.633–646 (an extensive discussion with examples).

Procedure

Evaluate how well your design attains Water and Waste Design Development Goal C—**The waste pipe system is designed and sized to carry the expected load.**

1. Identify the locations of all plumbing fixtures. Use either a building section diagram [*MEEB*, Figures 10.46 and 10.47, pp.643-644] or an axonometric plumbing wall diagram [*MEEB*, Figure 10.34, p.630].
2. Draw the waste pipe system serving the fixtures, minimizing pipe length wherever possible.
3. Size each fixture runout so runout size equals trap size [*MEEB*, Table 10.8, p.638]. Alternately, determine the drainage fixture unit value [*MEEB*, Table 10.9, p.639] and assign an appropriate runout pipe size [*MEEB*, Table 10.11, p.640]. Label the plumbing diagram accordingly.
4. Size fixture branches and vertical stacks according to the number of drainage fixture units each serves [*MEEB*, Table 10.11, p.640]. Label the plumbing diagram accordingly.
5. Determine the size and pitch of the building drain. What vertical drop is required for the drain to reach the septic tank at this pitch? Show how this is compatible with your site's topography.
6. Size the venting pipe system. For individual fixtures, vent size equals trap size. For collective vents, see *MEEB*, Table 10.12, p.641. Label the plumbing diagram accordingly.
7. Discuss changes necessary to meet Goal C.

DESIGN DEVELOPMENT

DRAINFIELD

Discussion

You have already sized your septic tank and drainfield to accommodate your expected waste load and soil type. Now you must layout your drainfield lines with the proper length (100' maximum) and spacing (6' o.c. minimum). You must also evaluate its position on your site: Ideally the septic tank lies downhill from the building, and the drainfield has adequate clearances from structures, vegetation, and bodies of water. When the drainfield is above the septic tank, a pump must be used to serve the drainfield.

Procedure

Evaluate how well your design attains Water and Waste Design Development Goal D—**The septic tank and drainfield are properly located on the site.**

1. Draw required code clearances for features on your site plan [*MEEB*, Table 10.5, p.608]. Identify areas that are suitable for septic tank placement and drainfield location.
2. Draw the location and layout of the septic tank and drainfield lines on your site plan. Drainlines must meet code [*MEEB*, Table 10.6, p.611].
3. Discuss whether or not your septic tank and drainfield meet code requirements. Indicate any design changes you deem necessary to meet Goal D.

E4 WATER AND WASTE

DESIGN DEVELOPMENT

E4

DOMESTIC HOT WATER (DHW) SYSTEM REQUIREMENTS

E4.6

Discussion

The temperature of stored hot water is higher than the desired temperature at the point-of-use. An accurate, hot-water-use estimate will account for this temperature difference and for the amount of cold water (entering the building at groundwater temperature) that is mixed with the hot water load.

Procedure

1. Determine your site's groundwater temperature [Figure E4.6.1].

2. Determine the average point-of-use water temperature for your building [*MEEB*, Table 9.7, p.538]. Assume your water heater thermostat setting to be 20° higher than this average (and at least 10° higher than the highest point-of-use temperature).

3. Calculate the percentage of stored hot water that goes into producing the total amount of hot water used.

   % hot = (T_{AV} – GWT) / (T_{therm} – GWT)

   where:

   T_{AV} = average point-of-use temperature (°F)
   GWT = site's groundwater temperature (°F)
   T_{therm} = water heater thermostat setting (°F)

4. Calculate the daily total use of stored hot water.

   S = (T) (% hot)

   where:

   S = total stored hot water (gallons/day)
   T = total hot water use in your building (gallons/day) [D4.4]

5. Determine the total heat required (Btu/day) to heat the stored water.

   TH_R = (S) (8.33 lb/gal) (T_{therm} – GWT)

DHW Requirements [suggested format]

GWT (°F)	T_{AV} (°F)	T_{therm} (°F)	Hot Water (%)	S (gal/day)	T (gal/day)	TH_R (Btu/day)
68	*90*	*120*	*42*	*504*	*1,200*	*218,400*

DOMESTIC HOT WATER (DHW) SYSTEM REQUIREMENTS (continued)

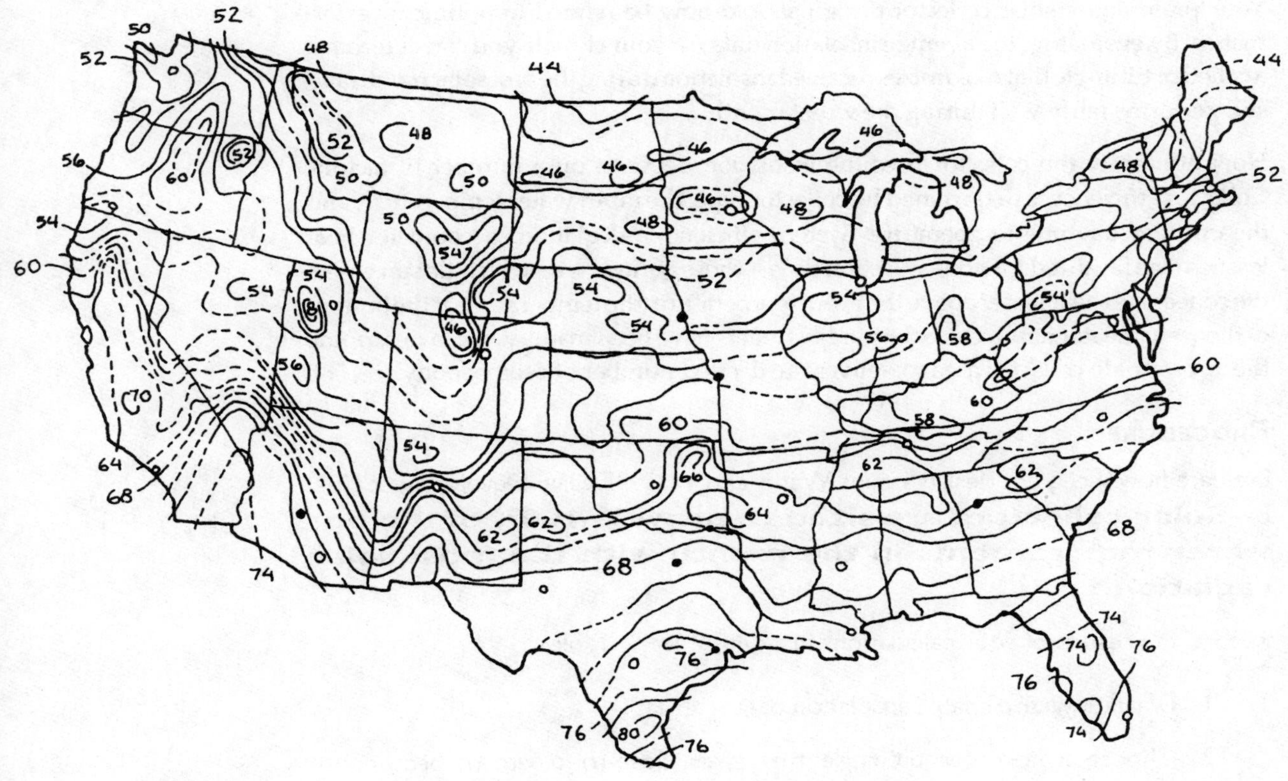

Figure E4.6.1 Groundwater Temperatures in the Continental United States. Reprinted, by permission, from the American Solar Energy Society, *Passive Solar Journal* 1 no.1, 5.

SOLAR DHW SYSTEM SIZING

Discussion

Your preliminary solar collector design should now be refined to optimize performance. By evaluating the average insolation data for your climate you can determine a collector tilt angle that maximizes received insolation during the best solar month and still performs fairly well during the worst month.

How efficiently the collector uses the insolation depends on a number of factors. Collector efficiency is determined by collector type, the input water temperature, and the outdoor daytime temperature. System efficiency calculations account for heat losses in the lines and for non-collection time—those periods when too little sun enters the collector to heat water above the input temperature (the temperature at the bottom of the water heater tank). Once these adjustments have been made, you can determine the appropriate collector area to deliver the desired number of Btus per day.

Procedure

Evaluate how well your design attains Water and Waste Design Development Goal E—**Solar collectors are sized to provide 100% of the hot water requirements in the month with the most solar radiation.**

Record the results of your calculations in a DHW Sizing Table.

1. Evaluate your climate's insolation data.
2. Choose a collector tilt angle that gives close-to-maximum best-month performance without suffering the worst-month performance.

Average Daily Solar Radiation Data (SOLMET) (Btu/ft²)

	Worst Month				Best Month			
Tilt angle:	0°	30°	60°	90°	0°	30°	60°	90°
Madison	389	429	525	477	1,933	1,364	1,010	505
Dodge City	731	857	1,047	952	2,294	1,565	1,188	447
Phoenix	931	1,104	1,214	1,104	2,737	1,765	1,073	346
Charleston	744	732	789	665	1,859	1,251	876	416
Spokane	255	383	434	409	2,357	2,415	2,121	1,368

3. Determine hourly insolation, I_o, for the best and worst months. Assume the hourly total is 18% of the daily insolation.

SOLAR DHW SYSTEM SIZING (continued)

4. Determine input water temperature, T_i. For the best month, the sun heats the water heater above the thermostat setting, so:

   $T_i = T_{therm} + 5°$.

   For the worst month, the collector is fed colder water from the bottom of the tank, so:

   $T_i = T_{therm} - 15°$.

5. Find average daytime outdoor temperature, T_A, for the best (hottest) and worst (coolest) months [*MEEB*, Table B.15, pp.1459–1463].

6. Calculate $(T_i - T_A)/I_o$ for "entry into" the nomograph in *MEEB*, Figure 9.31, p.563, to estimate collector efficiency for the best and worst months.

7. Calculate system efficiency, E, for the best and worst months.

   E = (0.8) (collector efficiency)

8. Determine the collector area that will deliver 100% of your hot water needs in the best month.

   area = TH_R / [(I) (E)]

   where:

   TH_R = total heat required (Btu/day) [E4.6]
   I = total daily insolation (Btu/dayft^2)
   E = best-month system efficiency

NOTE: If your collector is made of modular units, round the area to the next larger modular size (e.g., 40 ft^2 means 3 @ 3' x 5' panels).

9. Determine how much heat will be collected during the worst month. What is the percentage of the total heat required?

   TH_C = (area) (I) (E)

   where:

   I = total daily worst-day insolation (Btu/dayft^2)
   E = worst-month system efficiency

Solar DHW Sizing [suggested format]

	I (Btu/dayft^2)	I_o (Btuh/ft^2)	T_i (°F)	T_A (°F)	$(T_i - T_A)/I_o$	Collector Efficiency	System Efficiency	Collector Area (ft^2)	TH_R (Btu/day)	TH_C (Btu/day)
Best Month	1,250	225	125	80	0.20	.70	.56	25	17,500	17,500
Worst Month	728	131	105	49	0.43	.50	.40	25	17,500	7,280

SOLAR DHW SYSTEM SIZING (continued)

E4.7

10. Average your best- and worst-month performances to estimate yearly percentage of solar water heating.

11. How does the area of the solar collector compare with your previous estimate [D4.4]? How will this difference affect your design? Discuss whether or not your DHW system provides sufficient solar heated water. Indicate any design changes that are necessary to meet Goal E.

E4 WATER AND WASTE

DESIGN DEVELOPMENT

CISTERN SIZING

E4

E4.8

Discussion

Your cistern should be sized to supply enough water during your climate's dry season without falling below minimum capacity. Since the cistern outlet is located higher than the floor of the tank, the cistern can be as much as 10% full and still "run dry." A common sizing guideline is to accommodate at least minimum capacity plus the average rainy season rainfall. A month-by-month comparison of collected rainfall to actual water use will tell you if your cistern is adequate.

Procedure

Evaluate how well your design attains Water and Waste Design Development Goal F—**The cistern capacity is adequate to meet expected water needs year-round.**

1. Identify your climate's rainy season. Beginning with the first month after the rainy season, list the normal monthly rainfall [B4.1], in feet, for each month of the year.

NOTE: B4.1 and NOAA data are listed in inches.

2. Determine how much rain is collected during each month.

   rain collected = **(A) (R) (7.48 gal/ft³)**

   where:

   A = catchment area (three-fourths of roof area) (ft^2)

   R = monthly rainfall (ft) [B4.2]

3. Compare your cistern's capacity with the minimum sizing guideline. Does your cistern hold minimum capacity (10% total capacity) plus the average rainy season rainfall? If not, redesign or explain why you cannot meet this guideline.

4. Determine how much water your cistern holds at the end of each month. Begin with the holdings at 90% capacity after the rainy season ends, then for each month, add the rainfall collected and subtract the monthly water use. Does your cistern's capacity ever equal less than minimum capacity? If so, redesign and recalculate.

Cistern Design Scenario [suggested format]

Month	Monthly Rainfall (inches)	Rain Collected (gallons)	Monthly Use (gallons)	Cistern Holdings Before (gallons)	Cistern Holdings After (gallons)
#1 June	*1.36*	*2,543*	*3,200*	*4,550*	*3,898*
#2 July	*0.40*	*748*	*3,200*	*3,898*	*1,446*
carry forward to next month				*1,446*	

DESIGN DEVELOPMENT

DESIGN REVIEW

E4.9

Discussion

You have just completed design development of the major components of your building's water and waste system. The methods you have used should give you some insight for successfully designing the remainder of the water and waste system. This design review affords you the opportunity to view your water and waste design holistically and to discuss the conflicts and tradeoffs with other design considerations for the entire program.

Procedure

1. Compare the developed design with the schematic design [D4]. Discuss how it has evolved and why.
2. You have explored the use of water and waste systems, acoustic, daylighting, and thermal strategies. Comment on what tradeoffs have been made and what strategies worked well together.
3. Illustrate your discussion of points (1) and (2) with annotated, schematic diagrams of your water and waste design for the entire building and site.

APPENDICES

A–F

CONTENTS

A. Building Programs

B. Site Plans

C. NOAA Climatological Data

D. Sun Peg Shadow Plots

E. LOF Sun Charts

F. GDDM Footprints

APPENDIX A
BUILDING PROGRAMS

APP A
1–22

CONTENTS

1. A Beer Garden
2. A Bowling Alley
3. A Branch Bank
4. A Commercial Business School
5. A Community Center
6. A Corporate Skyscraper
7. A Design Office
8. An Elementary School
9. A Fire Station
10. A Fitness Center
11. A Garden Apartment Complex
12. A Linen Supply and Cleaner
13. A Preschool
14. A Public Library
15. A Regional Bus Terminal
16. A Retirement Home
17. A Rolls-Royce Showroom
18. A Shelter for the Homeless
19. A Supermarket
20. A 24-Hour Emergency Medical Clinic
21. A 24-Hour Restaurant
22. A Western Boot Factory

APPENDIX A

BUILDING PROGRAMS

A BEER GARDEN

INDOORS:

Bandstand
Cold-storage vault
Dance floor
Hall with seating for 150
Kitchen for burgers/snacks
Long bar
2 Restrooms with telephone lobby

OUTDOORS:

Bandstand
Dance floor (300 ft^2)
Enclosed patio with seating for 100

EQUIPMENT:

Amp/speakers for bands
6 Beer signs
Cash register
Chiller for cold storage
2 Commercial chillers @ bar
Ice chest @ bar
Jukebox
Short-order kitchen
2 Vending machines

PARKING:

Service dock
70 Spaces

OCCUPANCY:

11a.m.–2 a.m.
Peak hours: 5 p.m.–8 p.m.

A BOWLING ALLEY

Control desk

Manager's office

Pro shop (400 ft^2)

2 Public restrooms

16 Regulation lanes with spectator gallery

100 Rental lockers

Snack bar with outdoor eating area

EQUIPMENT:

16 Automatic pinsetters

Ball cleaner

3 Cash registers

Commercial chiller

Electric typewriter

16 Scoresheet projectors

Short-order kitchen

4 Vending machines

PARKING:

Drop-off area

Service dock to pro shop, snack bar

60 Spaces

OCCUPANCY:

10 a.m.–2 a.m.

Peak hours: September–May; weekdays; 6 p.m.–midnight

A BRANCH BANK

2 Bathrooms

Conference room

Drive-up teller

Lobby

6 Open-plan offices

3 Private offices

Staff lounge

6 Teller stations

24-Hour automated teller with appropriate shelter

Walk-in vault

Outdoor staff break area

EQUIPMENT:

Coffee urn

5 Electric typewriters

Mainframe computer

Refrigerator

15 Terminals, each with keyboard, display, and printer

2 Vending machines

PARKING:

Armored car parking

Drive-up area

10 Public spaces

15 Staff spaces

OCCUPANCY:

Office hours: Monday–Thursday; 8 a.m.–5 p.m.
Friday; 8 a.m.–7 p.m.

Drive-up only: Monday–Friday; 7 a.m.–8 a.m.; 5 p.m.–7 p.m.

APPENDIX A

BUILDING PROGRAMS

A COMMERCIAL BUSINESS SCHOOL 4

ADMINISTRATIVE OFFICES:

- Accountant
- 4 Instructors
- President
- Secretary/receptionist

CLASSROOMS:

- 2 Computer rooms with 12 stations each
- Lecture room with seating for 50
- Typing room with 12 desks

SERVICE:

- Janitor's closet
- Lunchroom
- 2 Staff bathrooms
- 2 Student bathrooms
- Patio

EQUIPMENT:

- 12 Electric typewriters
- Mainframe computer system
- 4 Overhead projectors
- 25 Personal computer systems
- 2 Photocopy machines
- Slide projector
- 6 Vending machines
- 2 Water cooler/drinking fountains

PARKING:

- 6 Employee spaces
- 30 Student spaces

OCCUPANCY:

- 2 Classes of 12 each + staff; Monday–Friday; 9 a.m.–5 p.m.
- 4 Classes of 12 each + staff; Monday–Friday; 7 p.m.–10 p.m.
- 4 Classes of 12 each + staff; Saturday; 8 a.m.–5 p.m.

APPENDIX A

BUILDING PROGRAMS

A COMMUNITY CENTER

Assembly room for 120 people

2 Bathrooms

3 Community services offices

Crafts room

Janitor's closet

Kitchen

Lounge/lobby

Neighborhood association office

Storage room

Deck/patio area(s) for socializing, eating, and crafts

Equipment:

Coffee maker

Electric kiln

Gas-fired kiln

Handheld power tools (several)

Lathe

Microwave oven

Oven/range

Potter's wheel

Refrigerator

2 Table saws

Parking:

Drop-off area

Dumpster access

50 Spaces

Occupancy:

10 Staff, 200 neighbors (maximum)

Open: 8 a.m.–11 p.m.

Peak hours: classes; 1 p.m.–5 p.m.
assemblies; 7 p.m.–10 p.m.

A CORPORATE SKYSCRAPER

Commercial kitchen

Corporate art gallery

Employee cafeteria

Executive penthouse floor

Lobby and reception area

Mechanical equipment floor

10 Office floors:
- 2 Bathrooms
- Office bay (2500 ft^2)
- Janitor's closet
- 2 Vertical circulation cores

Entry courtyard

Outdoor eating area

Solid-waste management center

EQUIPMENT:

Commercial kitchen

11 Electric coffee urns

2 Elevators

275 Personal computers

5 Photocopiers

5 Vending machines

11 Water coolers

PARKING:

Bus stop

Delivery space/loading dock

200 Spaces, some below grade

OCCUPANCY:

275 Employees; peak:	Monday–Friday; 8 a.m.–5 p.m.
27 Employees; off-peak:	Monday–Friday; 5 p.m.–midnight
	Saturday; 9 a.m.–6 p.m.

APPENDIX A

BUILDING PROGRAMS

A DESIGN OFFICE

- 2 Bathrooms
- Business office for 2 people
- Conference room for 12 people
- Drafting room with 6 stations, each with drawing board and PC
- Kitchenette area
- Principal's office
- Reception/display area
- Deck area for lunch and celebrations

Equipment:

- Blueprint machine
- Coffee maker
- Electric typewriter
- Hot plate
- 8 Personal computers
- Photocopier
- 2 Plotters
- Refrigerator

Parking:

- 6 Client spaces
- 10 Employee spaces

Occupancy:

- Monday–Friday; 8:30 a.m.–5 p.m.
- Occasional all-nighters and weekends

AN ELEMENTARY SCHOOL

2 Bathrooms

6 Classrooms for 24 students each (400 ft^2)

Gymnasium

Kindergarten classroom

Library

Lobby

2 Locker rooms

Lunchroom

Nurse's office

Principal's office

Storage area

Teachers' lounge

Outdoor playcourt for kindergarten class

Playground with softball field

Equipment:

Audio-visual equipment

Electric coffee urn

Electric typewriter

Micro-computer

Small photocopier

Small refrigerator

2 Vending machines

Parking:

Bus drop-off

2 Bus spaces

Service dock

10 Staff spaces

10 Visitor spaces

Occupancy:

Year-round; Monday-Friday; 8 a.m.–11 p.m.

Once a week; 7 p.m.–11 p.m.

A FIRE STATION

Business office

Dispatcher's office

2 Dorm bathrooms

3–12' x 36' Equipment bays

Fire chief's office

Hose tower

Kitchen/dining area/lounge

Lobby

Men's dorm with 15 beds

2 Public bathrooms

2 Shower rooms

Training classroom/exercise room

Women's dorm with 5 beds

Equipment wash area

Outdoor barbeque area/sports area

Equipment:

Audio-visual equipment

Coffee urn

Commercial kitchen

2 Electric typewriters

3 Micro-computers

Parking:

5 Public spaces

15 Staff spaces

Occupancy:

10 People; everyday; 24-hours/day

20 People at shift changes

APPENDIX A

BUILDING PROGRAMS

A FITNESS CENTER

Aerobics room

Indoor swimming pool (25 meter)

Janitor's closet

Lobby

2 Locker rooms/restrooms

4 Racquetball courts

Snack bar

Sports shop

Weight room

Deck/patio for eating/drinking

Deck/patio for exercising

4 Outdoor tennis courts

EQUIPMENT:

2 Cash registers

Main pool heater/filter/pump system

Small commercial kitchen

2 Sound systems

Walk-in refrigerator

2 Whirlpool baths with heaters and jets

PARKING:

Drop-off space (port cochère)

Dumpster access

25 Public spaces

6 Staff spaces

OCCUPANCY:

6 Staff

Peak hours: 50 patrons (maximum); 11 a.m.–1 p.m.; 5 p.m.–11 p.m.

Off peak: 20 patrons (maximum); 6 a.m.–11 a.m.; 1 p.m.–5 p.m.

APPENDIX A

BUILDING PROGRAMS

A GARDEN APARTMENT COMPLEX

Apartments:

- 2 One-bedroom
- 2 Three-bedroom
- 6 Two-bedroom

} each with deck or patio

Community Spaces:

- Laundry room
- Party room
- 2 Public restrooms
- 2 Public saunas
- Public deck/patio area
- Swimming pool, outdoor (50' x 20' minimum)

Equipment: (public spaces only)

- 3 Automatic washers
- 3 Electric driers
- Kitchen
- 2 Saunas
- Pool heater/pump/filter system

Parking:

- 10 Carports
- 2 Dumpsters
- 10 Guest spaces

Occupancy:

- 1.5 People/bedroom average

A LINEN SUPPLY AND CLEANER

2 Bathrooms

Business office for 3 people

Factory bay:

- Drying
- Folding
- Pressing
- Receiving (dirty laundry)
- Shipping (clean laundry)
- Washing

Lobby

Lounge with small kitchen

Manager's office

Receptionist

Outdoor lounge

Equipment:

Coffee urn

2 Conveyor systems with 1 motor-driven belt

2 Drum presses

12 Heavy-duty dryers

12 Heavy-duty washers

4 Micro-computers

4 Pressing machines

Small kitchen

2 Vending machines

Parking:

2 Loading docks

15 Staff spaces

Occupancy:

2 Shifts of 16 people each; Monday-Saturday; 4 a.m.–noon; 2 p.m.–10 p.m.

A PRESCHOOL

2 Adults' bathrooms

2 Children's bathrooms

Director's office

4 Instruction/play areas (500 ft^2 each)

Snack preparation area

Visitor reception area

Covered deck or patio (200 ft^2)

Play-yard (20,000 ft^2)

Equipment:

Coffee urn

Electric typewriter

Micro-computer

Refrigerator

Stove/range

Water heater

Parking:

Pick-up/drop-off area

6 Staff spaces

4 Visitor spaces

Occupancy:

60 Children, 6 staff; September–May; Monday–Friday; 7 a.m.–6 p.m.

40 Children, 4 staff; June–August

A PUBLIC LIBRARY

- Archives (1,000 ft^2)
- 2 Bathrooms
- Book bindery and workshop (2,000 ft^2)
- Card catalogue with microfiche files
- Circulation desk
- Lecture room for 50 people
- 3 Reading areas for adults, children, and periodicals (1,000 ft^2 each)
- Stack area (5,000 ft^2)
- Staff lounge
- Storage room (4,000 ft^2)
- Controlled outdoor reading area

Equipment:

- Audio-visual equipment
- Coffee urn
- 2 Electric typewriters
- 2 Microfiche viewers
- 2 Photocopiers

Parking:

- Automobile passenger drop-off area
- Bus space
- 10 Public spaces
- Service dock
- 5 Staff spaces

Occupancy:

Monday–Saturday; 10 a.m.–10 p.m.

A REGIONAL BUS TERMINAL

Business office

Freight office

Lobby/waiting room

Manager's office

2 Private bathrooms

2 Public bathrooms

Reception room

Snack bar

Ticket office

Outdoor waiting shelters for local buses

Equipment:

2 Cash registers

2 Coffee urns

2 Electric typewriters

3 Micro-computers

Public address system

Small commercial kitchen

6 Vending machines

Parking:

Automobile passenger drop-off area

2 Local bus stops

10 Public spaces

4 Regional bus spaces

Occupancy:

Average 20 people/hour waiting; 6 a.m.–1 a.m.

Peak hours: average 80 people/hour; 7 a.m.–9 a.m.; 5 p.m.–8 p.m.

A RETIREMENT HOME

Administration office

2 Communal bathrooms:

- Handicap shower each
- 2 Lavs each
- 2 Toilets each
- Tub/shower each
- Urinal (men's only)

Dining hall, seating for 30

Kitchen for cooking, preparation, and storage

Laundry room and linen storage

Nurses' station with medical records, lounge, and lockers

2 Public restrooms

8 Resident's rooms (150 ft^2 plus balcony each)

8 Resident's rooms (200 ft^2 plus balcony each)

6 Resident's rooms (200 ft^2 plus kitchenette, half-bath, and balcony each)

Patio or deck, partially covered

Equipment:

2 Clothes dryers

2 Clothes washers

2 Coffee urns

2 Electric typewriters

Elevator (if 2 or 3 stories)

Hot water heaters

Institutional kitchen

2 Microcomputers

Photocopier

22 Televisions

2 Vending machines

2 Water coolers

Parking:

16-seater-van space

10 Staff spaces

10 Visitor/resident spaces

Occupancy:

24 Hours

Visiting hours: Monday–Friday; 7:30 p.m.–10 p.m.
Weekends; all day

APPENDIX A

BUILDING PROGRAMS

A ROLLS-ROYCE SHOWROOM

Business offices for accountant, executive washroom, president, and secretary

Display floor with 2-car display and 4 sales "offices"

Service department:

- 2 Bathrooms
- 2-Bay service garage
- Parts counter and storage (1,000 ft^2)
- Waiting area

EQUIPMENT:

Coffee maker

2 Computerized tune-up diagnostic machines

4 Electric handtools

Electric typewriter

2 Hydraulic lifts

Personal computer

Pneumatic power tool compressor

PARKING:

6 Customer spaces

Delivery space

6 New or used car spaces

6 Service client car spaces

6 Staff spaces

OCCUPANCY:

Administration and Service; Monday–Friday; 8 a.m.–5 p.m.

Sales; Monday–Saturday; 8 a.m.–9 p.m.

APPENDIX A

BUILDING PROGRAMS

A SHELTER FOR THE HOMELESS

3 Bathrooms/shower rooms, one each for men, women, and families

Business office for director and secretary

Childcare room with toilet room

3 Counselers' offices

Dining hall, seats 40

Kitchen

Laundry room and linen storage

Men's dormitory with 20 beds

2 Public restrooms

Reception/waiting area

Recreation room

10 Rooms for families of 3 to 6

Women's dormitory with 10 beds

Outdoor dining and seating area

Outdoor play area

Equipment:

2 Coffee urns

2 Dryers

Institutional kitchen

4 Personal computers

3 Televisions

2 Washers

Parking:

6 Staff spaces

6 Visitor spaces

Occupancy:

24 Hours with peaks at dinner and breakfast

APPENDIX A

BUILDING PROGRAMS

A SUPERMARKET

Bakery

Butcher shop

Cart storage area

10 Checkout stations

Employee lounge

Floor manager's office

2 Private bathrooms

Shelving bay (7,000 ft^2)

Storage

Store manager's office

Walk-in cold storage room

Outdoor garden shop

Equipment:

Baker's oven

Beverage cooler (25 lineal feet)

Bucking bronco ride

10 Cash registers/computers

Coffee urn

Dairy cooler (25 lineal feet)

Electric typewriter

Freezer (50 lineal feet)

Meat cooler (25 lineal feet)

Meat cutter and slicer

Mini-computer

2 Video games

Parking:

2 Cart return enclosures

70 Patron spaces

Service dock

25 Staff spaces

Occupancy:

Staff; daily; 7 a.m.–8 a.m.; 11 p.m.–2 a.m.

Public; 8 a.m.–11 p.m.

A 24-HOUR EMERGENCY MEDICAL CLINIC

Business office

2 Doctors' offices

3 Examination rooms with half-baths

Janitor's closet

Nurses' office

Reception desk

Staff lounge with half-bath

Supply storage room

Waiting room

X-ray room

Outdoor patio associated with lounge

EQUIPMENT:

Coffee maker

3 Micro-computer systems

Range cook-top

2 Refrigerators, one each in storage area and lounge

Stereo and intercom system

4 Surgical lighting systems

X-ray equipment

PARKING:

Ambulance access

Drop-off area

10 Patient spaces

10 Staff spaces

OCCUPANCY:

1 Receptionist, 12 patients/hour, 4 nurses, 16 family members/hour average, 2 doctors, 1 bookkeeper; 8 a.m.–8 p.m.

2 Patients/hour, 2 nurses, 1 janitor, and 2 family members/hour average, 1 doctor; 8 p.m.–8 a.m.

A 24-HOUR RESTAURANT

- 2 Bathrooms
- Dining room for 80 people
- Food and supplies storage
- Hosting station
- Kitchen and grill
- Lobby and waiting area
- Lounge for 20 people
- Lunch counter for 12 people
- Manager/bookkeeper office
- Patio dining area for 40 people

EQUIPMENT:

- 2 Cash registers
- Commercial coffee maker
- Commercial dishwasher
- 2 Commercial ranges
- Electric typewriter
- Exhaust fan
- Ice maker
- 4 Infrared warming lamps
- Large freezer
- Large grill
- Micro-computer
- 2 Microwave ovens
- 2 Ovens
- Television
- 2 Video games
- Walk-in cooler

PARKING:

- 2 Bus spaces
- 50 Patron spaces
- Service dock
- 12 Staff spaces

OCCUPANCY:

- 24 Hours/day
- Peak hours: 7 a.m.–9 a.m.; 11:30 a.m.–1:30 p.m.; 5 p.m.–8 p.m.

A WESTERN BOOT FACTORY

2 Bathrooms

Business office for bookkeeper, manager, and secretary

Factory bay with 10 stations

Janitor's room

Lunchroom and patio

Warehouse with "workspace" for 4 people (2,000 ft^2)

EQUIPMENT:

Coffee maker

Electric typewriter

10 Industrial sewing machines

2 Vending machines

PARKING:

Dumpster

18 Employee spaces

5 Public spaces

Service dock with 2 bays

OCCUPANCY:

Management (5 people); Monday, Wednesday, Friday; 8 a.m.–6 p.m.

Labor (14 people); Monday, Wednesday, Friday; 9 a.m.–5 p.m.

APPENDIX B

SITE PLANS

APP B

1–4

CONTENTS

1. SITE A PLAN
   SITE A MATRIX PLANS
2. SITE B PLAN
   SITE B MATRIX PLANS
3. SITE C PLAN
   SITE C MATRIX PLANS
4. SITE D PLAN
   SITE D MATRIX PLANS

APPENDIX B

SITE PLANS

SITE A PLAN

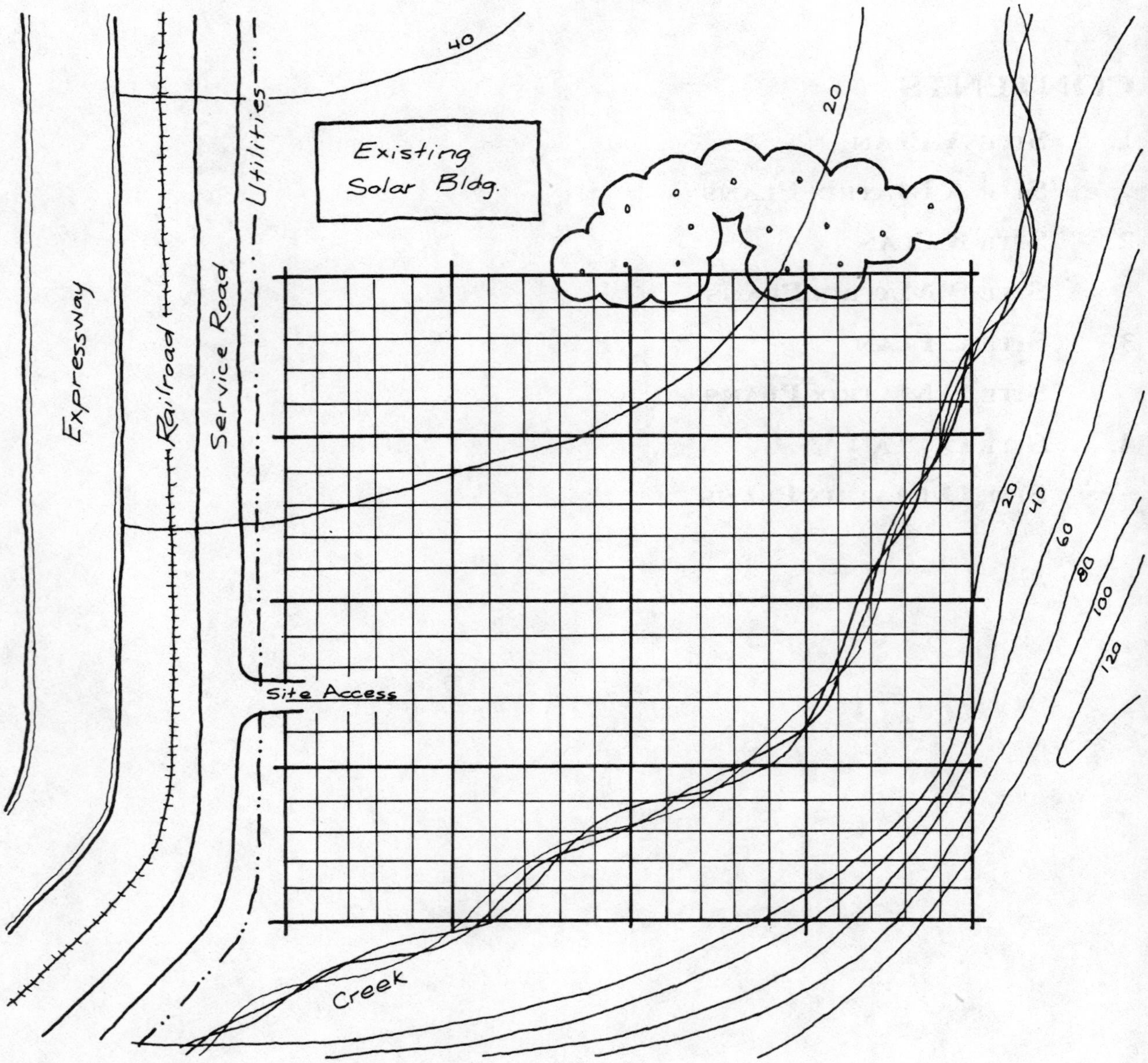

Expressway
Railroad
Service Road
Utilities
40
Existing Solar Bldg.
20
Site Access
20
40
60
80
100
120
Creek

Indigenous Vegetation

Madison, Salem, Spokane	– Fir
Dodge City	– Oak
Charleston	– Pine
Phoenix	– Olive

grid is 20' x 20'

SITE A MATRIX PLANS

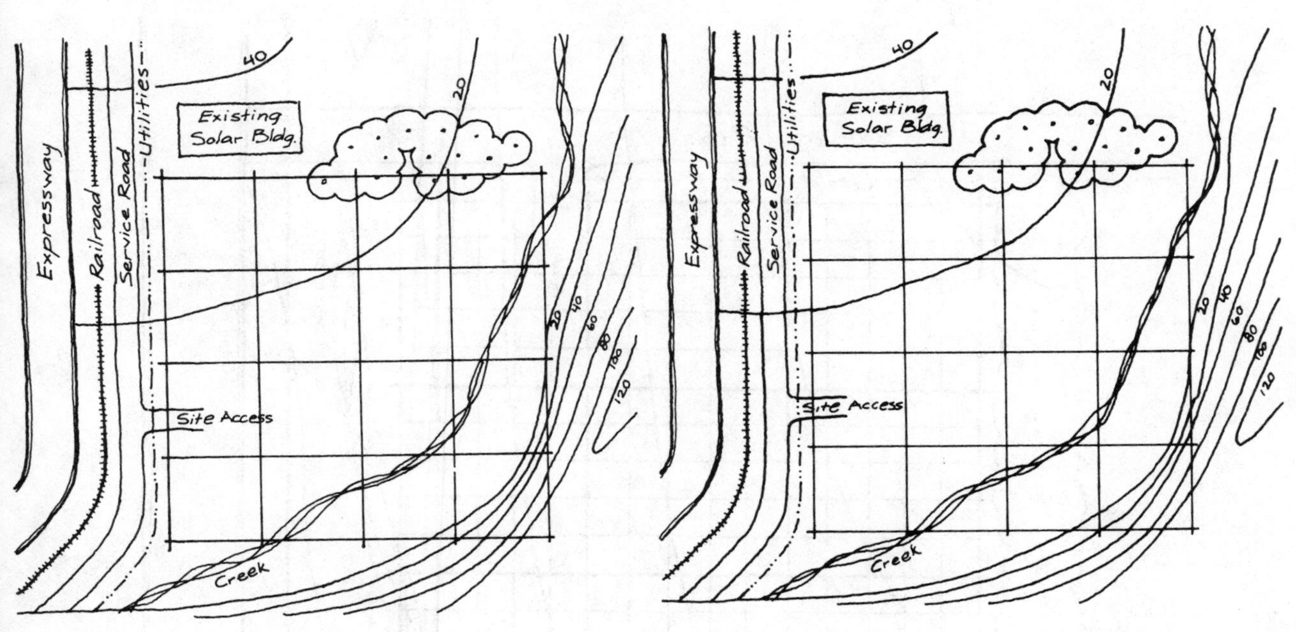

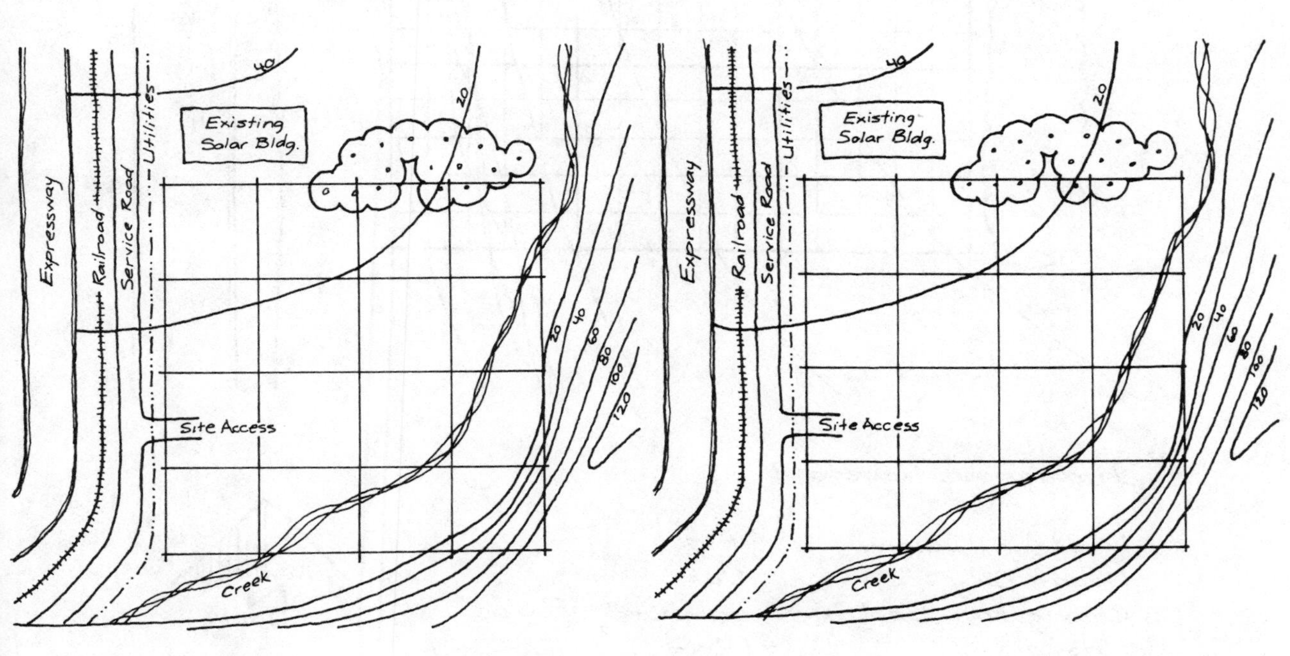

APPENDIX B

SITE PLANS

SITE B PLAN

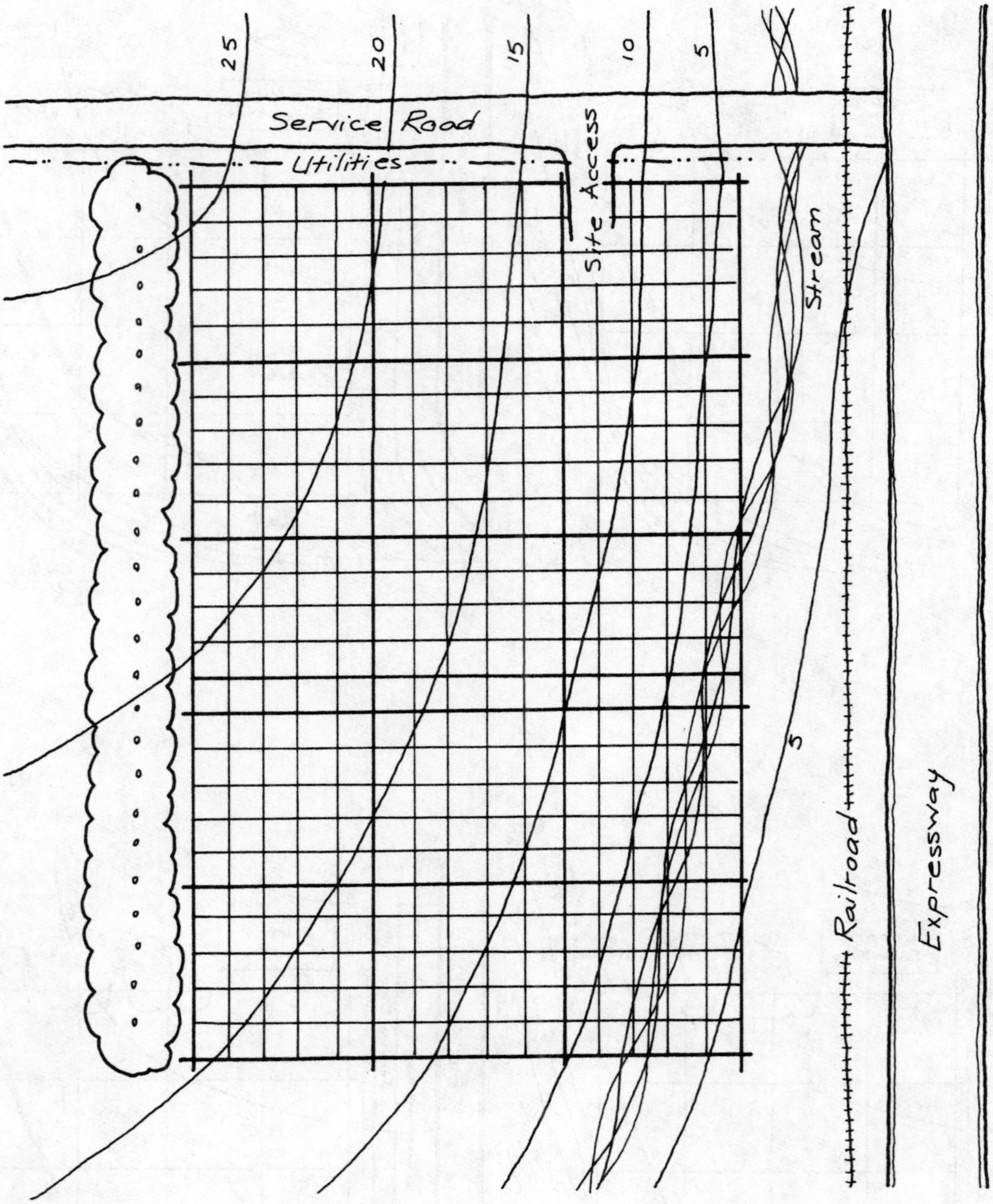

25 20 15 10 5

Service Road

Utilities

Site Access

Stream

5

Railroad

Expressway

Indigenous Vegetation

Madison, Salem - Fir
Dodge City - Oak
Charleston, Spokane - Lombardy Poplar
Phoenix - Olive

grid is 20'×20'

APPENDIX B

SITE PLANS

SITE B MATRIX PLANS

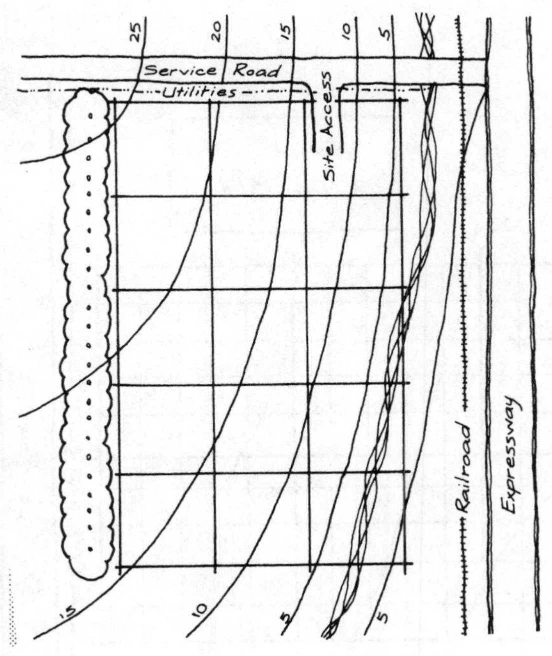

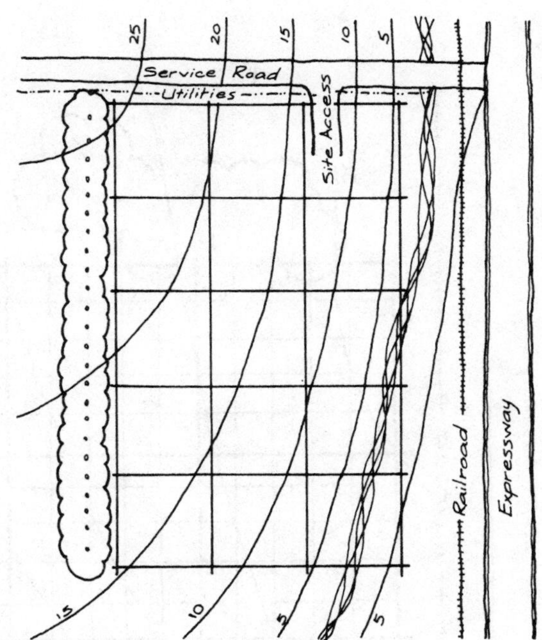

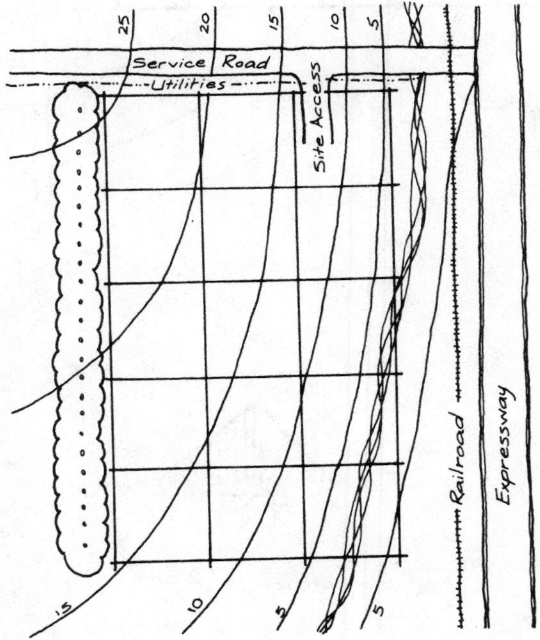

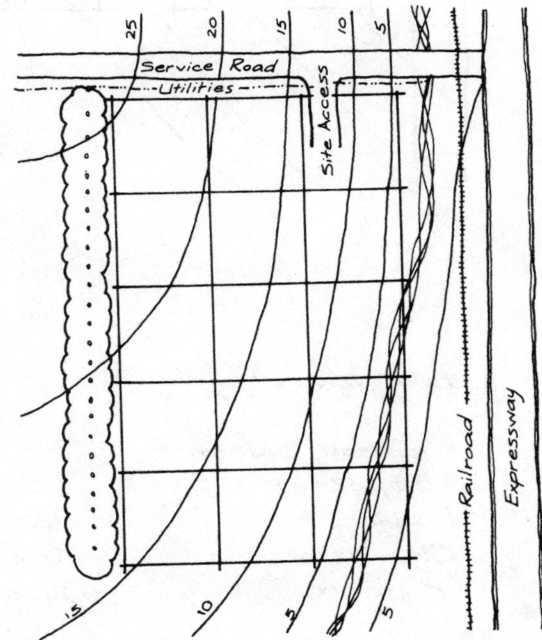

SITE C PLAN

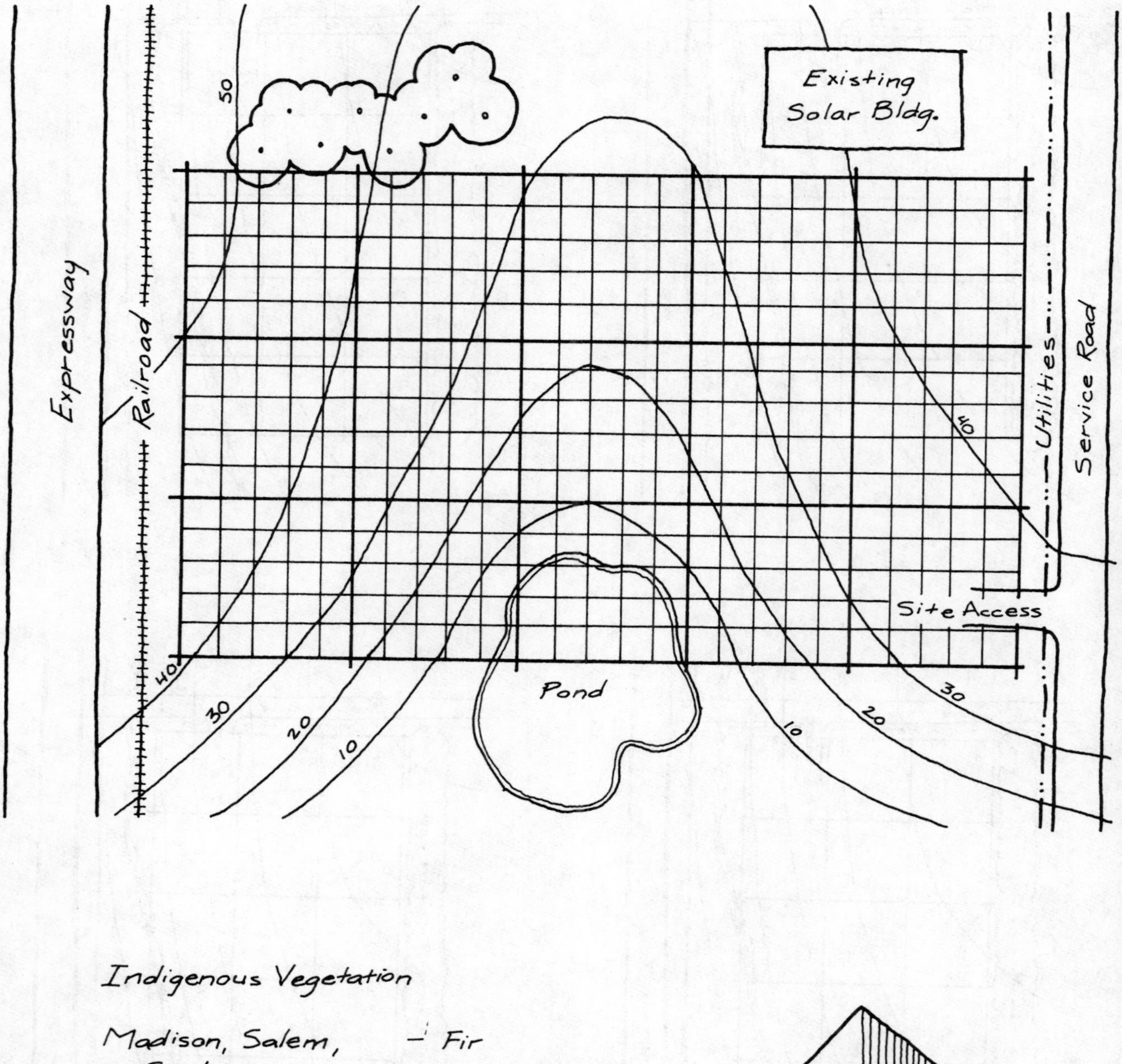

Indigenous Vegetation

Madison, Salem, Spokane	– Fir
Dodge City	– Oak
Charleston	– Pine
Phoenix	– Olive

grid is 20'×20'

SITE C MATRIX PLANS

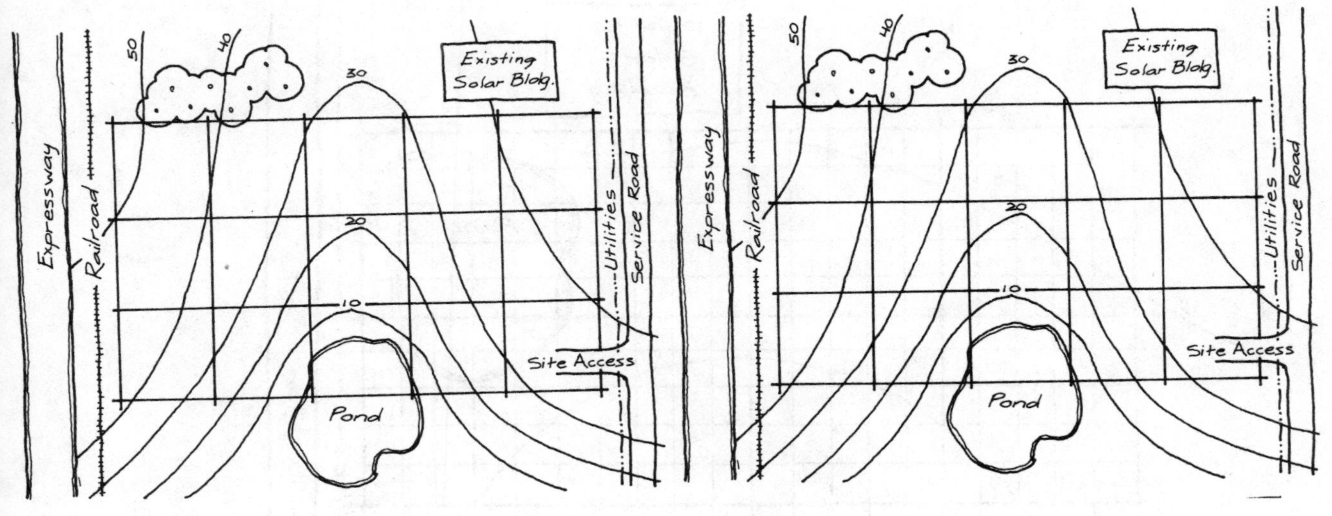

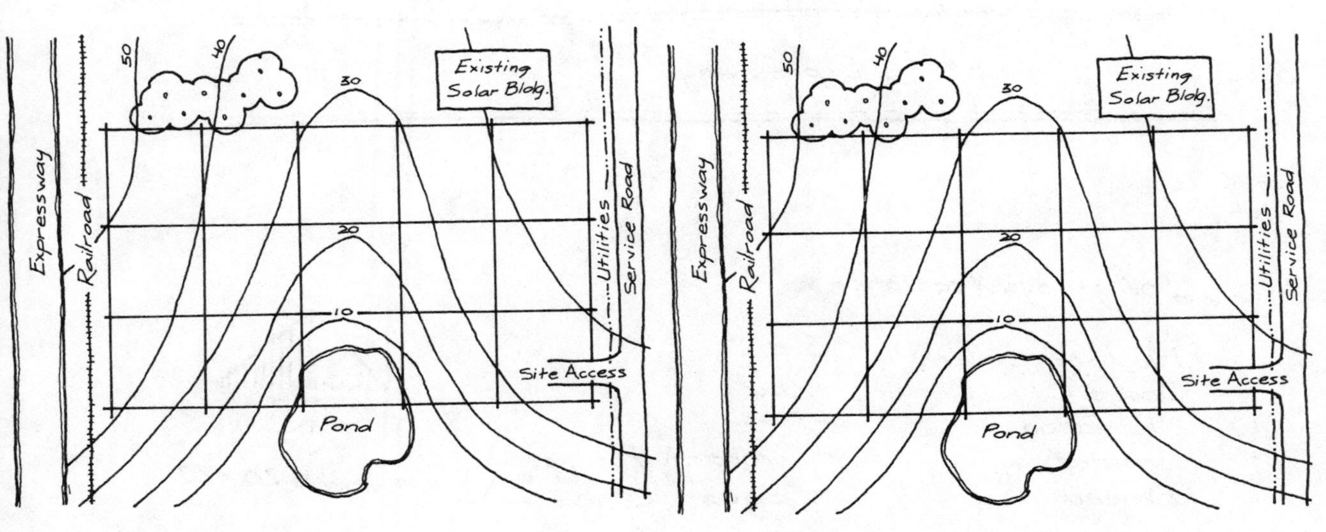

SITE D PLAN

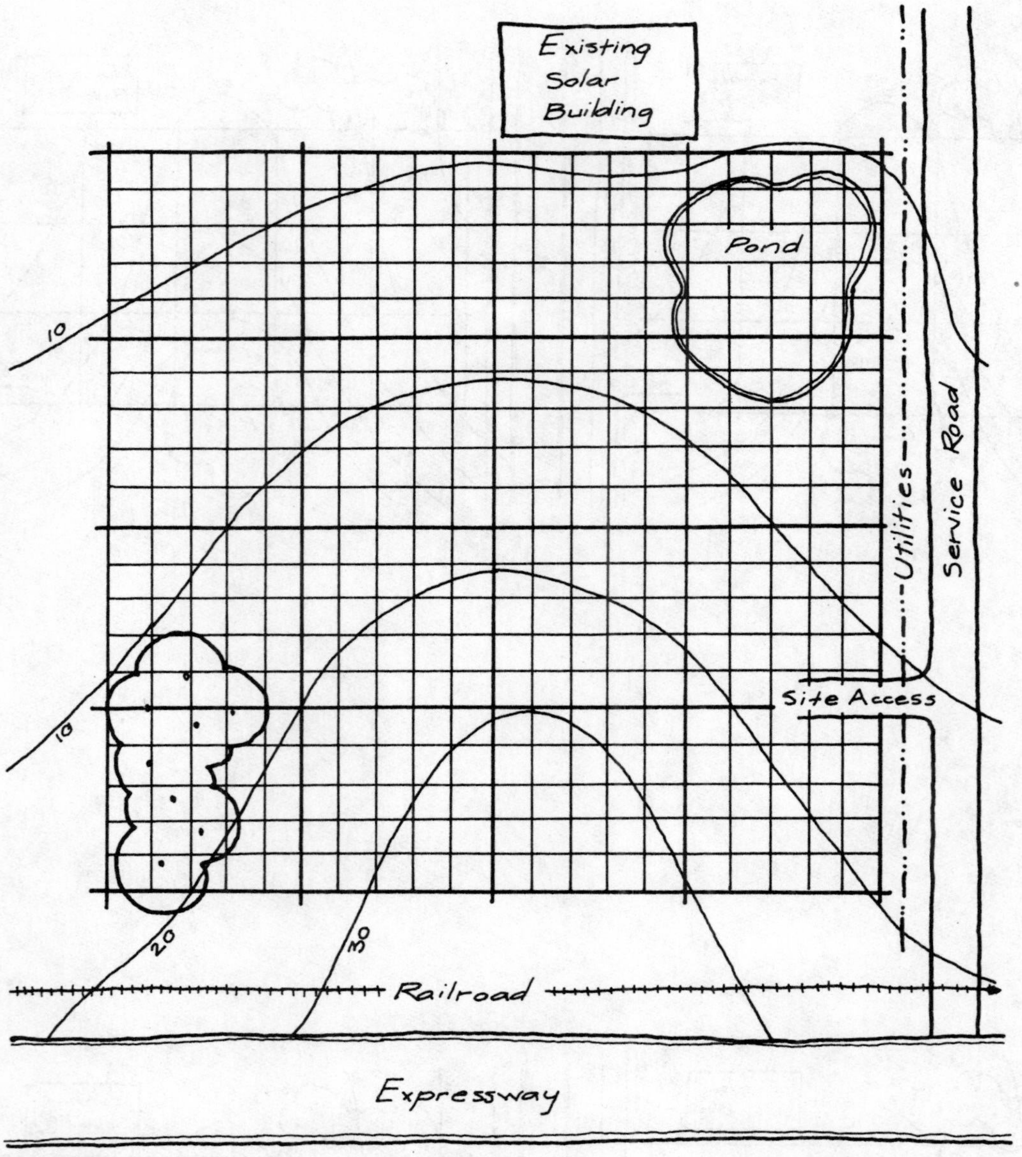

Existing Solar Building

Pond

10

10

20

30

Utilities

Service Road

Site Access

Railroad

Expressway

Indigenous Vegetation

Madison, Salem	– Fir
Dodge City	– Oak
Charleston	– Pine
Spokane	– Lombardy Poplar
Phoenix	– Saguaro Cactus

grid is 20' × 20'

APPENDIX B

SITE PLANS

SITE D MATRIX PLANS

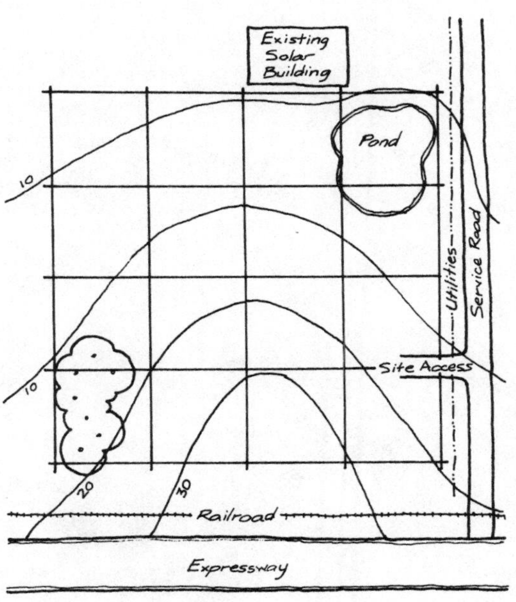

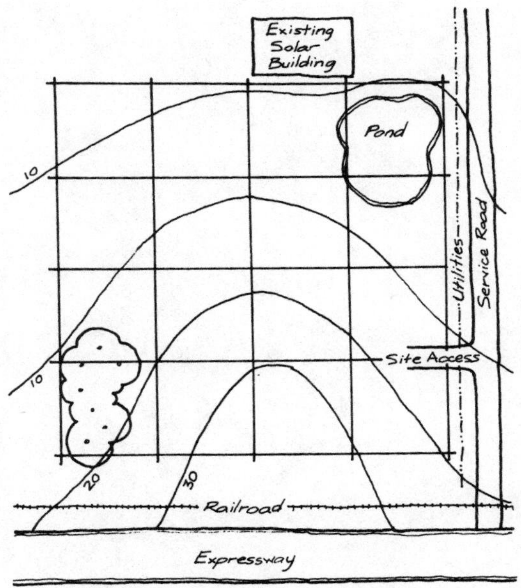

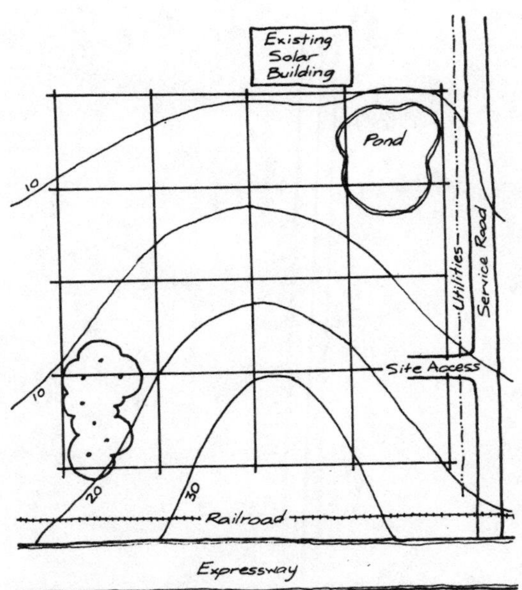

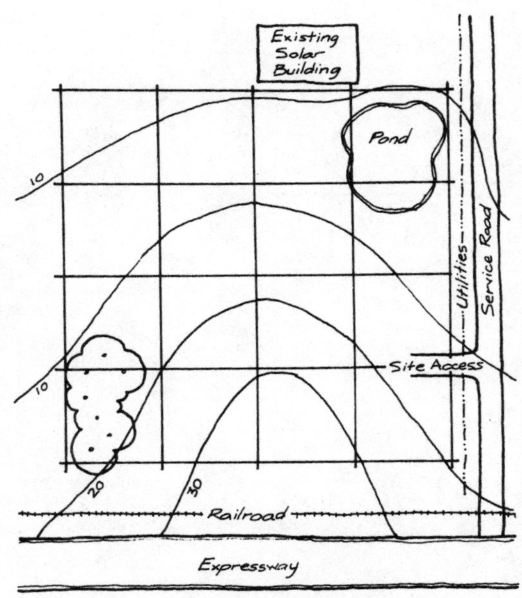

APPENDIX C

NOAA CLIMATOLOGICAL DATA

1–11

CONTENTS

CLIMATOLOGICAL DATA

Narrative Climatological Summary and Normals, Means, and Extremes for the Following Cities:

1. Charleston, South Carolina
2. Dodge City, Kansas
3. Madison, Wisconsin
4. Phoenix, Arizona
5. Spokane, Washington

Data for locations throughout the U.S. are available from U.S. National Oceanic and Atmospheric Administration, Environmental Data and Information Service, National Climatic Center, Asheville, NC 28801–2696, or check with your local U.S. government document depository.

WIND ROSES

These wind roses have been redrawn to isolate the five example climates/cities:

6. January and February
7. March and April
8. May and June
9. July and August
10. September and October
11. November and December

Wind rose data for locations throughout the U.S. are available in the U.S. Department of Commerce, *Climatic Atlas of the United States.*

APPENDIX C

NOAA CLIMATOLOGICAL DATA CHARLESTON, SOUTH CAROLINA

Narrative Climatological Summary

Charleston is a peninsula city bounded on the west and south by the Ashley River, on the east by the Cooper River, and on the southeast by a spacious harbor. Weather records for the airport are from a site some 10 miles inland. The terrain is generally level, ranging in elevation from sea level to 20 feet on the peninsula, with gradual increases in elevation toward inland areas. The soil is sandy to sandy loam with lesser amounts of loam. The drainage varies from good to poor. Because of the very low elevation, a considerable portion of this community and the nearby coastal islands are vulnerable to tidal flooding.

The climate is temperate, modified considerably by the nearness to the ocean. The marine influence is noticeable during winter when the low temperatures are sometimes 10–15°F higher on the peninsula than at the airport. By the same token, high temperatures are generally a few degrees lower on the peninsula. The prevailing winds are northerly in the fall and winter, southerly in the spring and summer.

Summer is warm and humid. Temperatures of 100°F or more are infrequent. High temperatures are generally several degrees lower along the coast than inland due to the cooling effect of the sea breeze. Summer is the rainiest season with 41% of the annual total. The rain, except during occasional tropical storms, generally occurs as showers or thunderstorms.

The fall season passes through the warm Indian summer period to the pre-winter cold spells which begin late in November. From late September to early November the weather is mostly sunny and temperature extremes are rare. Late summer through early fall is the period of maximum threat to the South Carolina coast from hurricanes.

The winter months, December through February, are mild with periods of rain. However, the winter rainfall is generally of a more uniform type. There is some chance of a snow flurry, with the best probability of its occurrence in January, but a significant amount is rarely measured. An average winter would experience less than one cold wave and severe freeze. Temperatures of 20°F or less on the peninsula and along the coast are very unusual.

The most spectacular time of the year, weatherwise, is spring with its rapid changes from windy and cold in March to warm and pleasant in May. Severe local storms are more likely to occur in spring than in summer.

The average occurrence of the first freeze in the fall is early December, and the average last freeze is late February, giving an average growing season of about 294 days.

APPENDIX C

NOAA CLIMATOLOGICAL DATA
CHARLESTON, SOUTH CAROLINA
(continued)

1

NORMALS, MEANS, AND EXTREMES													
Latitude: 32°54'N Longitude: 80°02'W Elevation: 40 feet Time Zone: Eastern													
	Jan	Feb	Mar	Apr	May	Jun	Jul	Aug	Sep	Oct	Nov	Dec	Year
TEMPERATURE °F													
Normals													
Daily Maximum	58.8	61.2	68.0	76.0	82.9	87.0	89.4	88.8	84.6	76.8	68.7	61.4	75.3
Daily Minimum	36.9	38.4	45.3	52.5	61.4	68.0	71.6	71.2	66.7	54.7	44.6	38.5	54.2
Monthly	47.9	49.8	56.7	64.3	72.2	77.6	80.5	80.0	75.7	65.8	56.7	50.0	64.8
Extremes													
Record High	83	87	90	94	98	103	104	102	99	94	88	83	104
Record Low	6	12	15	29	36	50	58	56	42	27	15	8	6
NORMAL DEGREE DAYS													
Heating (base 65°F)	543	434	286	69	6	0	0	0	0	76	262	471	2147
Cooling (base 65°F)	13	9	29	48	229	378	481	465	321	101	13	6	2093
% POSSIBLE SUNSHINE	57	60	65	71	69	65	66	63	60	63	59	56	63
MEAN NUMBER OF DAYS													
Sunrise to Sunset													
Clear	9.1	8.8	9.2	11.4	8.1	6.1	5.0	5.6	6.9	11.6	12.1	9.6	103
Partly Cloudy	6.5	6.4	8.0	7.8	11.0	11.3	12.3	13.0	10.4	8.4	6.5	7.0	109
Cloudy	15.5	13.0	13.8	10.8	11.9	12.6	13.7	12.4	12.8	11.0	11.4	14.4	153
Precipitation	9.7	8.9	10.1	7.4	8.9	10.9	13.3	12.6	9.4	6.0	7.0	8.4	112
Snow, Ice Pellets	0.0	0.1	0.0	0.0	0.0	0.0	0.0	0.0	0.0	0.0	0.0	0.1	0.2
RELATIVE HUMIDITY (%)													
1 a.m.	80	79	81	83	88	89	90	91	90	87	85	82	85
7 a.m.	83	81	83	83	85	86	88	90	90	88	86	83	86
1 p.m.	55	52	51	49	53	58	62	63	62	55	53	55	56
7 p.m.	71	67	66	65	70	74	77	79	80	78	76	73	73
NORMAL PRECIPITATION (in.)	3.33	3.37	4.38	2.58	4.41	6.54	7.33	6.50	4.94	2.92	2.18	3.11	51.6
WIND													
Mean Speed (mph)	9.1	9.9	10.0	9.7	8.7	8.4	7.9	7.4	7.8	8.1	8.1	8.5	8.6
Prevailing Direction	SW	NNE	SSW	SSW	S	S	SW	SW	NNE	NNE	N	NNE	NNE

APPENDIX C

NOAA CLIMATOLOGICAL DATA DODGE CITY, KANSAS

Narrative Climatological Summary

The climate of Dodge City and southwestern Kansas is classified as semi-arid. Dodge City is nearly 300 miles east of the Rocky Mountains, but the weather reflects their influence. The mountains form a barricade against all but high altitude moisture from the southwest, west, and northwest. Chinook winds occur occasionally, but with less frequency and effect than at stations farther to the west. Relatively dry air predominating with an abundance of sunshine contribute to broad diurnal temperature ranges.

Thunderstorms during the growing season contribute most of the moisture. In general, the thunderstorms are widely scattered, occurring during the late afternoons and evenings. They are occasionally accompanied by hail and strong winds, but due to the local nature of the storms, damage to crops and buildings is spotty and variable. Winter is the dry season. However, the moisture accumulated during the winter months is important for the hard, winter wheat. The duration of snow cover is generally brief due to mild temperatures and an abundance of sunshine. The exception results from the occasional blizzard that spreads across the flat, treeless prairies of the high plains.

Afternoon temperatures in the 90s prevail during the summer months. Temperatures above 100°F are the exception. Due to low humidity and a continual breeze, these high temperatures are moderated. Temperatures normally drop sharply after sunset, allowing cool, comfortable nights. During the winter months, large temperature changes are frequent, but the duration of extreme cold spells is brief.

The visibility at Dodge City is generally unrestricted as the terrain is favorable for unrestricted movement of air and air masses. Western Kansas is noted for clear skies and an abundance of sunshine.

Based on the 1951–1980 period, the average first occurrence of 32°F in the fall is October 23 and the average last occurrence in the spring is April 21.

APPENDIX C

NOAA CLIMATOLOGICAL DATA
DODGE CITY, KANSAS (continued)

NORMALS, MEANS, AND EXTREMES													
Latitude: 37°46'N	Longitude: 99°58'W				Elevation: 2,582 feet				Time Zone: Central				
	Jan	Feb	Mar	Apr	May	Jun	Jul	Aug	Sep	Oct	Nov	Dec	Year
TEMPERATURE °F													
Normals													
Daily Maximum	41.1	47.2	55.0	67.4	76.2	87.2	92.5	90.8	81.5	71.0	54.5	45.3	67.5
Daily Minimum	17.9	22.7	29.2	41.1	52.0	62.0	67.4	65.7	56.6	44.4	30.4	22.1	42.6
Monthly	29.5	35.0	42.1	54.3	64.1	74.6	80.0	78.3	69.1	57.8	42.5	33.7	55.1
Extremes													
Record High	80	85	93	100	102	108	109	107	106	96	91	86	109
Record Low	-13	-15	-15	15	26	41	46	47	29	20	0	-21	-21
NORMAL DEGREE DAYS													
Heating (base 65°F)	1101	840	710	331	124	14	0	0	43	251	675	970	5059
Cooling (base 65°F)	0	0	0	10	96	302	465	412	166	28	0	0	1479
% POSSIBLE SUNSHINE	67	64	65	68	67	75	79	77	74	73	66	65	70
MEAN NUMBER OF DAYS													
Sunrise to Sunset													
Clear	11.0	9.0	9.2	9.3	8.9	11.6	12.7	13.6	13.9	14.7	11.7	11.4	137
Partly Cloudy	7.6	7.6	8.2	8.9	10.0	10.5	11.8	10.1	7.3	7.5	7.6	7.3	104
Cloudy	12.4	11.6	13.6	11.9	12.1	7.9	6.6	7.3	8.8	8.8	10.8	12.4	124
Precipitation	4.6	4.9	6.9	6.9	10.1	8.5	8.4	8.2	6.3	4.8	4.1	4.3	78
Snow, Ice Pellets	1.5	1.2	1.5	0.3	0.0	0.0	0.0	0.0	0.0	0.1	0.7	1.0	6.2
RELATIVE HUMIDITY (%)													
midnight	72	72	68	67	73	69	63	67	70	67	71	72	69
6 a.m.	76	76	76	75	81	78	75	78	79	74	76	76	77
noon	58	56	50	46	51	47	43	47	49	46	52	57	50
6 p.m.	59	53	46	42	48	43	38	42	45	47	56	61	48
NORMAL PRECIPITATION (in.)	0.45	0.57	1.47	1.84	3.28	3.02	3.08	2.54	1.86	1.27	0.76	0.52	20.7
WIND													
Mean Speed (mph)	13.6	14.0	15.7	15.5	14.6	14.2	13.1	12.7	13.7	13.5	13.8	13.5	14.0
Prevailing Direction	S	N	N	SSE	S	S	S	S	S	S	S	N	S

APPENDIX C

NOAA CLIMATOLOGICAL DATA MADISON, WISCONSIN

Narrative Climatological Summary

Madison is set on a narrow isthmus of land between Lakes Mendota and Monona. Lake Mendota (15 square miles) lies northwest of Lake Monona (5 square miles), and the lakes are only two-thirds of a mile apart at one point. Drainage at Madison is southeast through two other lakes into the Rock River, which flows south into Illinois, and then west to the Mississippi. The westward flowing Wisconsin River is only 20 miles northwest of Madison. Madison lakes are normally frozen from mid-December to early April.

Madison has the typical continental climate of interior North America with a large annual temperature range and with frequent short-period temperature changes. The range of extreme temperatures is from about 110°F to –40°F. Winter temperatures (December–February) average near 20°F and the summer average (June–August) is in the upper 60s. Daily temperatures average below 32°F about 120 days and above 40°F for about 210 days of the year.

Madison lies in the path of the frequent cyclones and anticyclones that move eastward over the area during fall, winter, and spring. In summer, the cyclones have diminished intensity and tend to pass farther north. The most frequent air masses are of polar origin. Occasional outbreaks of arctic air affect this area during the winter months. Although northward-moving, tropical air masses contribute considerable cloudiness and precipitation, the true Gulf air mass does not reach this area in winter and only occasionally at other seasons. Summers are pleasant with only occasional periods of extreme heat or high humidity.

There are no dry or wet seasons, but about 60% of the annual precipitation falls in the five months of May–September. Cold season precipitation is lighter, but lasts longer. Soil moisture is usually adequate in the first part of the growing season. During July, August, and September, the crops depend on current rainfall, which is mostly from thunderstorms and tends to be erratic and variable. Average occurrence of thunderstorms is just under 7 days per month during this period.

March and November are the windiest months. Tornadoes are infrequent. Dane County has about one tornado in every three to five years.

The ground is covered with 1 inch or more of snow about 60% of the time from about December 10 to near February 25 in an average winter. The soil is usually frozen from the first of December through most of March with an average frost penetration of 25–30 inches. The growing season averages 175 days.

Farming is diversified with the main emphasis on dairying. Field crops are mainly corn, oats, clover, and alfalfa, but barley, wheat, rye, and tobacco are also raised. Canning factories pack peas, sweet corn, and lima beans. Fruits are mainly apples, strawberries, and raspberries.

APPENDIX C

NOAA CLIMATOLOGICAL DATA
MADISON, WISCONSIN (continued)

3

NORMALS, MEANS, AND EXTREMES													
Latitude: 43°08'N	Longitude: 89°20'W				Elevation: 858 feet				Time Zone: Central				
	Jan	Feb	Mar	Apr	May	Jun	Jul	Aug	Sep	Oct	Nov	Dec	Year
TEMPERATURE °F													
Normals													
Daily Maximum	24.5	30.0	40.8	57.5	69.8	78.8	82.8	80.6	72.3	61.1	44.1	30.6	56.1
Daily Minimum	6.7	11.0	21.5	34.1	44.2	53.8	58.3	56.3	47.8	37.8	26.0	14.1	34.3
Monthly	15.6	20.5	31.2	45.8	57.0	66.3	70.6	68.5	60.1	49.5	35.1	22.4	45.2
Extremes													
Record High	56	61	82	94	93	101	104	102	99	90	76	62	104
Record Low	-37	-28	-29	0	19	31	36	35	25	13	-11	-25	-37
NORMAL DEGREE DAYS													
Heating (base 65°F)	1531	1246	1048	576	273	58	12	29	161	490	897	1321	7642
Cooling (base 65°F)	0	0	0	0	25	97	185	137	14	9	0	0	467
% POSSIBLE SUNSHINE	48	52	53	52	58	64	68	65	60	54	40	40	55
MEAN NUMBER OF DAYS													
Sunrise to Sunset													
Clear	7.5	7.2	6.3	6.4	6.9	7.3	9.1	9.3	9.6	9.4	5.7	6.4	91
Partly Cloudy	6.6	6.0	7.5	7.8	9.3	10.0	11.1	10.4	8.2	7.5	6.1	6.1	97
Cloudy	16.9	15.0	17.2	15.8	14.8	12.7	10.8	11.3	12.2	14.1	18.3	18.5	177
Precipitation	10.0	8.0	10.8	11.4	11.4	10.4	9.5	9.6	9.2	8.8	9.5	9.9	119
Snow, Ice Pellets	2.8	2.4	2.5	0.6	0.0	0.0	0.0	0.0	0.0	0.0	1.3	3.4	13.0
RELATIVE HUMIDITY (%)													
midnight	78	78	78	75	77	80	84	87	88	82	82	82	81
6 a.m.	79	80	82	81	80	82	86	91	92	86	85	83	84
noon	69	66	62	55	54	56	57	59	61	59	67	72	61
6 p.m.	73	70	65	56	54	56	58	62	69	68	75	77	65
NORMAL PRECIPITATION (in.)	1.11	1.02	2.15	3.10	3.34	3.89	3.75	3.82	3.06	2.24	1.83	1.53	30.8
WIND													
Mean Speed (mph)	10.6	10.4	11.2	11.4	10.1	9.2	8.1	8.0	8.7	9.6	10.8	10.3	9.9
Prevailing Direction	wnw	wnw	nw	nw	s	s	s	s	s	s	s	w	s

APPENDIX C

NOAA CLIMATOLOGICAL DATA PHOENIX, ARIZONA

Narrative Climatological Summary

Phoenix is located in the Salt River Valley at an elevation of about 1,100 feet. The valley is oval shaped and flat except for scattered precipitous mountains rising a few hundred to as much as 1,500 feet above the valley floor. Sky Harbor Airport, where the weather observations are taken, is in the southern part of the city. Six miles to the south of the airport are the South Mountains rising to 2,500 feet. Eighteen miles southwest, the Estrella Mountains rise to 4,500 feet, and 30 miles to the west are the White Tank Mountains rising to 4,100 feet. The Superstition Mountains, over 30 miles to the east, rise to as much as 5,000 feet. The valley, though located in the Sonora Desert, supports large acreages of cotton, citrus, and other agriculture along with one of the largest urban populations in the United States. The water supply for this complex, desert community is partly from reservoirs on the impounded Salt and Verde Rivers and partly from a large, underground water table.

Temperatures range from very hot in summer to mild in winter. Many winter days reach over 70°F and typical high temperatures in the middle of the winter are in the 60s. The climate becomes less attractive in the summer. The normal high temperature is over 90°F from early May through early October and over 100°F from early June through early September. Many days each summer will exceed 110°F in the afternoon and remain above 85°F all night. When temperatures are extremely high, the low humidity does not provide much comfort.

Indeed, the climate is very dry. Annual precipitation is only about 7 inches, and afternoon humidities range from about 30% in winter to only about 10% in June. Rain comes mostly in two seasons—from about Thanksgiving to early April there are periodic rains from Pacific storms; moisture from the south and southeast results in a summer thunderstorm peak during July and August. Usually the break from extreme dryness in June to the onset of thunderstorms in early July is very abrupt. Afternoon humidities suddenly double to about 20%, which with the great heat give a feeling of mugginess. Fog is rare, occurring about once per winter, and is unknown during the other seasons.

The valley is characterized by light winds. High winds associated with thunderstorms occur periodically in the summer. These occasionally create dust storms that traverse long distances across the deserts. Strong thunderstorm winds occur any month of the year but are rare outside the summer months. Persistent strong winds of 30 mph or more are rare except for two or three events in an average spring due to Pacific storms. Winter storms rarely bring high winds due to the relatively stable air in the valley during that season.

Based on the 1951–1980 period, the average first occurrence of 32°F in the fall is December 13 and the average last occurrence in the spring is February 7.

APPENDIX C

NOAA CLIMATOLOGICAL DATA
PHOENIX, ARIZONA(continued)

4

NORMALS, MEANS, AND EXTREMES													
Latitude: 33°26'N	Longitude: 112°01'W				Elevation: 1,110 feet				Time Zone: Mountain				
	Jan	Feb	Mar	Apr	May	Jun	Jul	Aug	Sep	Oct	Nov	Dec	Year
TEMPERATURE °F													
Normals													
Daily Maximum	65.2	69.7	74.5	83.1	92.4	102	105	102	98.2	87.7	74.3	66.4	85.1
Daily Minimum	39.4	42.5	46.7	53.0	61.5	70.6	79.5	77.5	70.9	59.1	46.9	40.2	57.3
Monthly	52.3	56.1	60.6	68.0	77.0	86.5	92.3	89.9	84.6	73.4	60.6	53.3	71.2
Extremes													
Record High	88	92	100	105	113	122	118	116	118	107	93	88	122
Record Low	17	22	25	32	40	50	61	60	47	34	25	22	17
NORMAL DEGREE DAYS													
Heating (base 65°F)	394	269	187	52	0	0	0	0	0	13	159	368	1442
Cooling (base 65°F)	0	20	51	142	376	645	846	772	588	273	27	6	3746
% POSSIBLE SUNSHINE	78	80	84	88	93	94	85	85	89	88	84	78	86
MEAN NUMBER OF DAYS													
Sunrise to Sunset													
Clear	14.0	12.7	14.6	17.1	21.0	23.2	16.3	17.7	21.6	20.4	17.7	15.3	212
Partly Cloudy	7.0	6.8	8.0	7.3	6.5	4.6	10.4	9.5	5.4	6.2	6.2	6.3	84
Cloudy	10.0	8.7	8.4	5.7	3.5	2.2	4.2	3.8	3.0	4.5	6.1	9.4	70
Precipitation	3.9	3.8	3.5	1.8	0.9	0.7	4.4	4.8	2.9	2.7	2.5	3.8	36
Snow, Ice Pellets	0	0	0	0	0	0	0	0	0	0	0	0	0
RELATIVE HUMIDITY (%)													
5 a.m.	66	60	56	43	35	31	45	51	50	51	58	66	51
11 a.m.	45	39	34	23	18	16	28	33	31	31	37	46	32
5 p.m.	32	27	24	16	13	11	20	23	23	23	27	34	23
11 p.m.	56	48	42	29	22	20	33	38	38	41	49	57	39
NORMAL PRECIPITATION (in.)	0.73	0.59	0.81	0.27	0.14	0.17	0.74	1.02	0.64	0.63	0.54	0.83	7.11
WIND													
Mean Speed (mph)	5.3	5.9	6.7	7.0	7.1	6.8	7.2	6.7	6.3	5.8	5.4	5.1	6.3
Prevailing Direction	E	E	E	E	E	E	W	E	E	E	E	E	E

APPENDIX C

NOAA CLIMATOLOGICAL DATA
SPOKANE, WASHINGTON

Narrative Climatological Summary

Spokane is located on the upper plateau area of the Columbia Basin in eastern Washington where the long, gradual slope from the Columbia River meets the sharp rise of the Rocky Mountain Ranges.

Much of the urban area of Spokane lies along both sides of the Spokane River at an elevation of approximately 2,000 feet, while the residential areas have spread across the plateaus with elevations up to 2,500 feet above sea level. Spokane International Airport is situated on a plateau 6 miles west-southwest and some 400 feet higher than the downtown business district.

The climate of Spokane combines some of the characteristics of damp, coastal weather and arid, interior conditions. Most of the air masses that reach Spokane are brought by the prevailing westerly and southwesterly wind flows. Frequently much of the moisture in the storms that move eastward and southeastward from the Gulf of Alaska and the eastern Pacific Ocean is precipitated as the storms are lifted across the Coast and Cascade Ranges. Annual precipitation totals in the Spokane area are generally less than 20 inches and less than 50% of the amounts received west of the Cascades. However, the precipitation and total cloudiness in the Spokane vicinity are greater than that of the desert areas of south-central Washington. The lifting action of the air masses as they move up the east slope of the Columbia Basin frequently produces the cooling and condensation necessary for formation of clouds and precipitation.

Infrequently, the Spokane area comes under the influence of dry, continental air masses from the north or east. When these air masses penetrate eastern Washington the results are high temperatures and very low humidity in the summer and sub-zero temperatures in the winter. In the winter most of the severe arctic outbursts of cold air move southward on the east side of the Continental Divide and do not affect Spokane.

In general, Spokane weather has the characteristics of a mild, arid climate during the summer months and a cold, coastal climate in the winter. Approximately 70% of the total annual precipitation falls between the first of October and the end of March and about half of that falls as snow. The growing season usually extends over nearly six months from mid-April to mid-October. Irrigation is required for all crops except dry-land grains. The summer weather is ideal for full enjoyment of the many mountain and lake recreational areas in the vicinity. Winter weather includes many cloudy or foggy days and below freezing temperatures with occasional snowfall of several inches. Sub-zero temperatures and traffic-stopping snowfalls are infrequent.

Based on the 1951–1980 period, the average first occurrence of 32°F in the fall is October 6 and the average last occurrence in the spring is May 4.

SPOKANE, WASHINGTON(continued)

NORMALS, MEANS, AND EXTREMES

Latitude: 47°38'N Longitude: 117°32'W Elevation: 2,357 feet Time Zone: Pacific

	Jan	Feb	Mar	Apr	May	Jun	Jul	Aug	Sep	Oct	Nov	Dec	Year
TEMPERATURE °F													
Normals													
Daily Maximum	31.3	39.0	46.2	56.7	66.1	74.0	84.0	81.7	72.4	58.3	41.4	34.2	57.1
Daily Minimum	20.0	25.7	29.0	34.9	42.5	49.3	55.3	54.3	46.5	36.7	28.5	23.7	37.2
Monthly	25.7	32.4	37.6	45.8	54.3	61.7	69.7	68.1	59.4	47.6	34.9	29.0	47.2
Extremes													
Record High	59	61	71	90	96	100	103	108	98	86	67	56	108
Record Low	-22	-17	-7	17	24	33	37	35	24	11	-21	-25	-25
NORMAL DEGREE DAYS													
Heating (base 65°F)	1218	913	849	576	339	140	17	63	209	539	903	1116	6882
Cooling (base 65°F)	0	0	0	0	8	41	162	159	41	0	0	0	411
% POSSIBLE SUNSHINE	27	40	54	61	63	66	80	77	71	55	28	22	54
MEAN NUMBER OF DAYS													
Sunrise to Sunset													
Clear	3.0	3.3	4.2	4.5	5.5	7.3	16.5	15.2	12.3	8.0	3.2	2.8	86
Partly Cloudy	4.3	5.0	7.8	8.3	10.1	10.3	8.3	8.4	8.1	7.7	5.0	3.9	87
Cloudy	23.7	20.0	19.0	17.2	15.4	12.4	6.1	7.4	9.6	15.3	21.8	24.3	192
Precipitation	14.2	11.4	11.5	8.6	9.4	7.7	4.3	5.0	5.7	7.6	12.6	15.0	113
Snow, Ice Pellets	5.3	2.9	1.6	0.2	0.0	0.0	0.0	0.0	0.0	0.1	2.0	5.0	17
RELATIVE HUMIDITY (%)													
4 a.m.	85	84	81	77	77	74	64	63	71	79	87	87	77
10 a.m.	83	80	69	57	53	49	40	43	51	66	83	86	63
4 p.m.	78	69	55	44	41	36	27	28	34	49	76	83	52
10 p.m.	84	81	74	65	63	58	45	46	56	70	85	87	68
NORMAL PRECIPITATION (in.)	2.47	1.61	1.36	1.08	1.38	1.23	0.50	0.74	0.71	1.08	2.06	2.49	16.7
WIND													
Mean Speed (mph)	8.8	9.3	9.7	10.0	9.2	9.2	8.6	8.2	8.3	8.2	8.7	8.6	8.9
Prevailing Direction	NE	SSW	SSW	SW	SSW	SSW	SW	SW	NE	SSW	NE	NE	SSW

APPENDIX C

NOAA CLIMATOLOGICAL DATA

JANUARY WIND ROSES

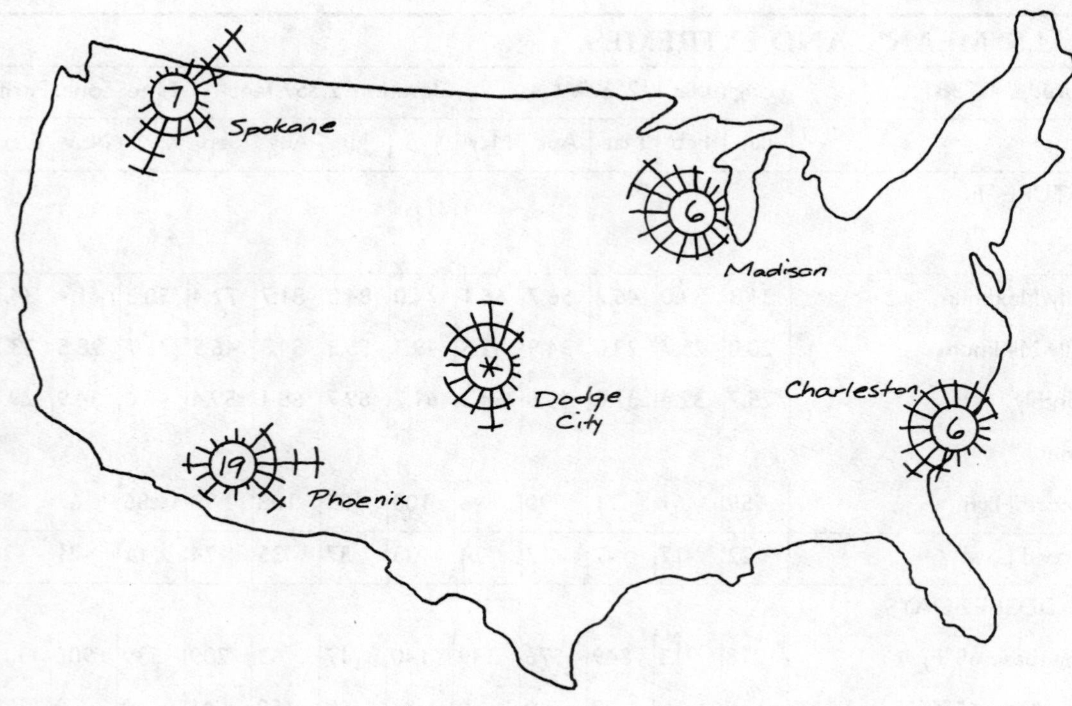

FEBRUARY WIND ROSES

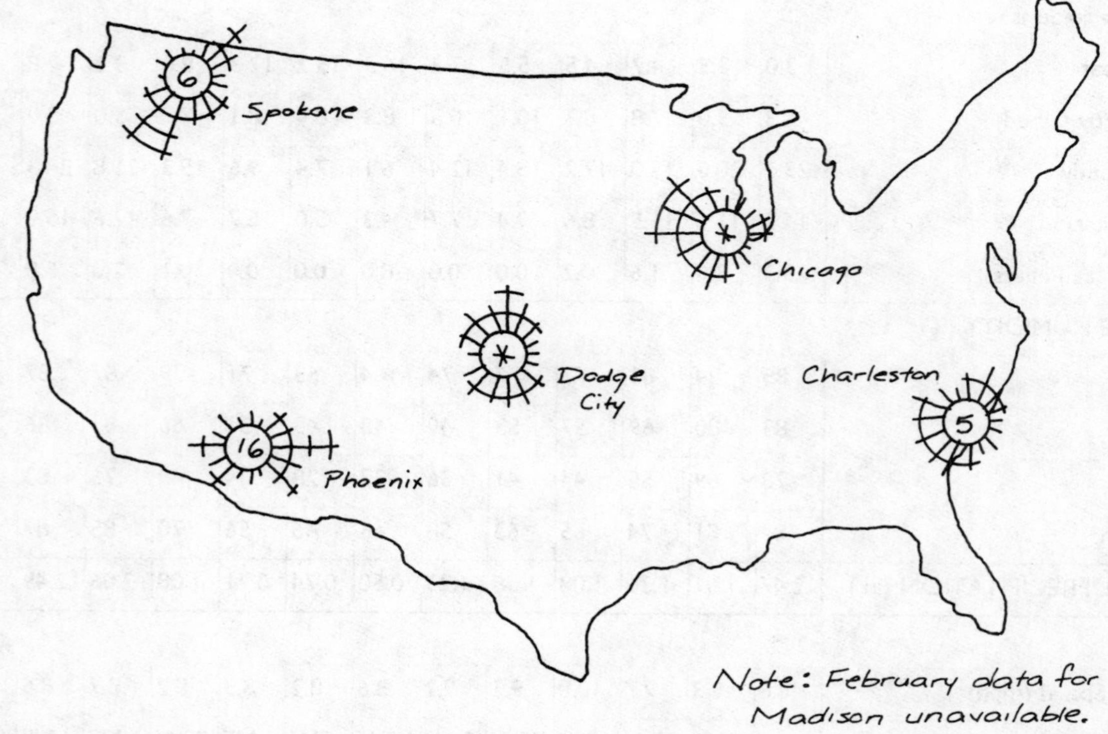

Note: February data for Madison unavailable.

APPENDIX C

NOAA CLIMATOLOGICAL DATA

MARCH WIND ROSES

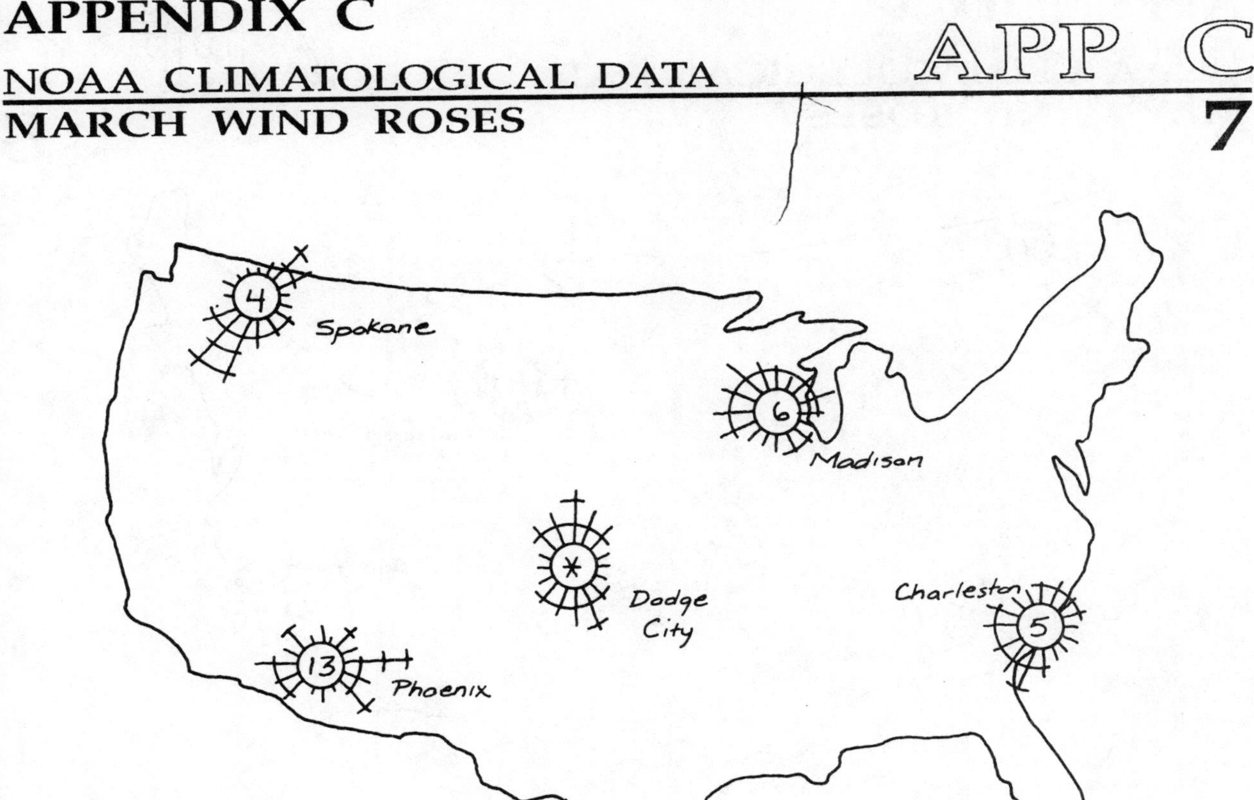

APRIL WIND ROSES

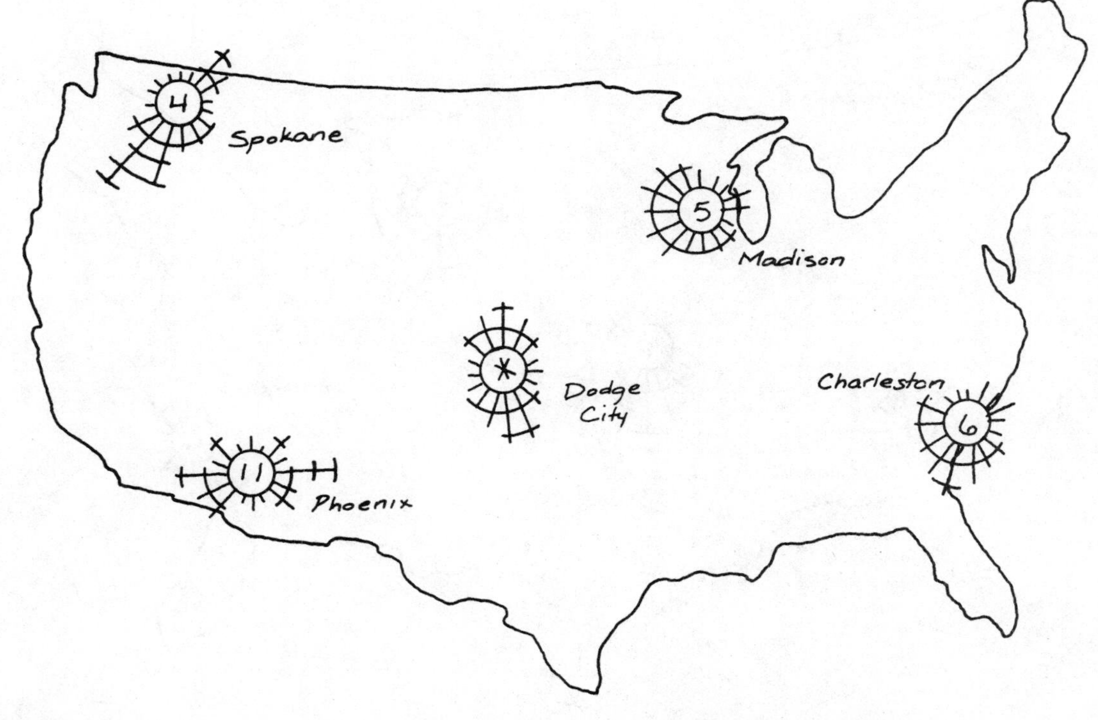

APPENDIX C

NOAA CLIMATOLOGICAL DATA

MAY WIND ROSES

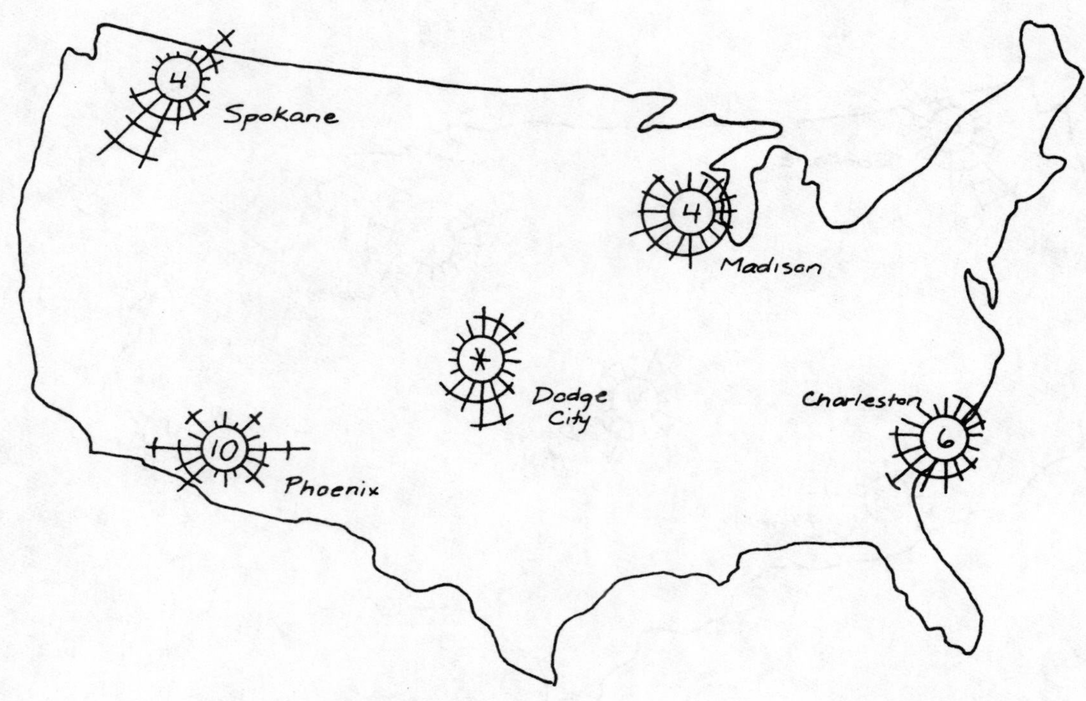

JUNE WIND ROSES

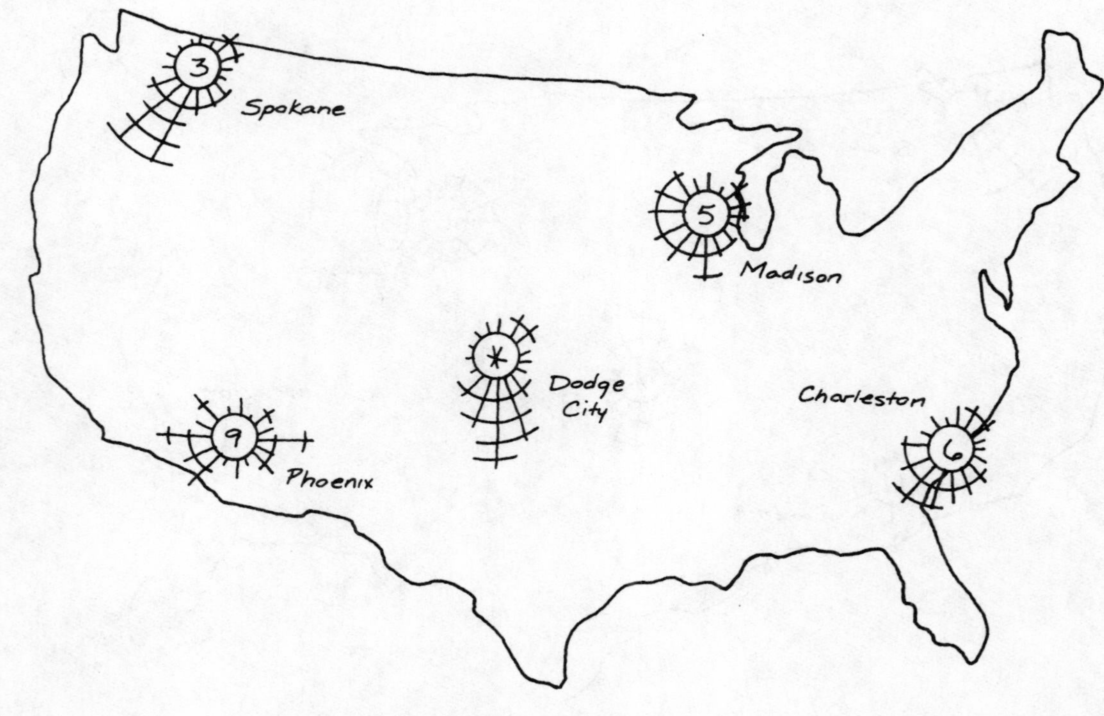

APPENDIX C

NOAA CLIMATOLOGICAL DATA

JULY WIND ROSES

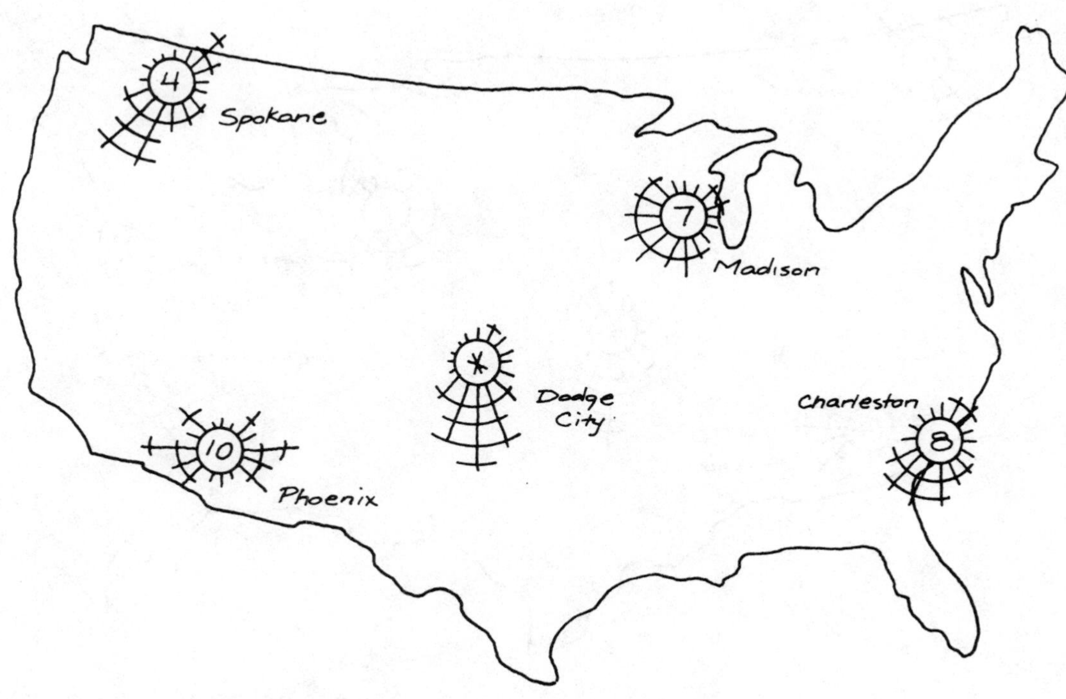

AUGUST WIND ROSES

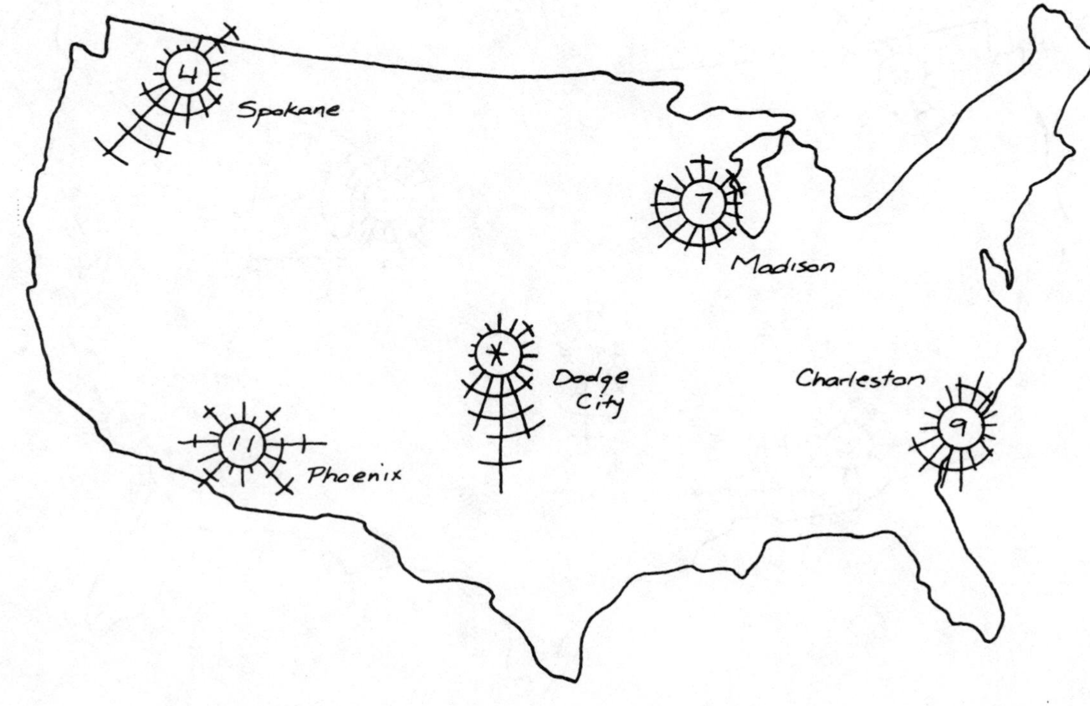

SEPTEMBER WIND ROSES

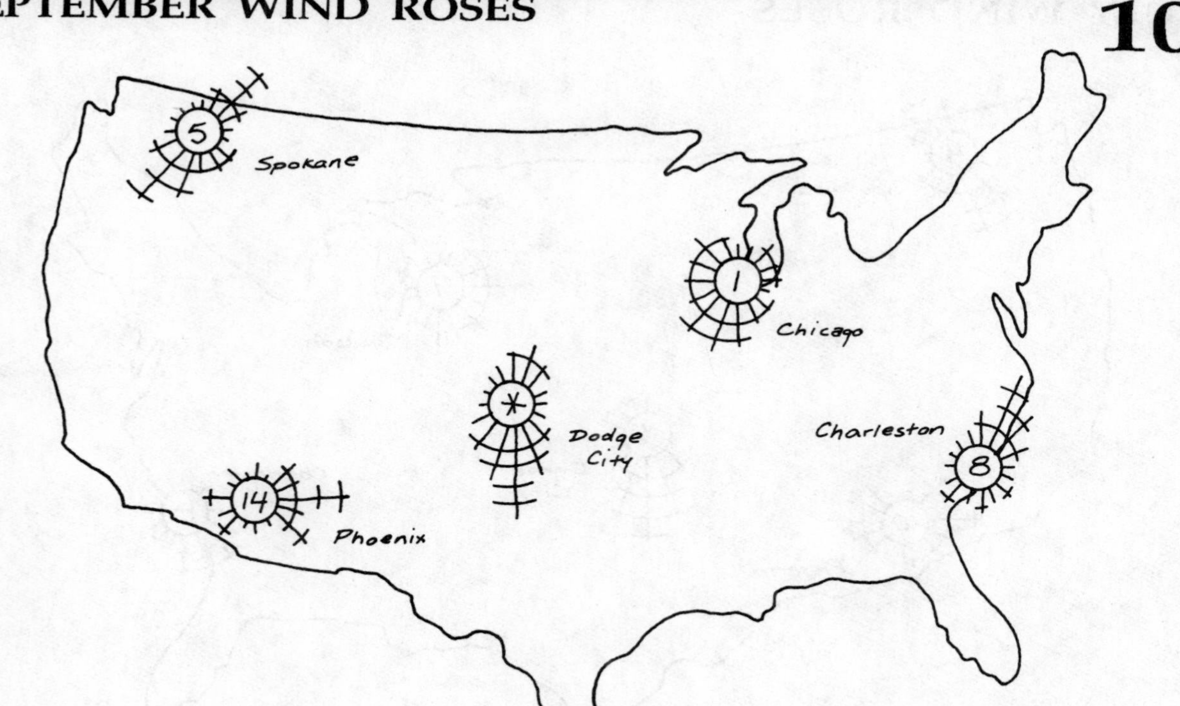

Note: September data for Madison unavailable.

OCTOBER WIND ROSES

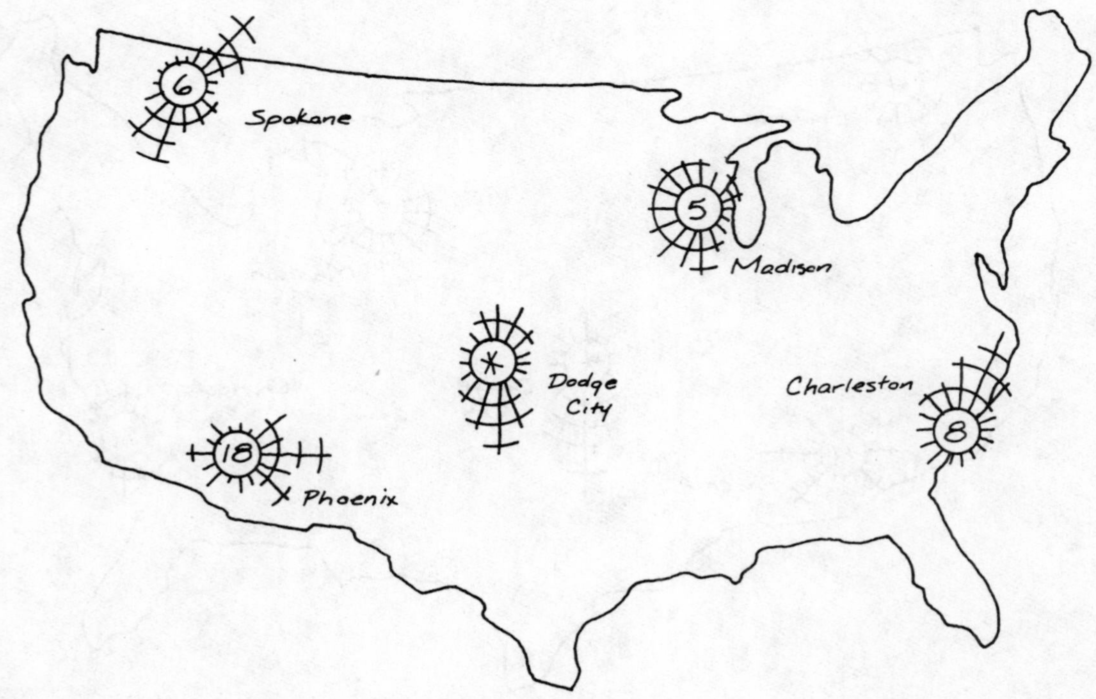

NOVEMBER WIND ROSES

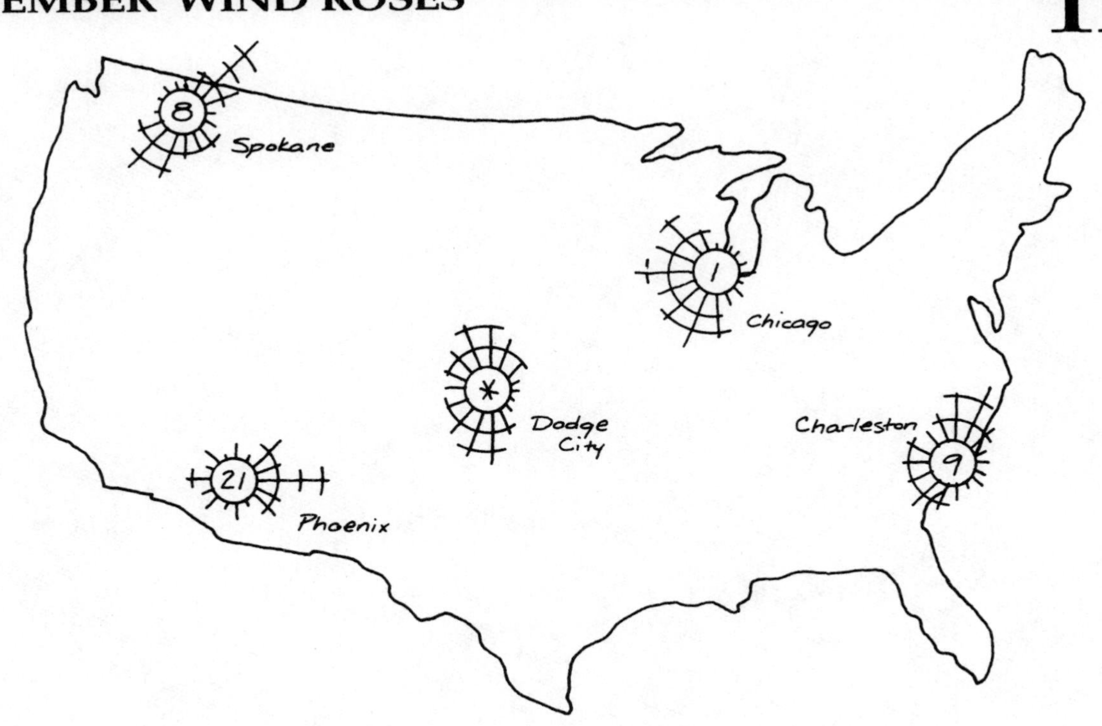

Note: November data for Madison unavailable.

DECEMBER WIND ROSES

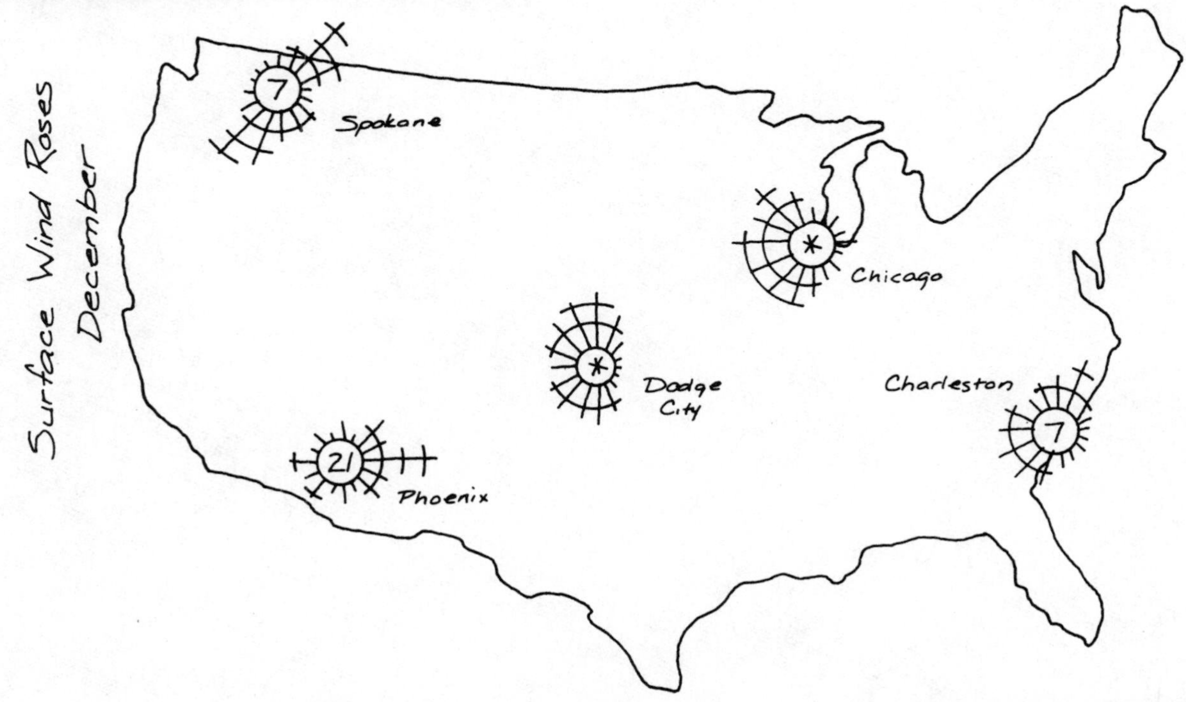

Note: December data for Madison unavailable.

APPENDIX D

SUN PEG SHADOW PLOTS

1–3

CONTENTS

1. 28° North Latitude
   32° North Latitude
2. 36° North Latitude
   40° North Latitude
3. 44° North Latitude
   48° North Latitude

APPENDIX D

SUN PEG SHADOW PLOTS

28° NORTH LATITUDE

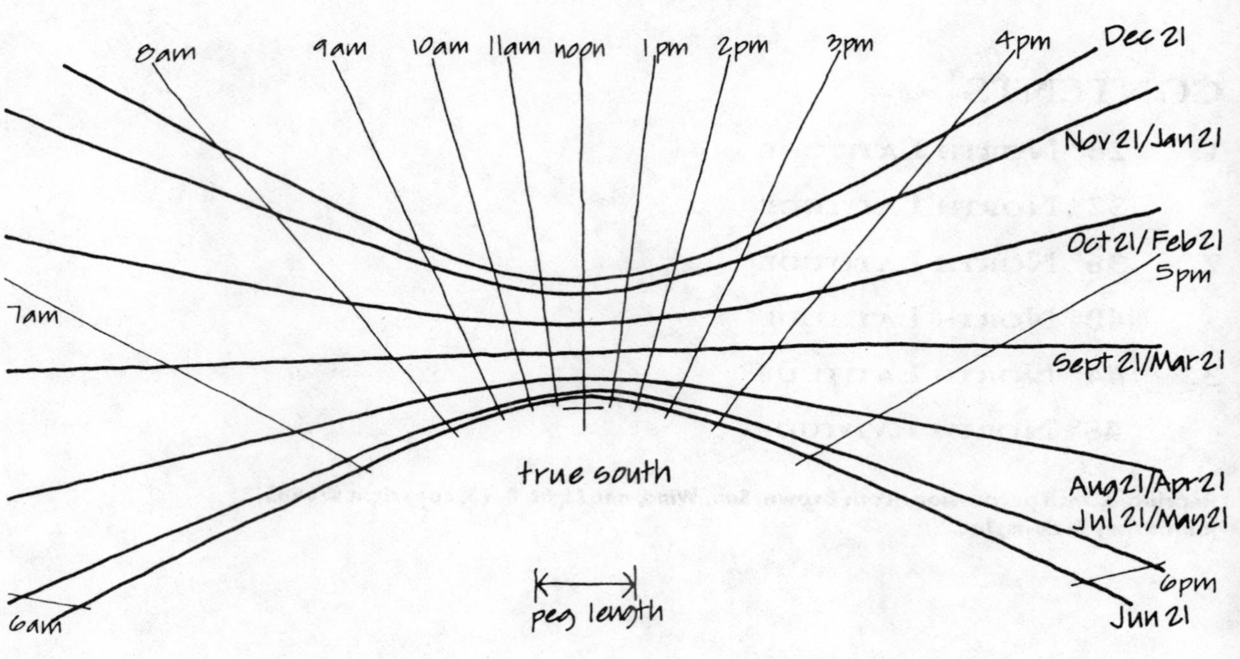

32° NORTH LATITUDE

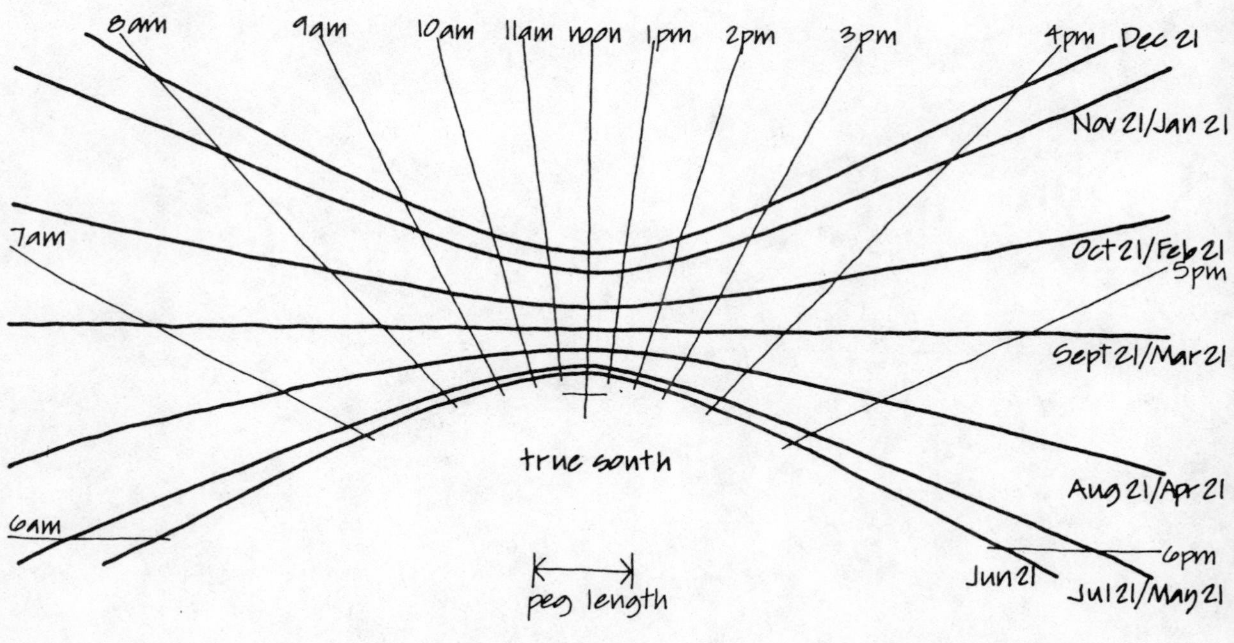

APPENDIX D

SUN PEG SHADOW PLOTS

36° NORTH LATITUDE

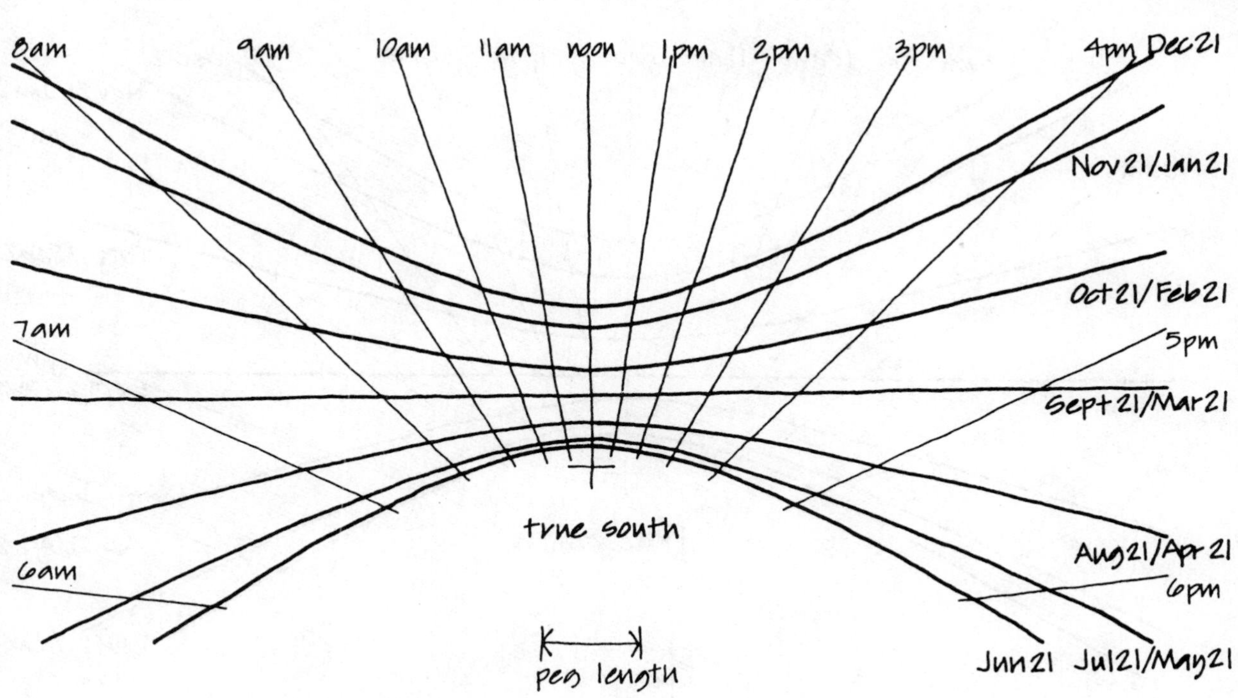

40° NORTH LATITUDE

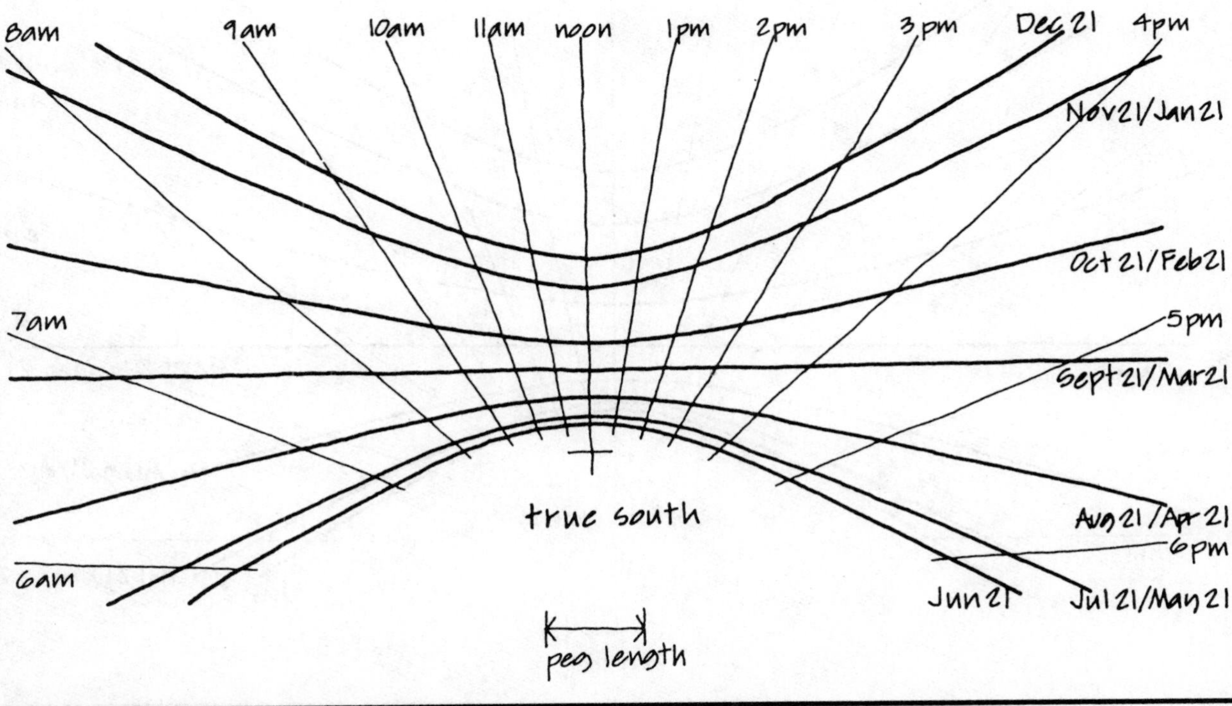

APPENDIX D

SUN PEG SHADOW PLOTS

44° NORTH LATITUDE

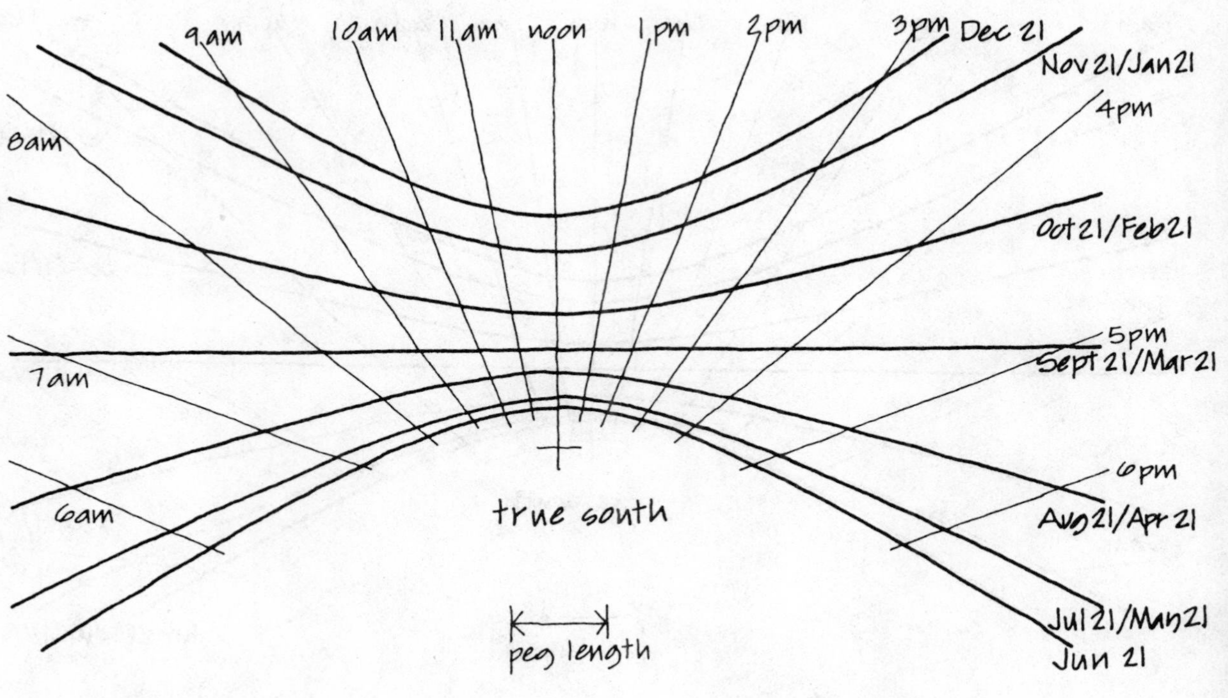

48° NORTH LATITUDE

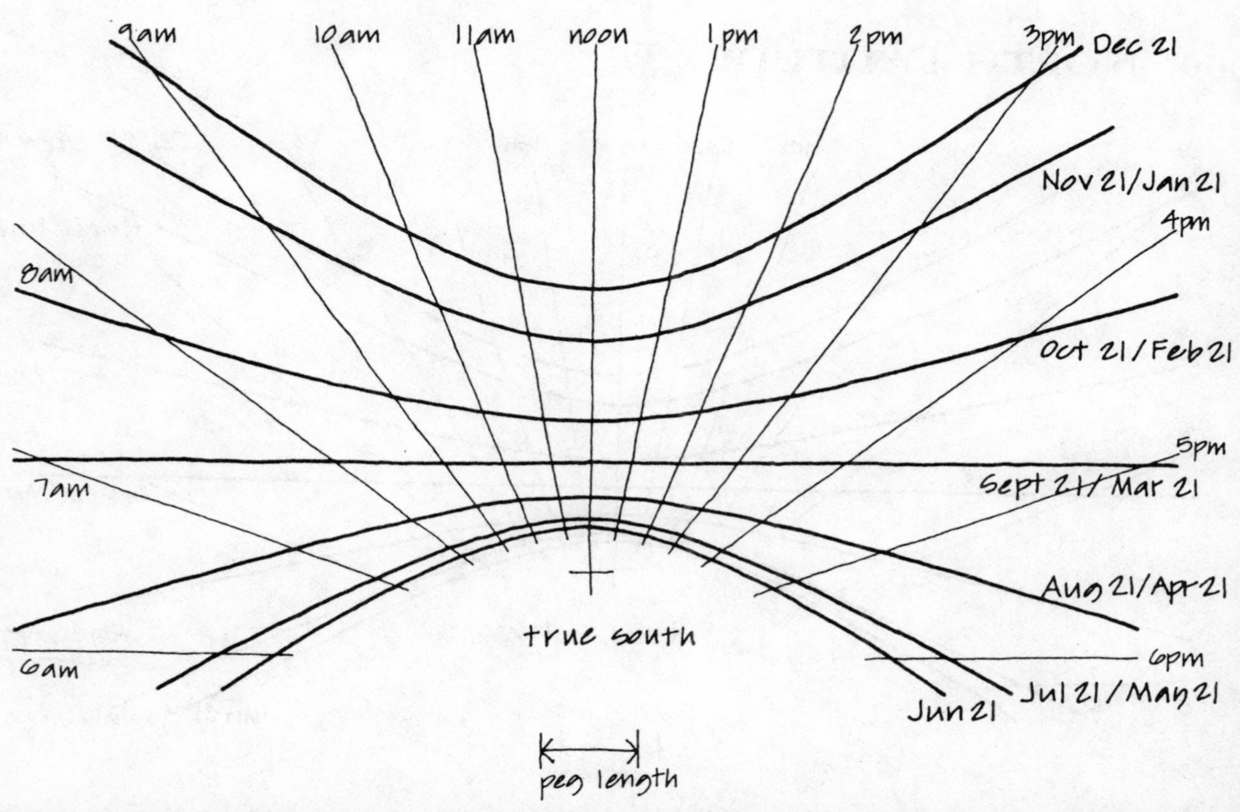

APPENDIX E

LOF SUN CHARTS

APP E

1–6

CONTENTS

1. 32° North Latitude Sun Chart
2. 36° North Latitude Sun Chart
3. 40° North Latitude Sun Chart
4. 44° North Latitude Sun Chart
5. 48° North Latitude Sun Chart
6. LOF Profile Angle Overlay

Reprinted, by permission, from Libbey–Owens–Ford Company, *Sun Angle Calculator*.

APPENDIX E

LOF SUN CHARTS

32° NORTH LATITUDE SUN CHART

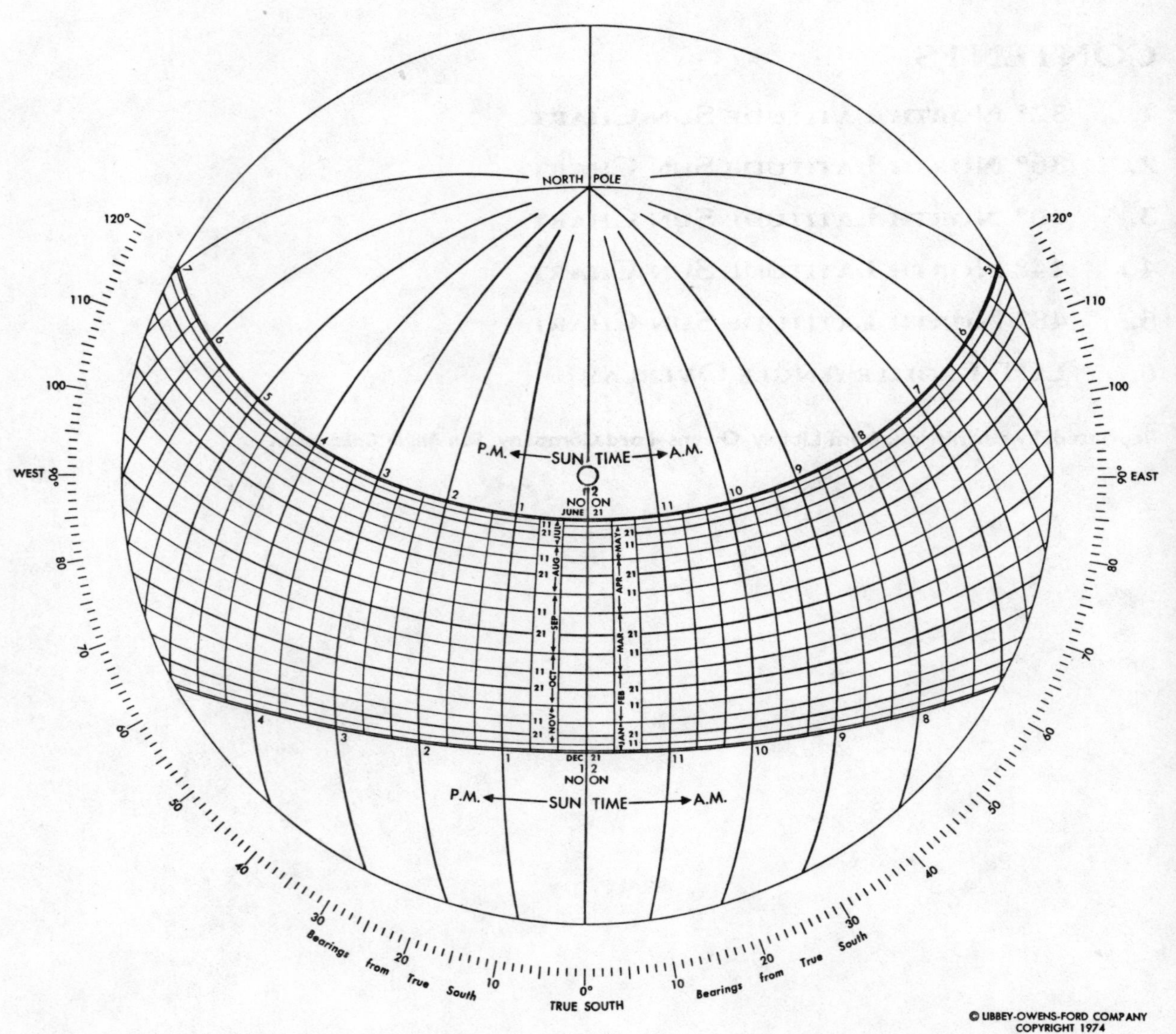

APPENDIX E

LOF SUN CHARTS

36° NORTH LATITUDE SUN CHART

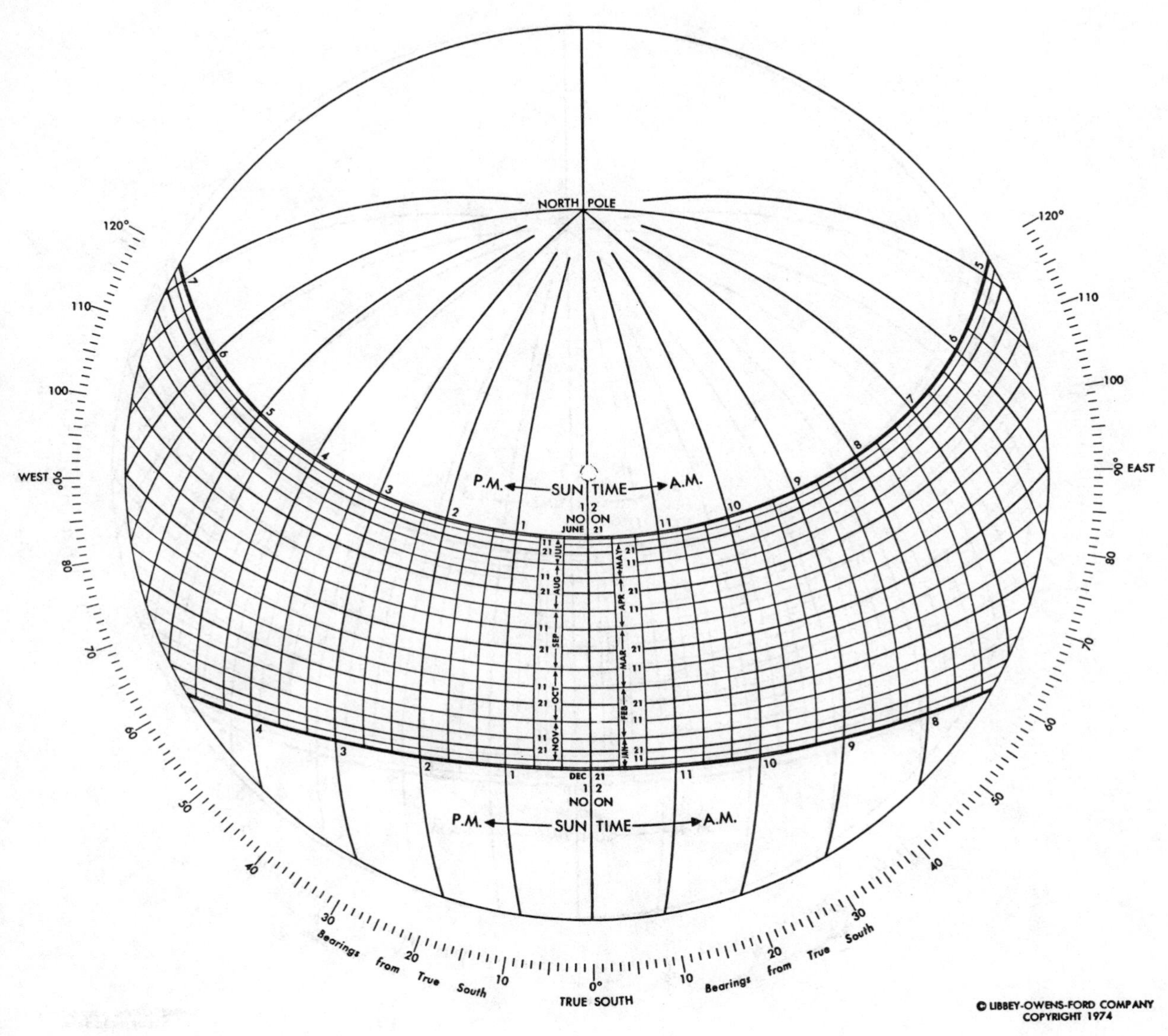

APPENDIX E

LOF SUN CHARTS

40° NORTH LATITUDE SUN CHART

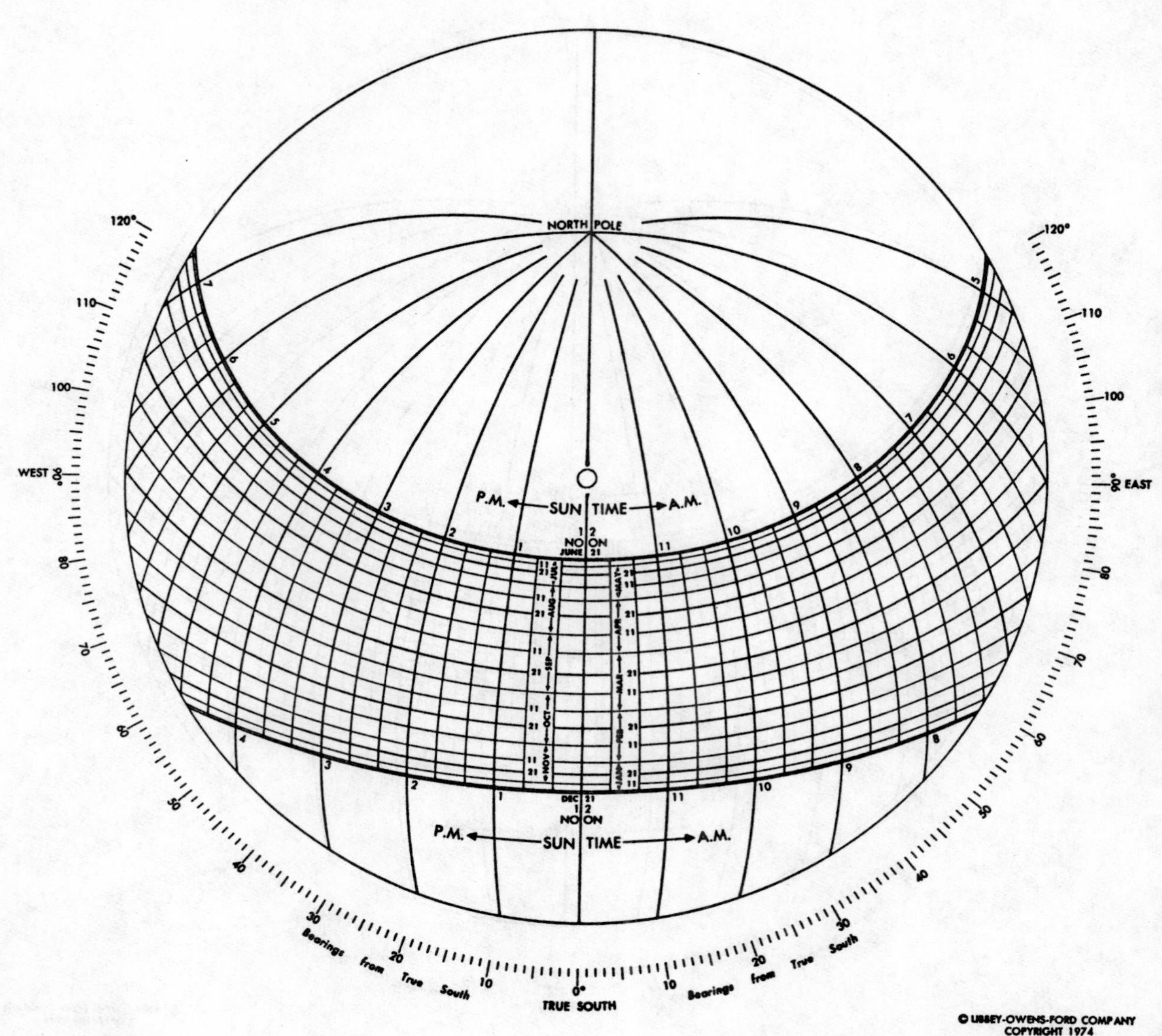

APPENDIX E

LOF SUN CHARTS

44° NORTH LATITUDE SUN CHART

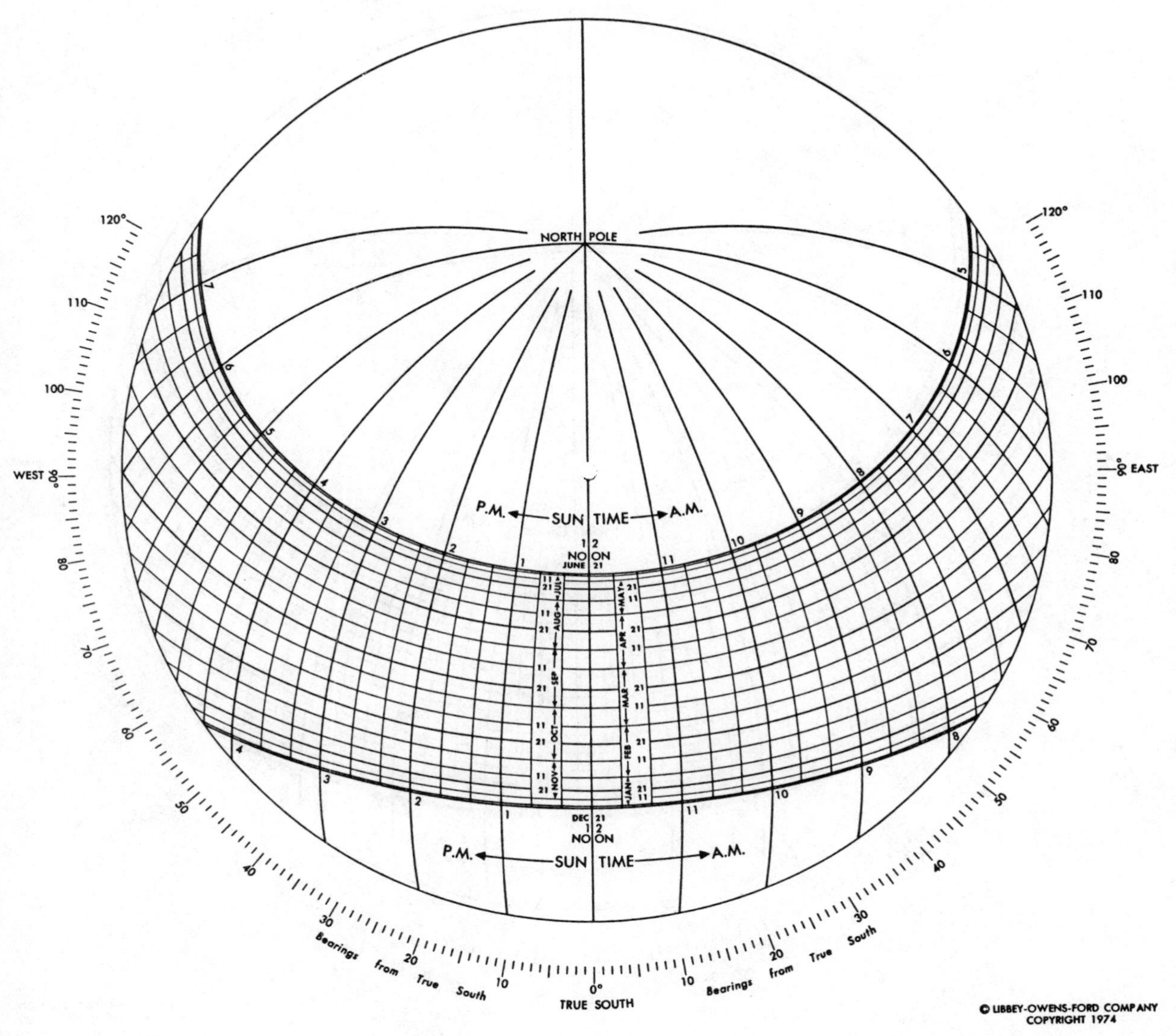

APPENDIX E

LOF SUN CHARTS

48° NORTH LATITUDE SUN CHART

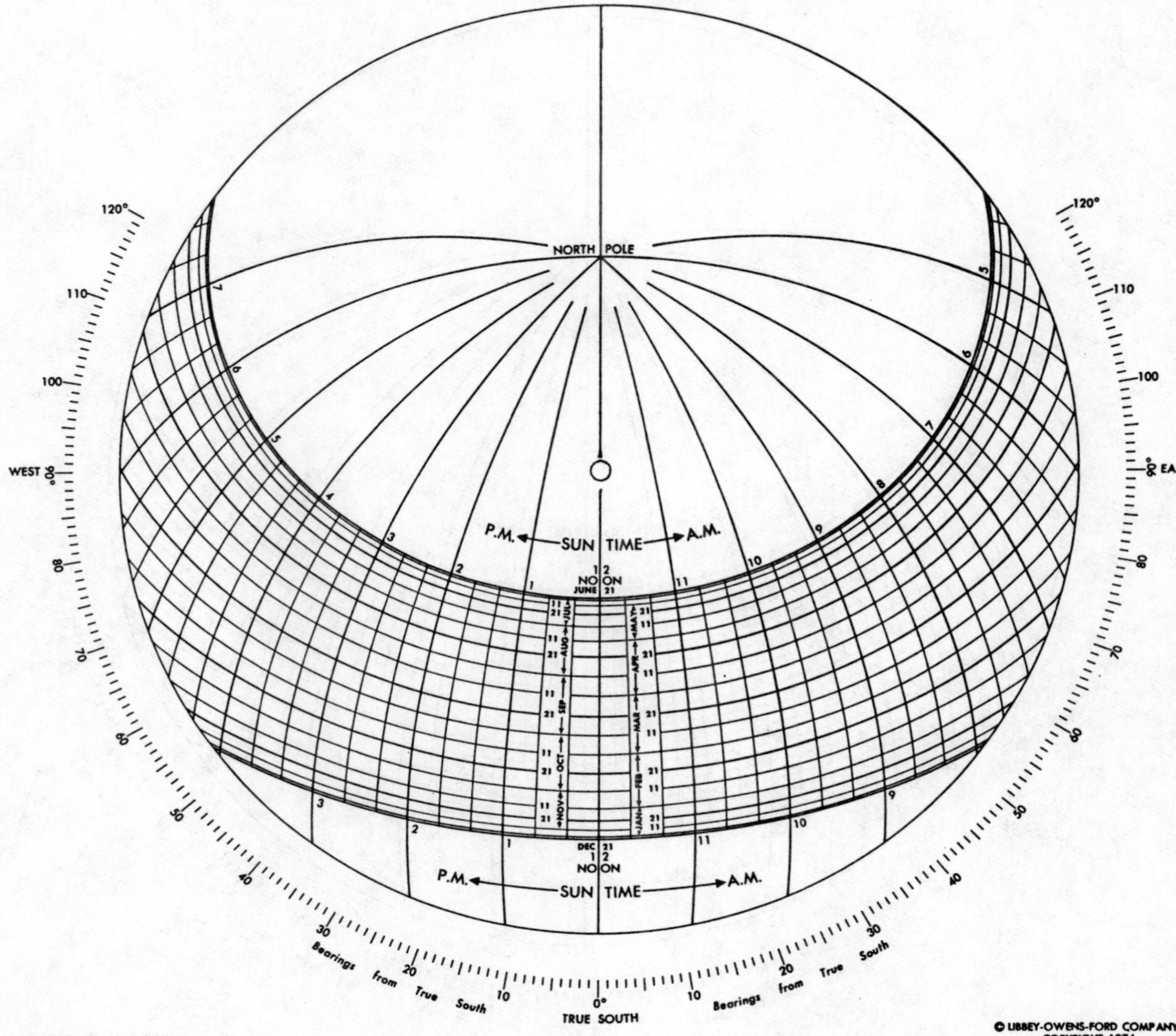

APPENDIX E

LOF SUN CHARTS
PROFILE ANGLE OVERLAY

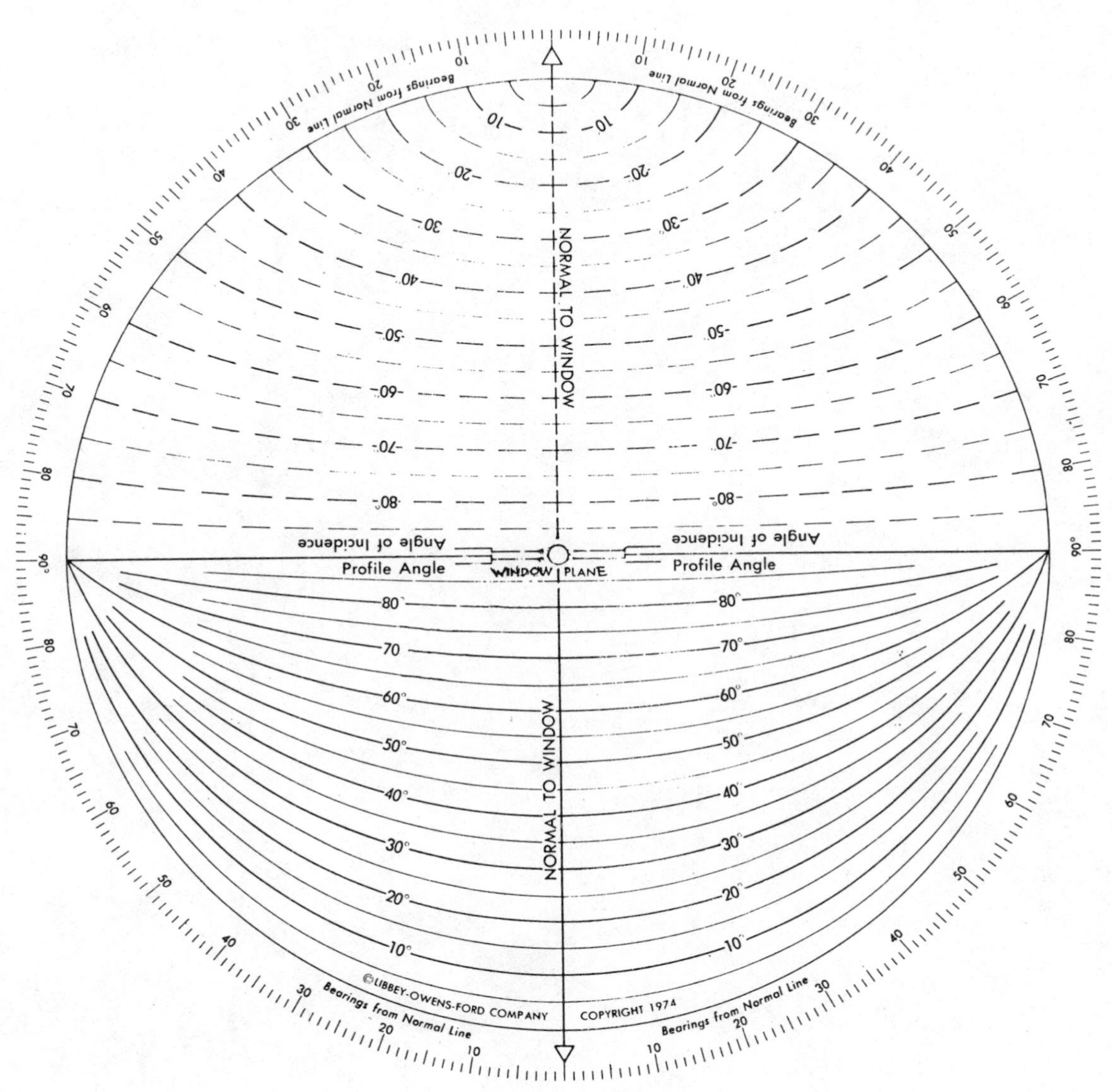

APPENDIX F

GDDM FOOTPRINTS

APP F

1–20

CONTENTS

1. A–7 AND A–8
2. A–9 AND A–10
3. A–11 AND A–12
4. A–13
5. A–14
6. A–15
7. A–16
8. A–17
9. A–18
10. A–19 AND A–20
11. A–21 AND A–22
12. A–23 AND A–24
13. A–30
14. A–31
15. A–32
16. A–33
17. A–34
18. A–35
19. A–36
20. A–37

Reprinted, with permission, from Millet and Bedrick, "Manual: Graphic Daylight Design Method."

A–7 H/W = 0.5; S/H = 0.0

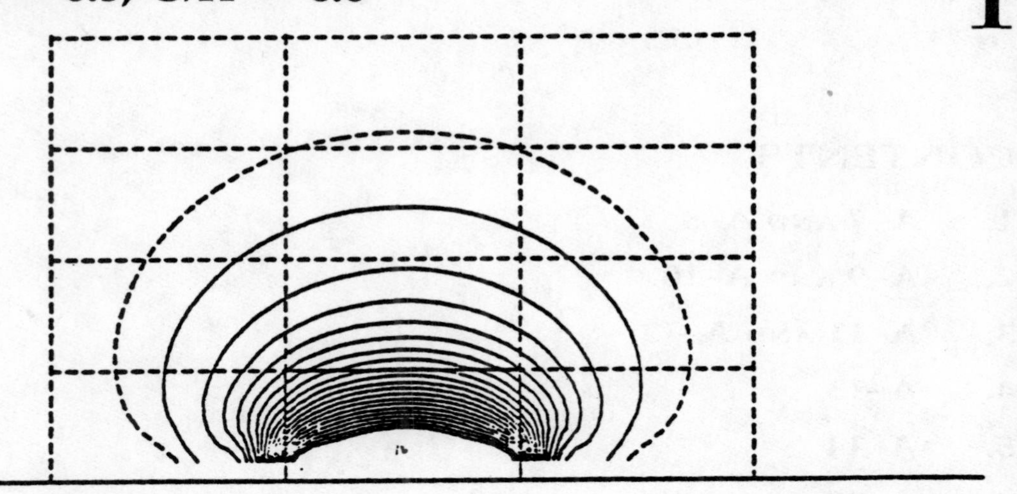

A–8 H/W = 0.5; S/H = 0.5

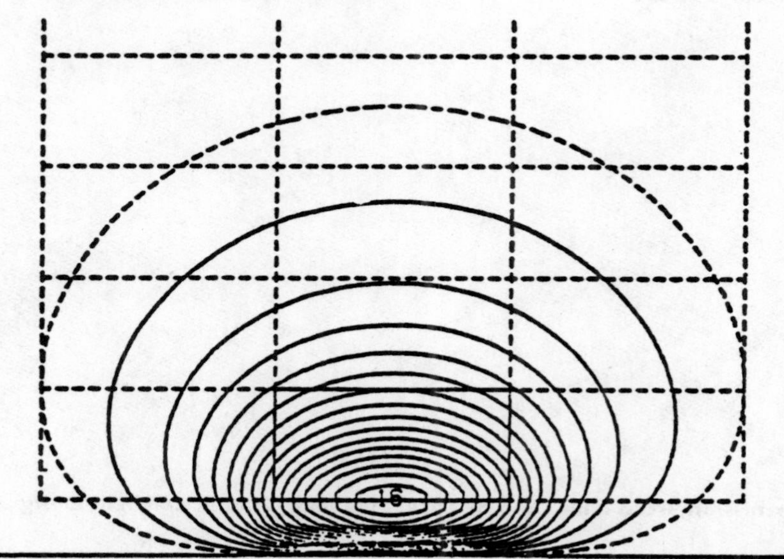

A–9 H/W = 0.5; S/H = 1.0

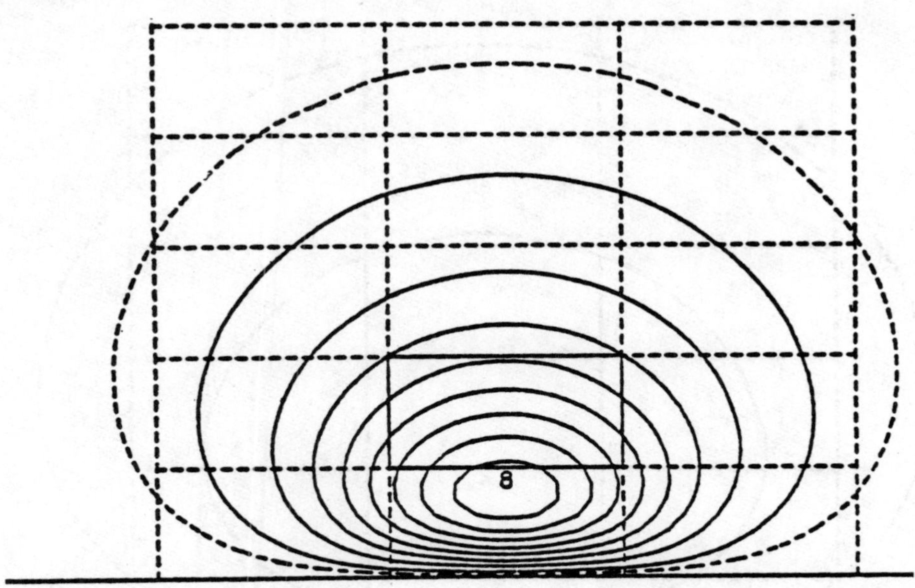

A–10 H/W = 0.5; S/H = 2.0

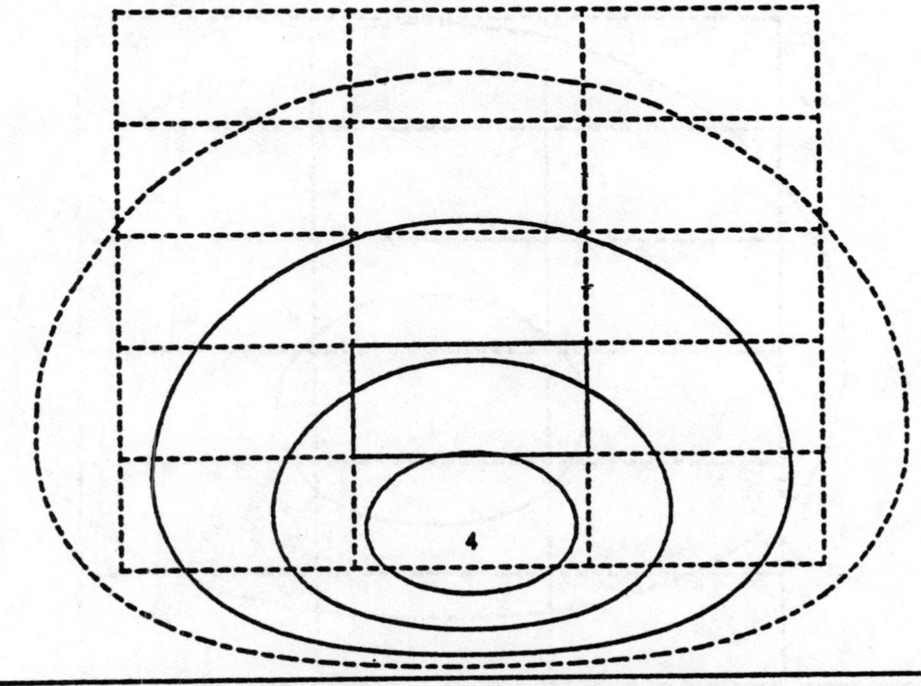

A–11 H/W = 0.5; S/H = 3.0

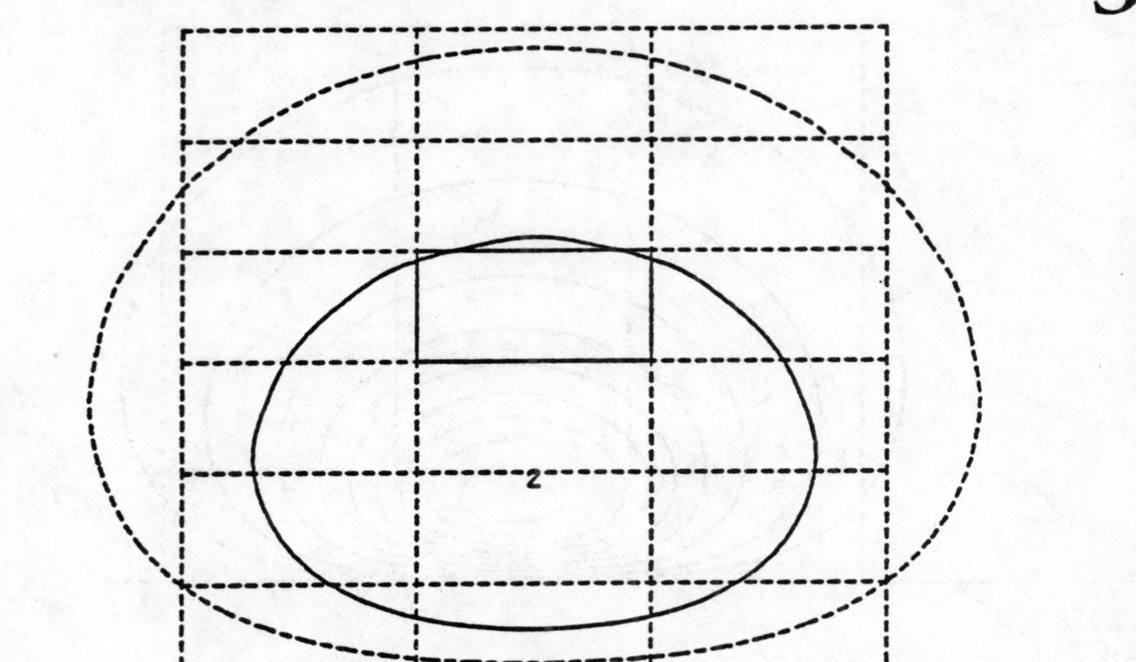

A–12 H/W = 0.5; S/H = 4.0

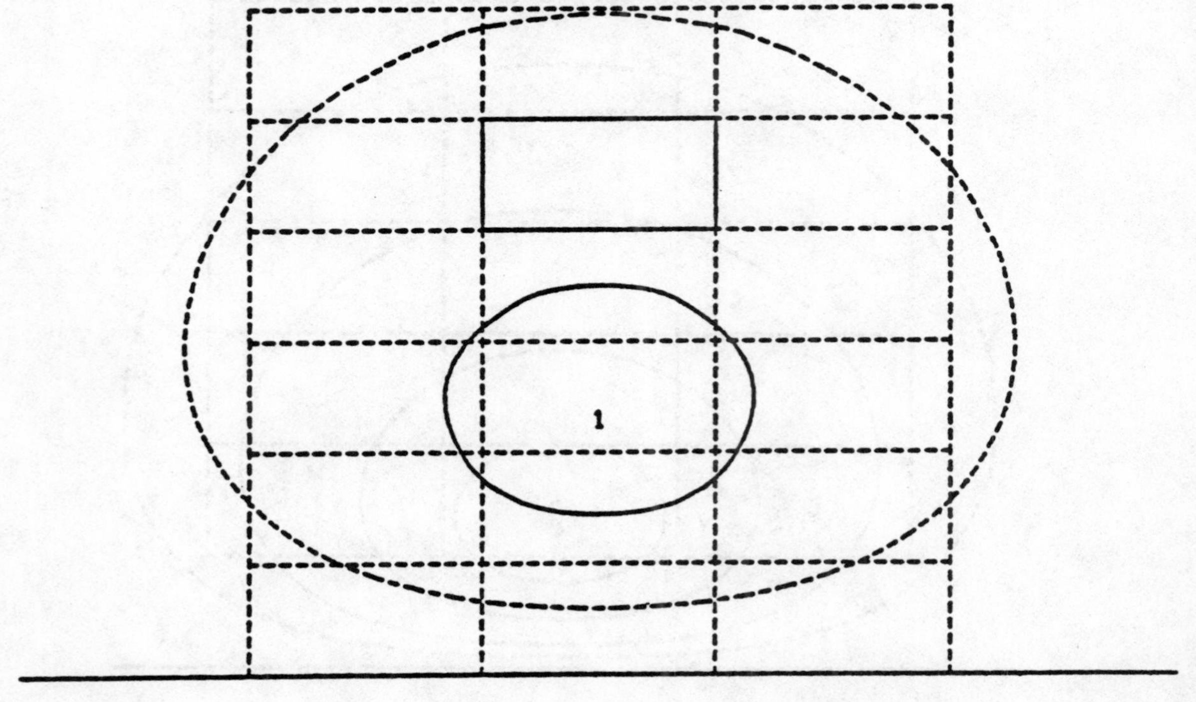

A–13 H/W = 1.0; S/H = 0.0

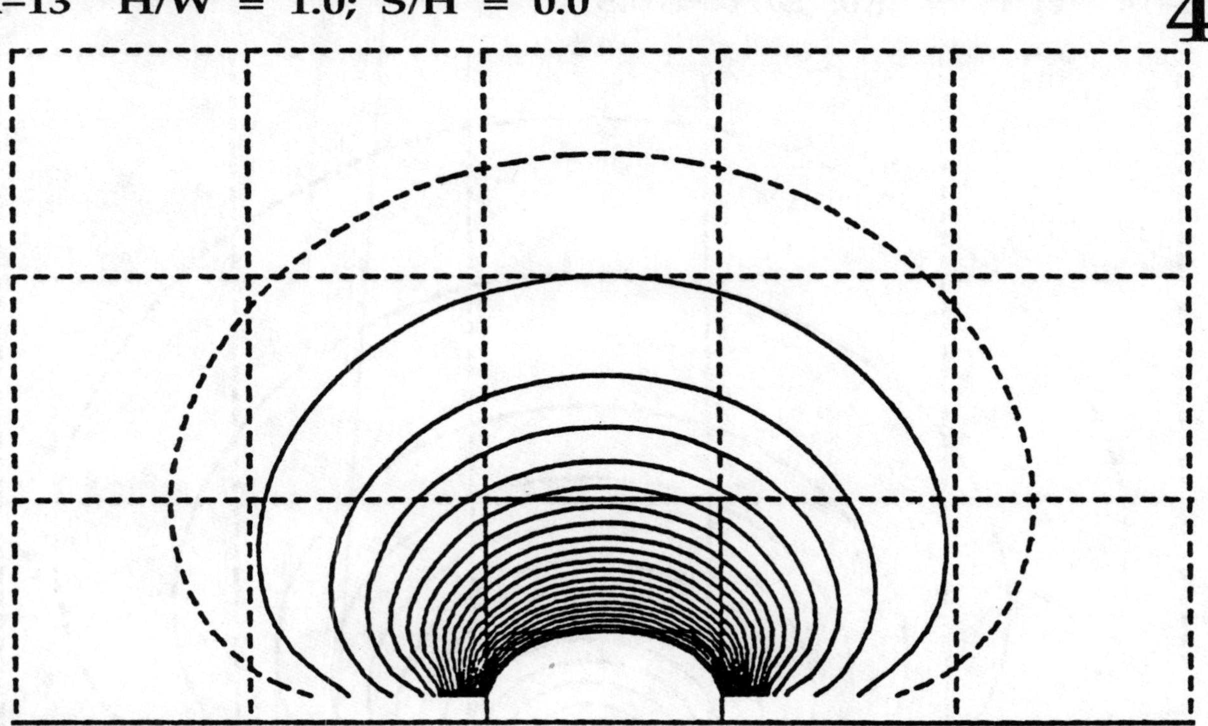

A–14 H/W = 1.0; S/H = 0.5

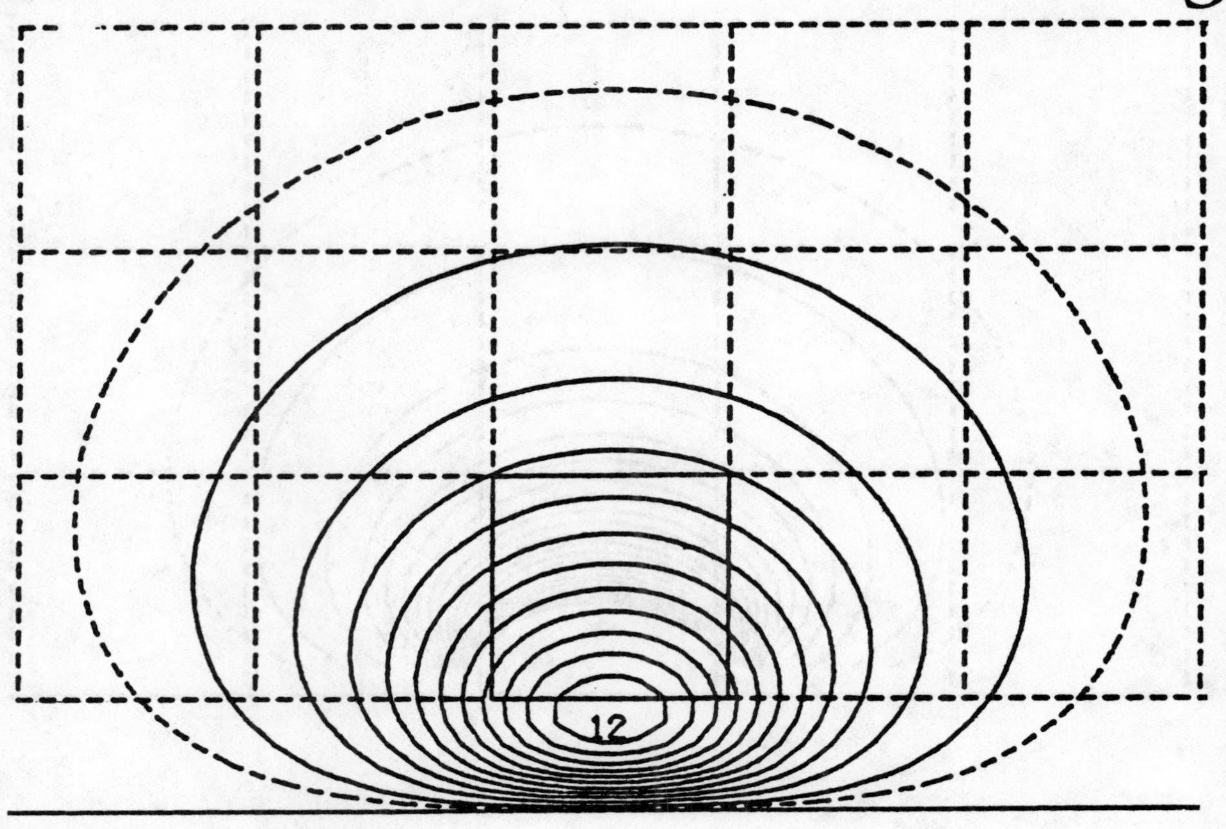

A–15 H/W = 1.0; S/H = 1.0

5

A–16 H/W = 1.0; S/H = 2.0

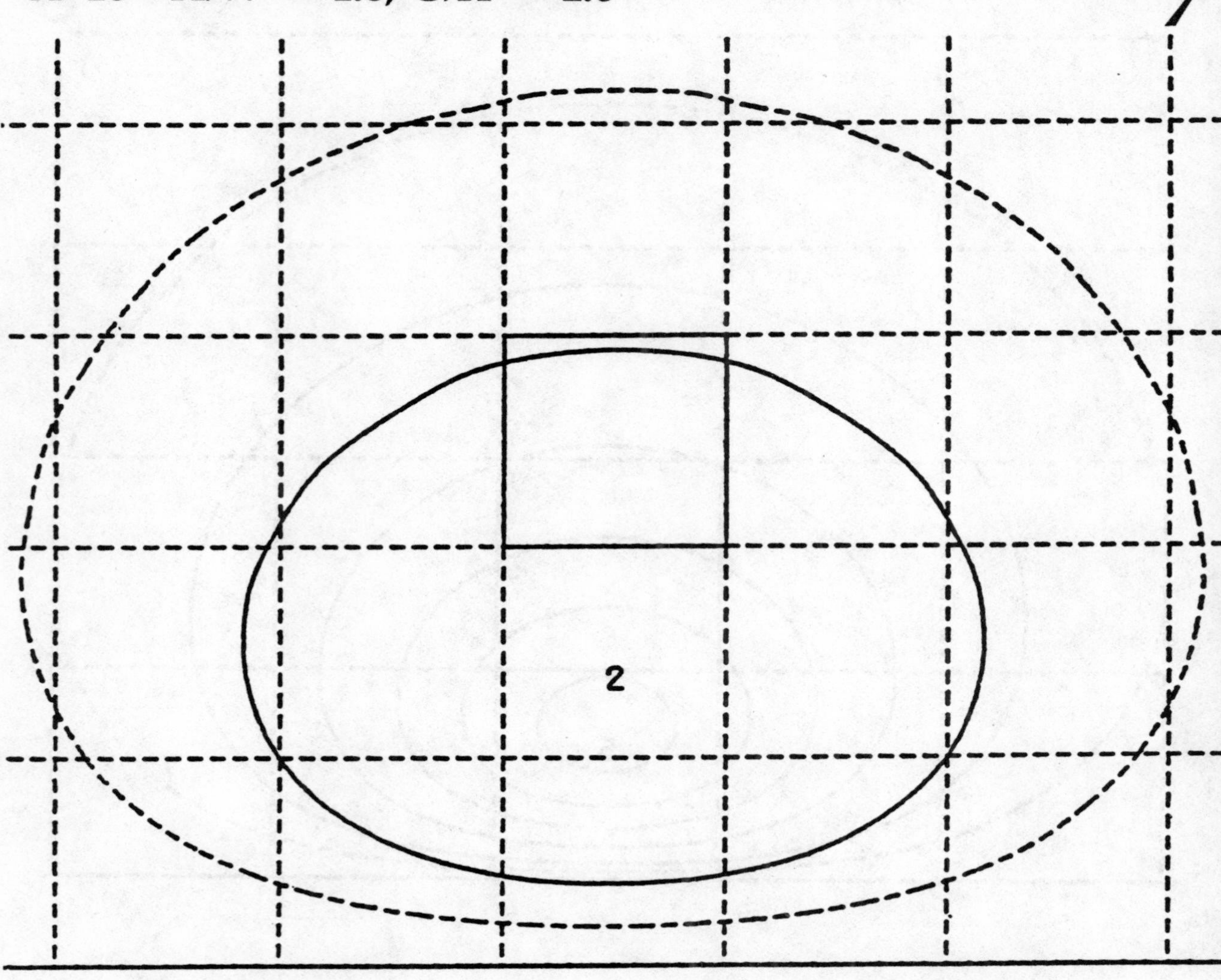

A–17 H/W = 1.0; S/H = 3.0

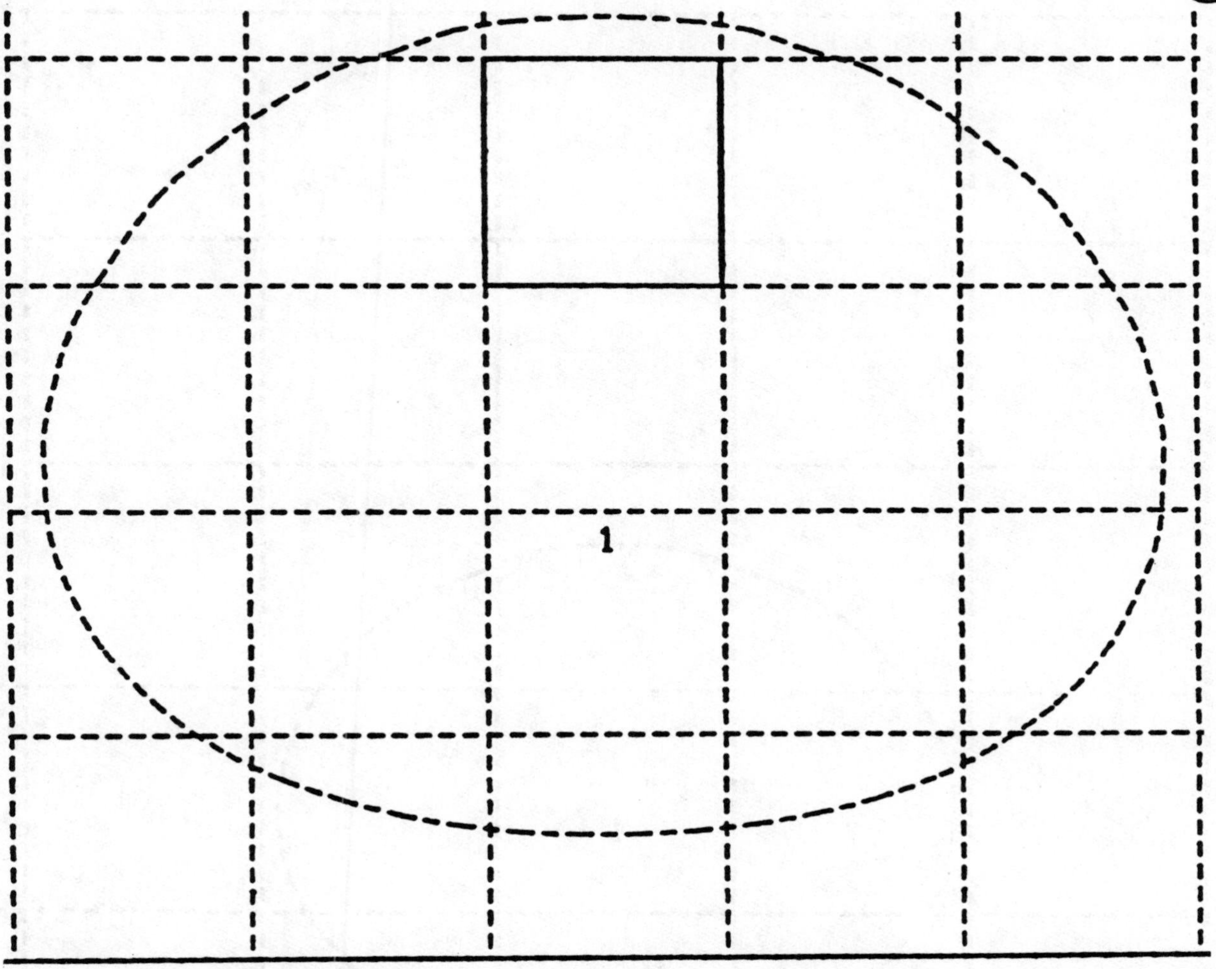

1

A–18 H/W = 1.0; S/H = 4.0

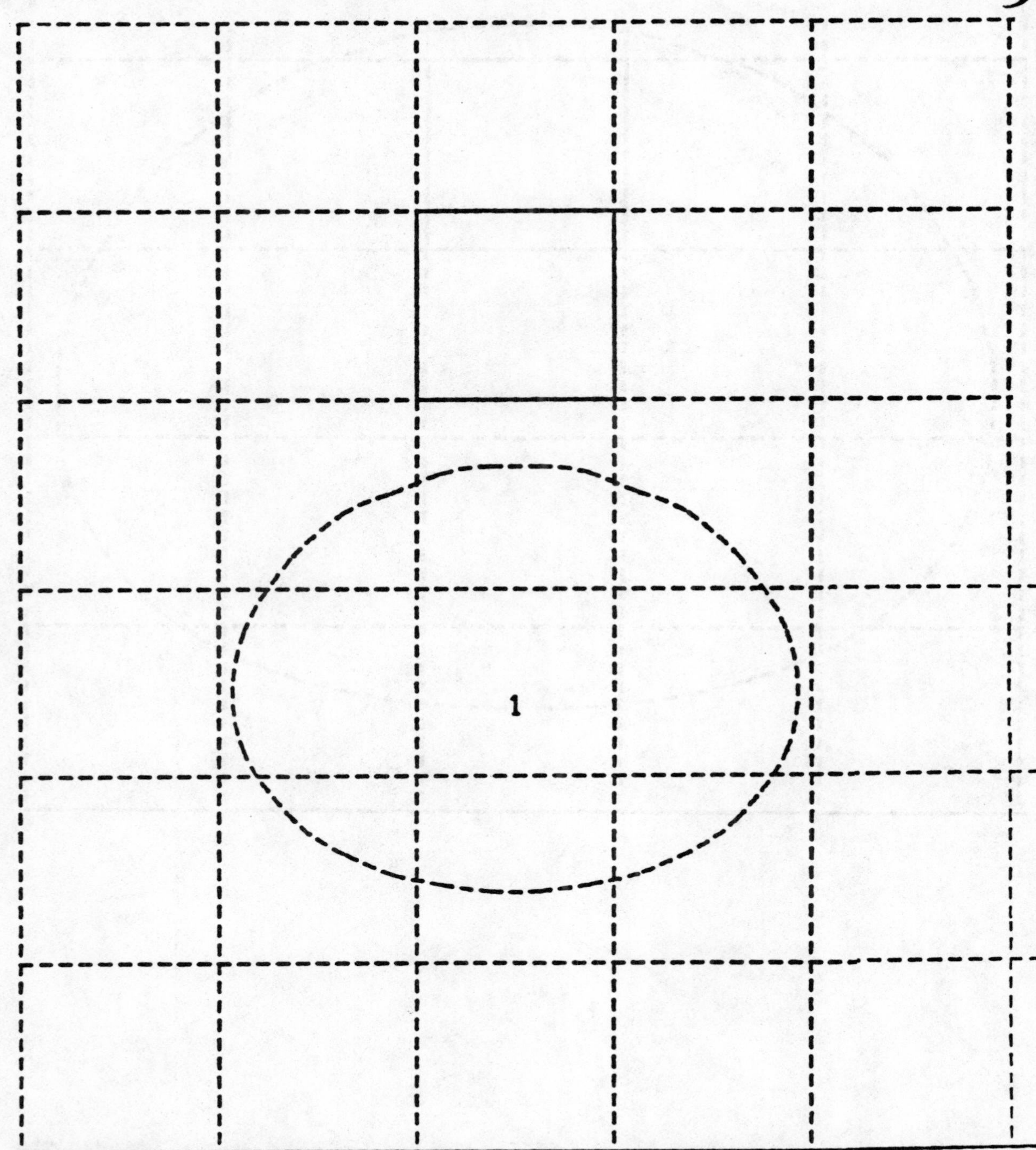

A–19 H/W = 2.0; S/H = 0.0

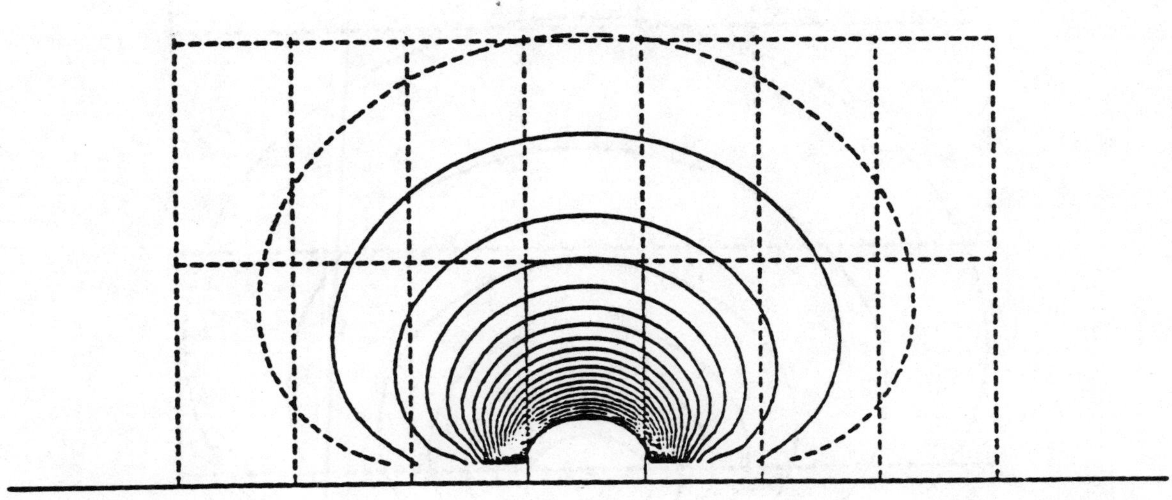

A–20 H/W = 2.0; S/H = 0.25

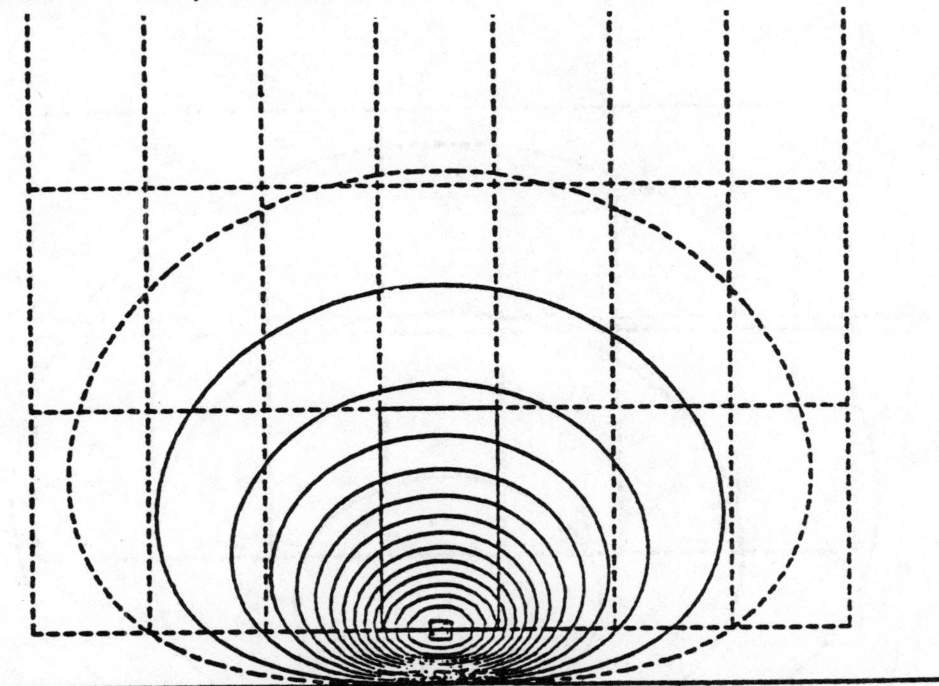

A–21 H/W = 2.0; S/H = 0.5

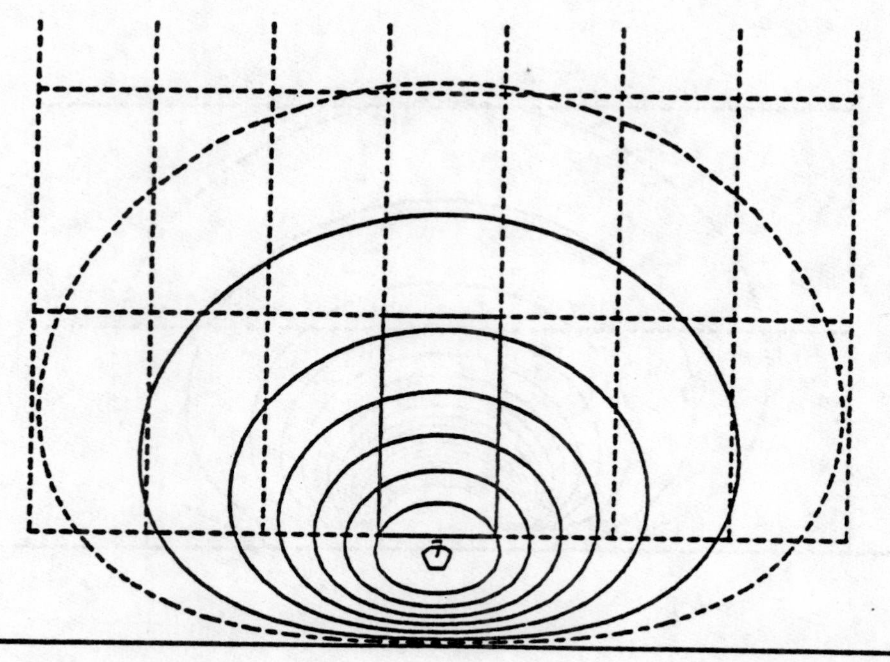

A–22 H/W = 2.0; S/H = 1.0

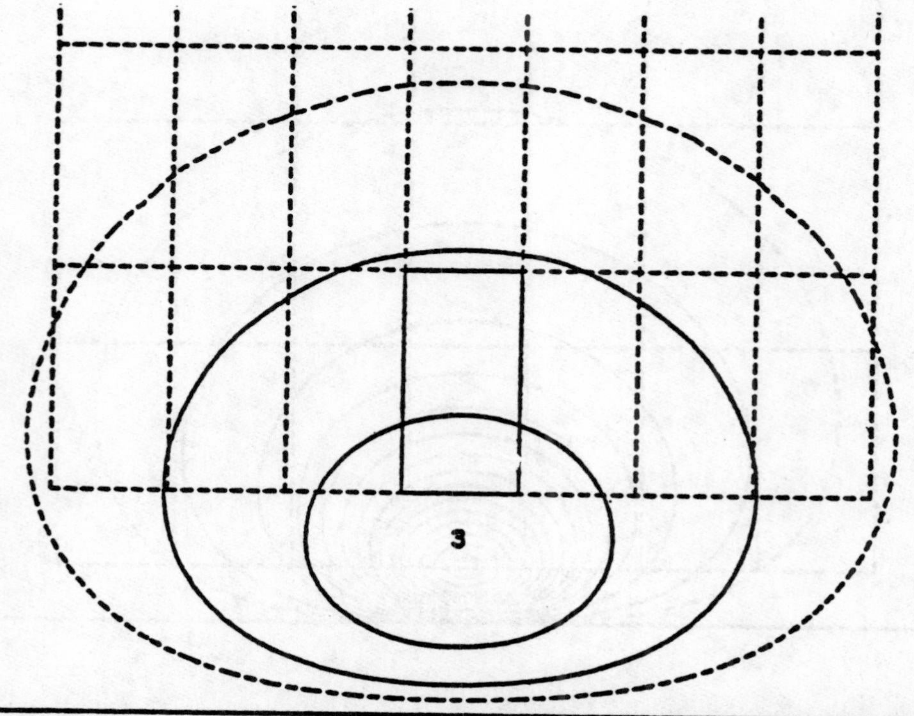

A–23 H/W = 2.0; S/H = 1.5

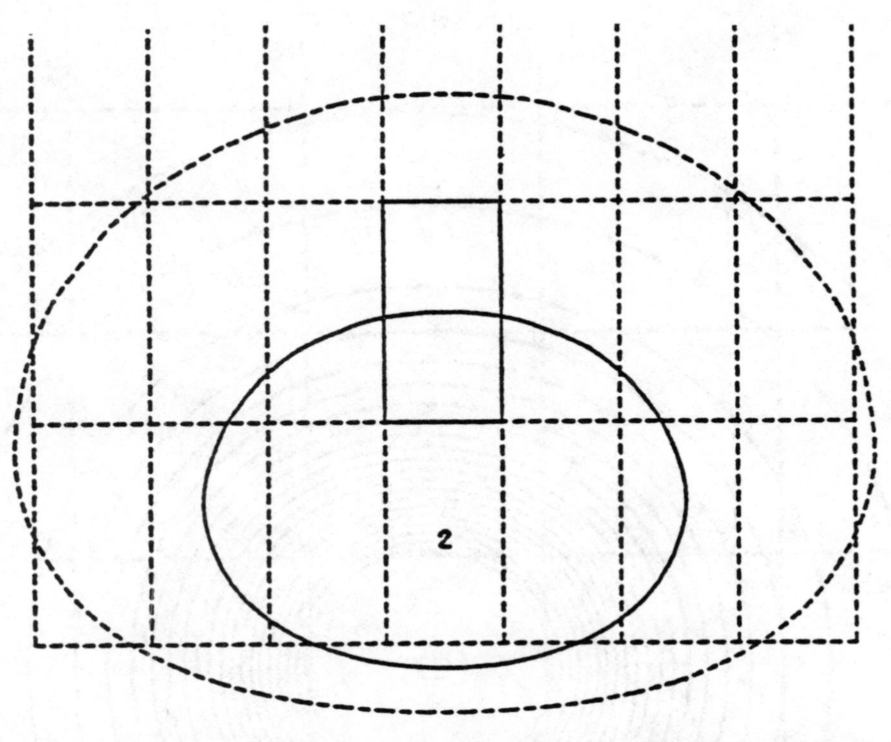

A–24 H/W = 2.0; S/H = 2.0

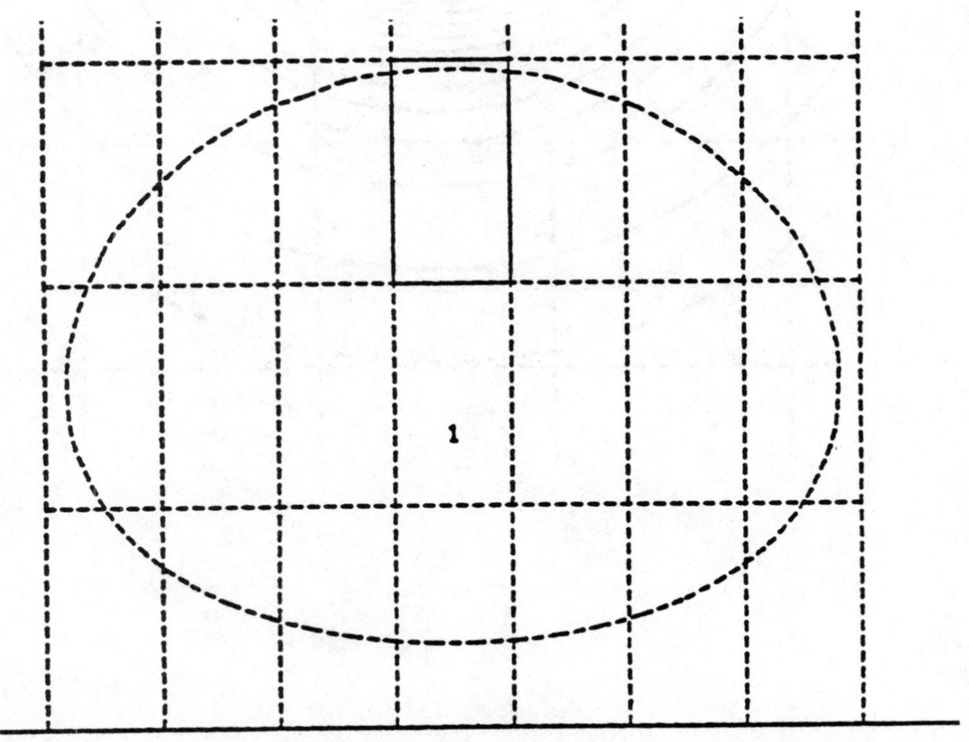

A–30 L/W = 1.0; S/W = 1.0

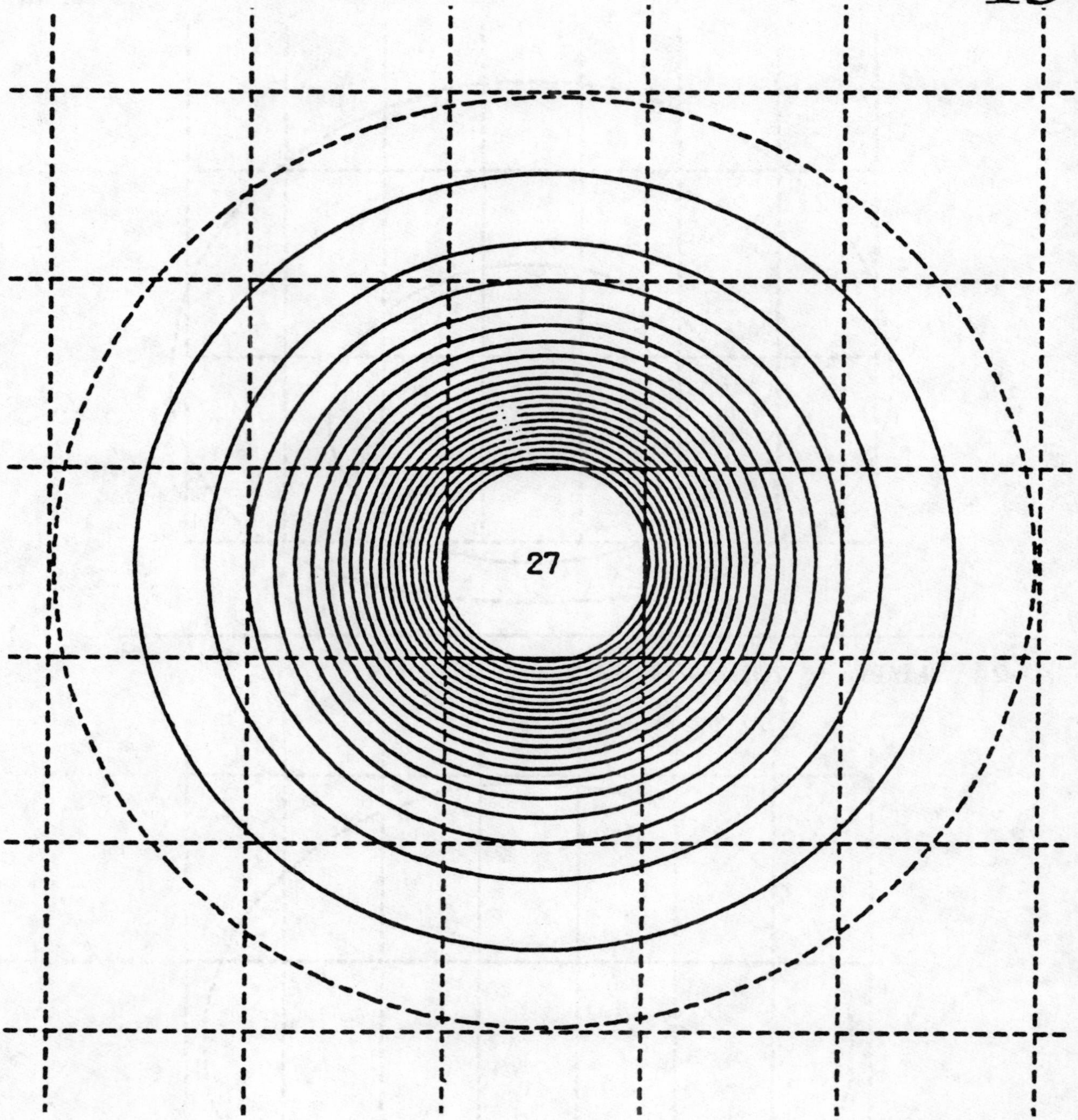

27

GDDM FOOTPRINTS

A–31 L/W = 1.0; S/W = 2.0

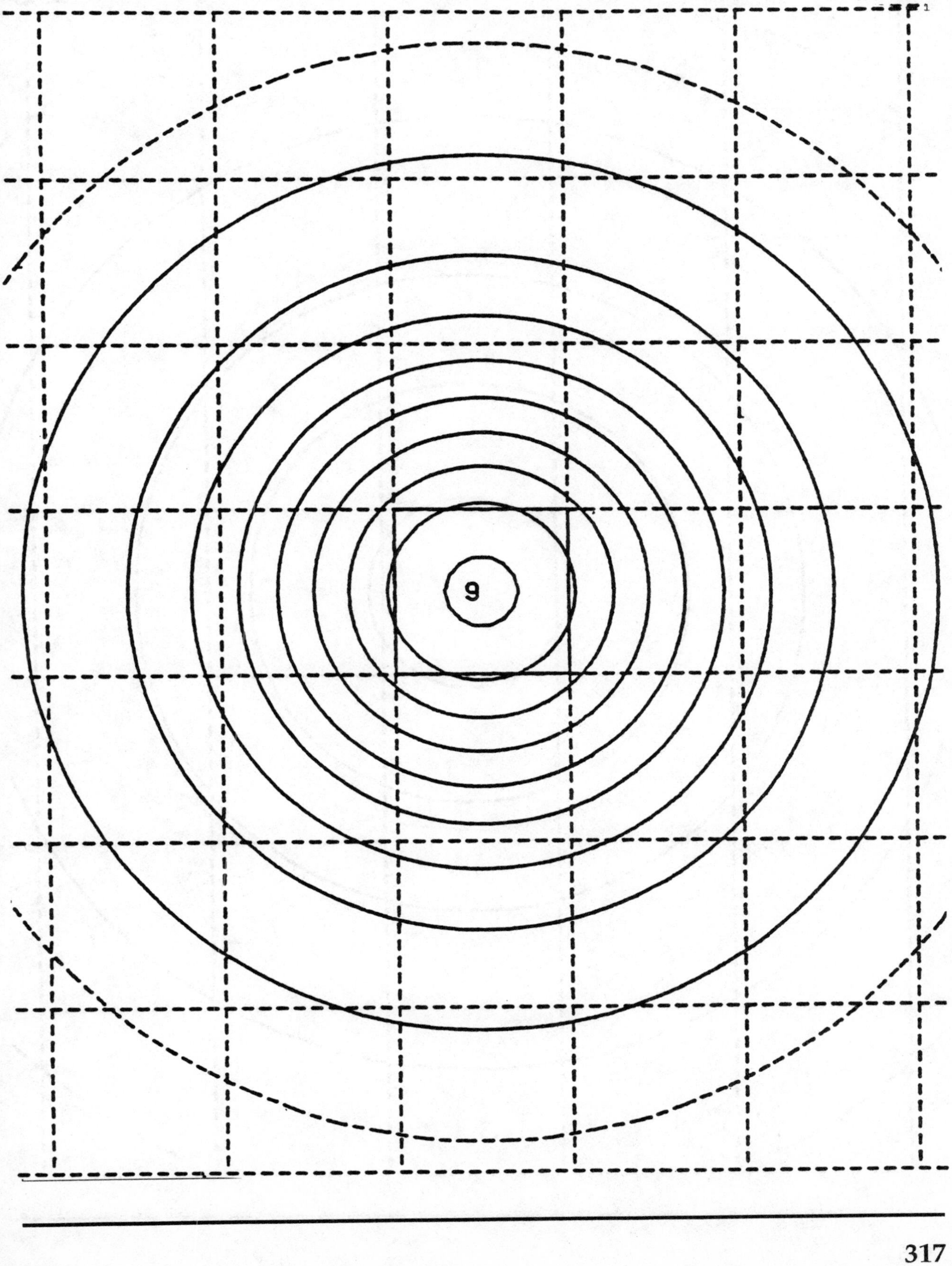

9

A–32 L/W = 1.0; S/W = 3.0

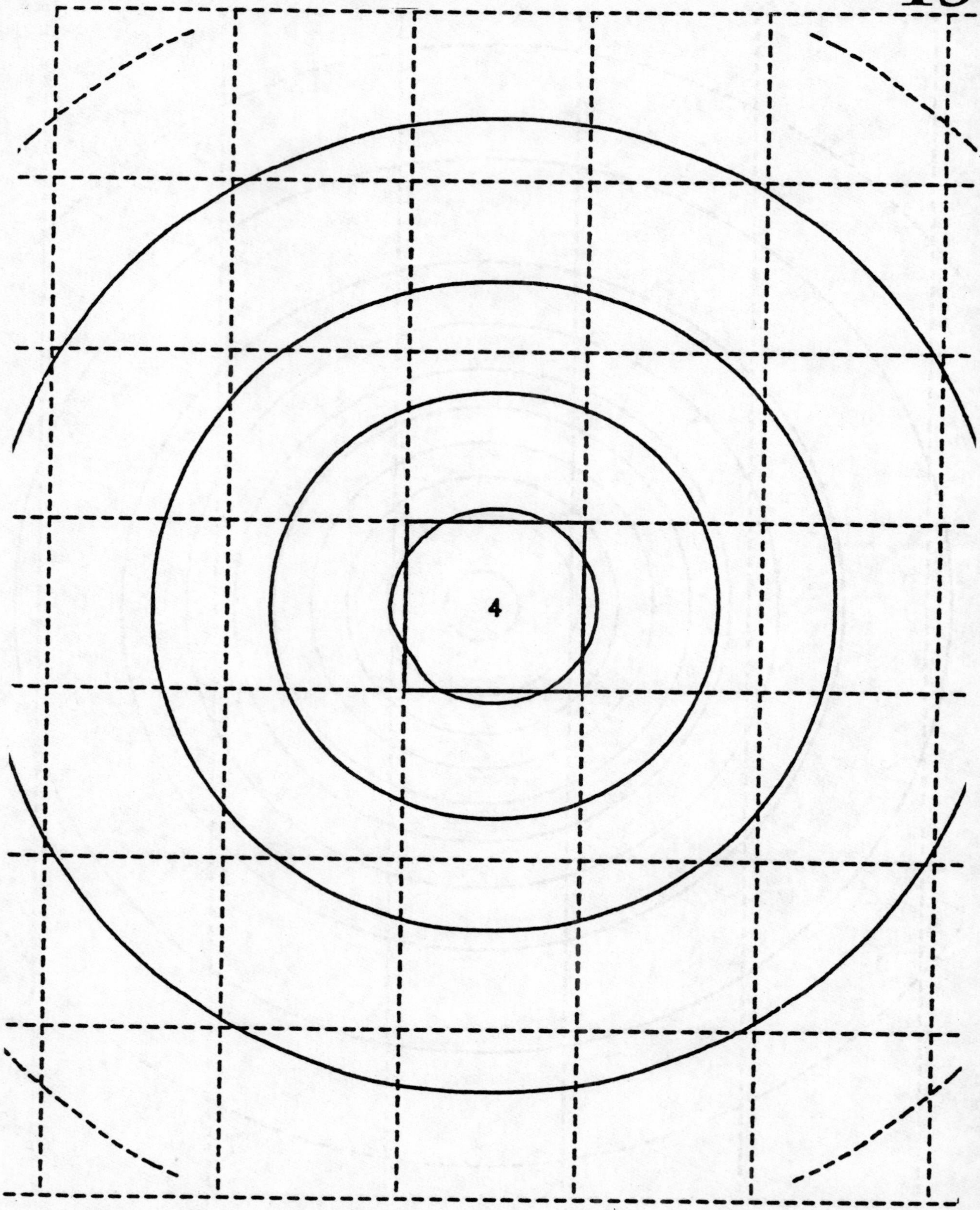

4

A–33 L/W = 1.0; S/W = 4.0

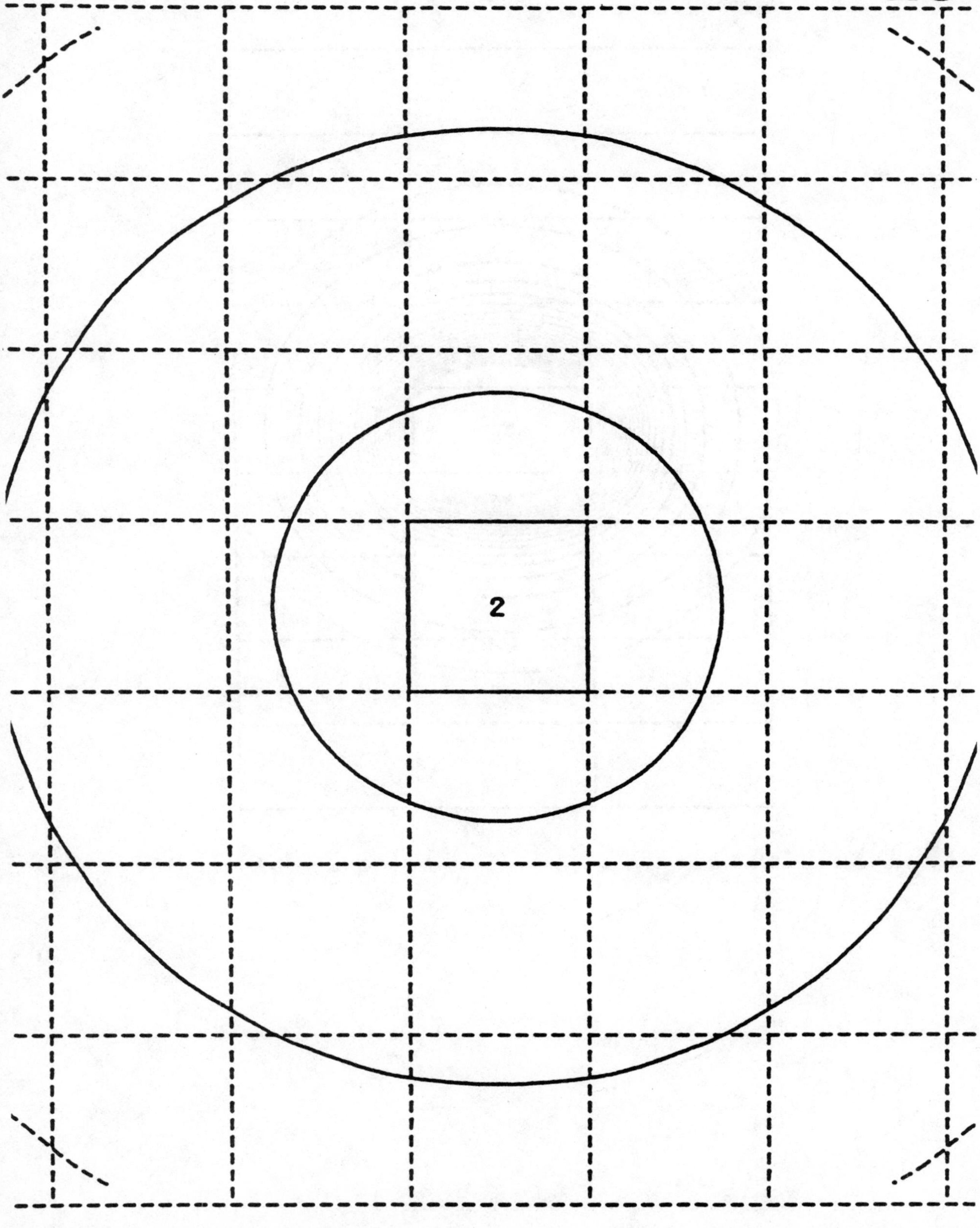

2

A–34 L/W = 2.0; S/W = 1.0

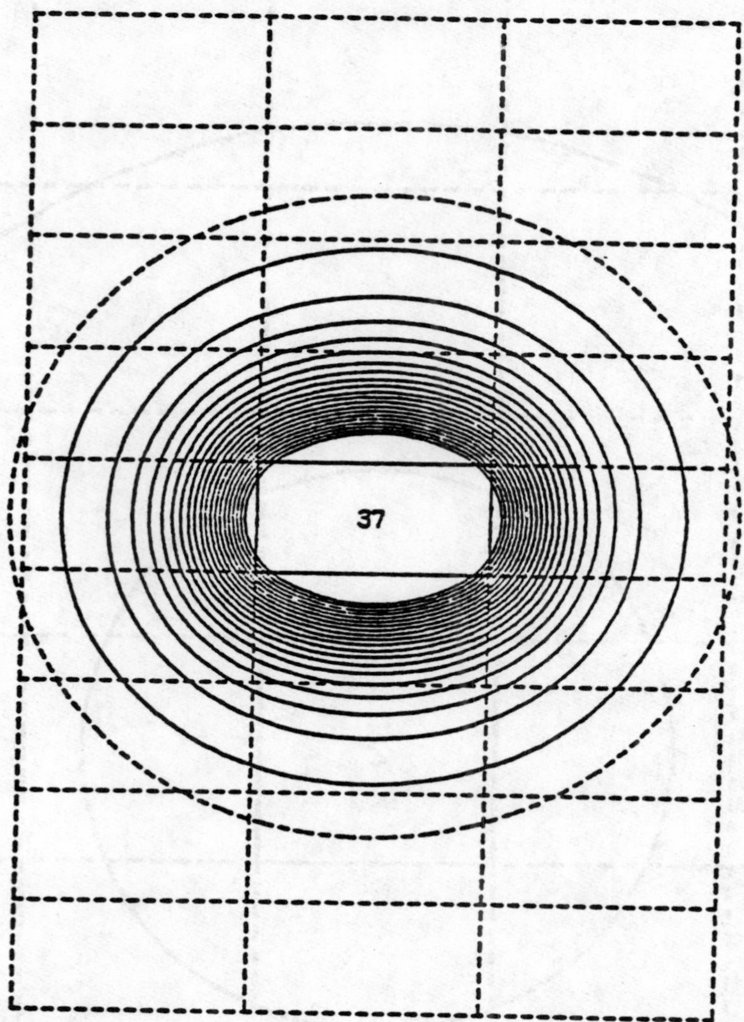

A–35 L/W = 2.0; S/W = 2.0

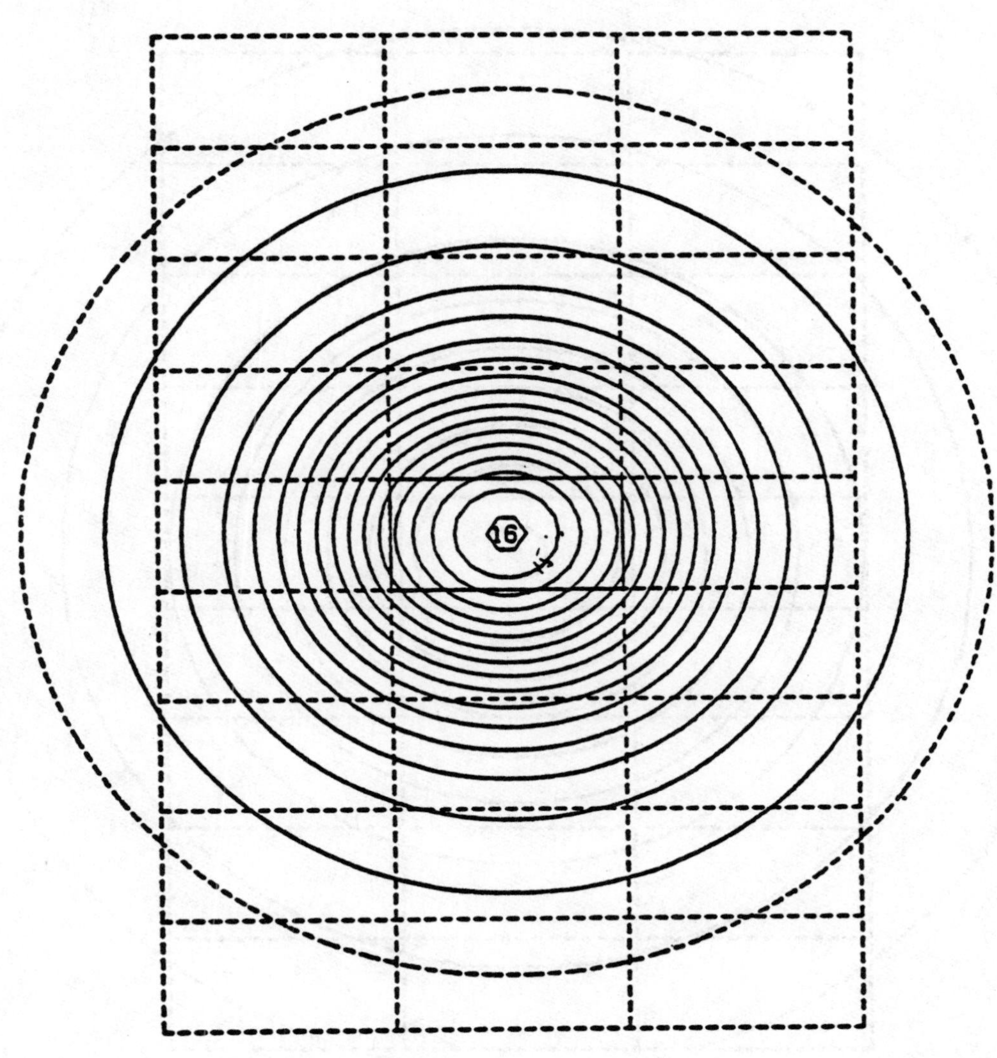

A–36 L/W = 2.0; S/W = 3.0

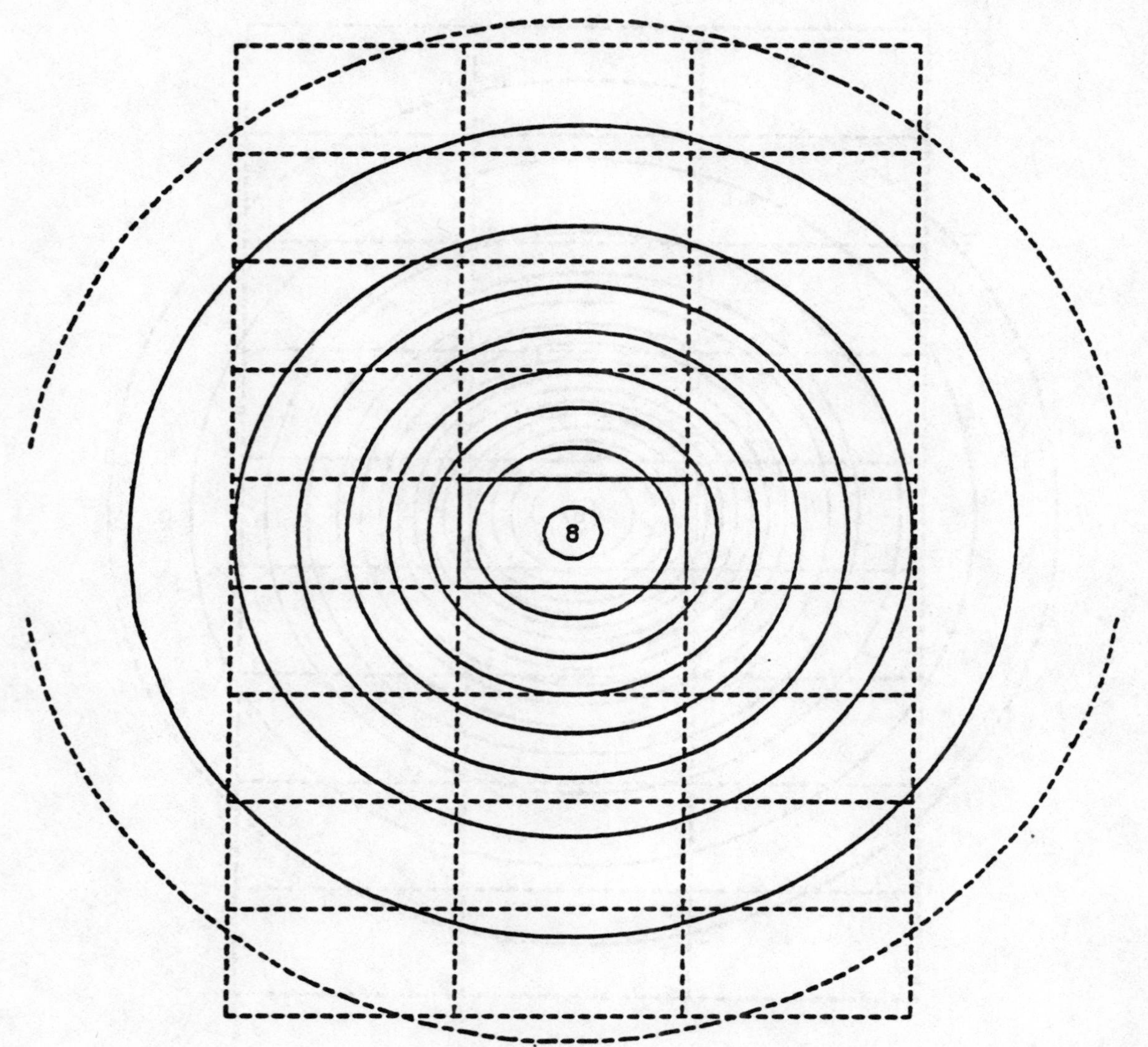

A–37 L/W = 2.0; S/W = 4.0

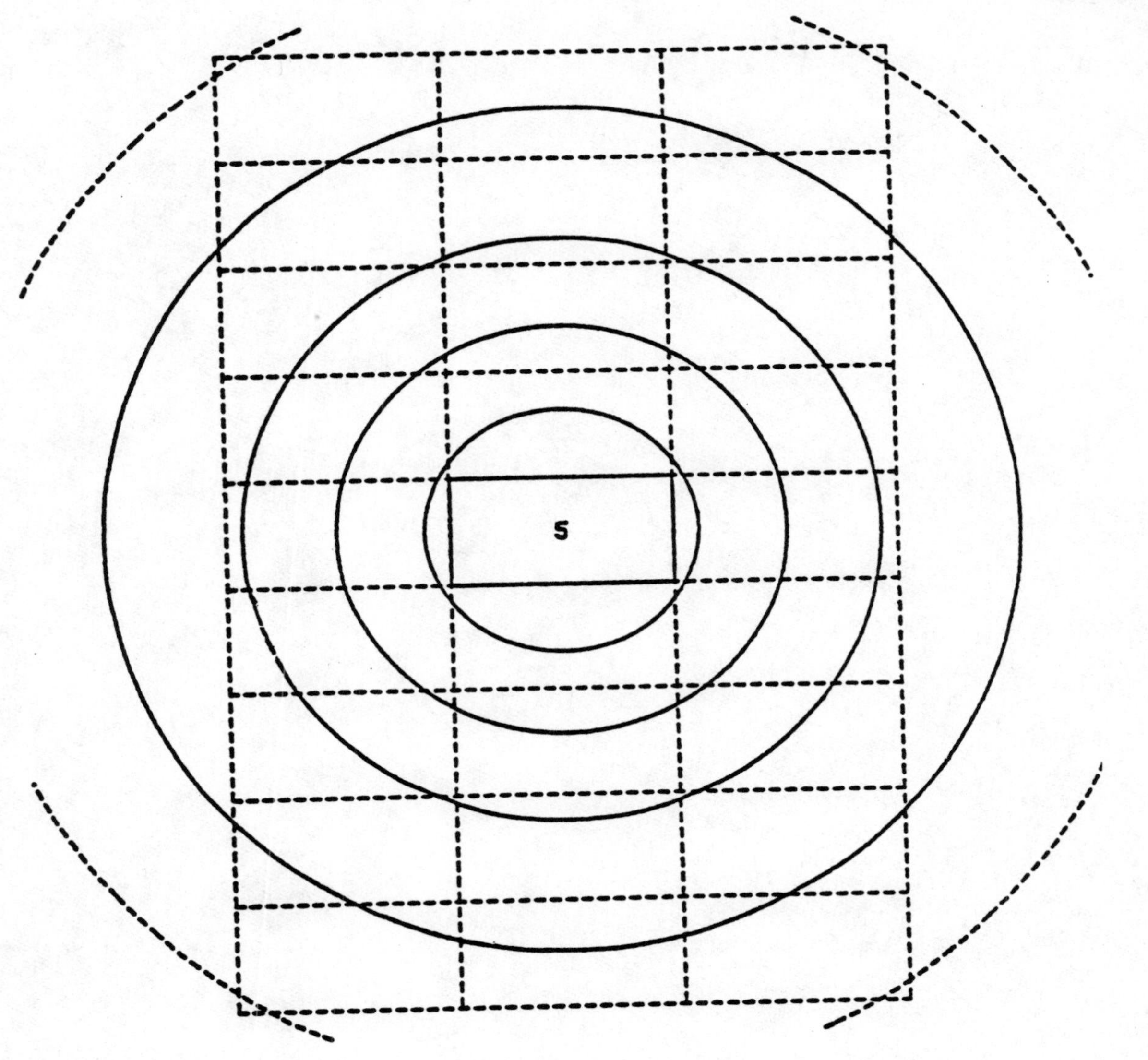

BIBLIOGRAPHY

Allen, Edward, and Joseph Iano. *The Architect's Studio Companion: Technical Guidelines for Prelminary Design*. New York: John Wiley & Sons, 1989.

American Society of Heating, Refrigerating and Air-Conditioning Engineers, Inc. (ASHRAE). "Energy Efficient Design of New Buildings Except Low-Rise Residential Buildings." *ASHRAE/IES Standard 90.1-1989*. Atlanta: ASHRAE Co-Sponsored by Illuminating Engineering Society of North America, 1989.

Balcomb, J. Douglas, et al. *Passive Solar Analysis*, Atlanta: ASHRAE, 1984.

Balcomb, J. Douglas, Dennis Barley, Robert McFarland, Joseph Perry, Jr., William Wray, and Scott Noll. *Passive Solar Design Handbook*. Vol. 2, *Passive Solar Design Analysis*. Washington: U.S. Government Printing Office, 1980.

Balcomb, J. Douglas, Robert W. Jones (ed.), Claudia E. Kosiewicz, Gloria S. Lazarus, Robert D. McFarland, and William O. Wray. *Passive Solar Design Handbook*. Vol. 3, *Passive Solar Design Analysis and Supplement*. Boulder: American Solar Energy Society, Inc., 1983.

Brown, G. Z. *Sun, Wind, and Light: Architectural Design Strategies*. New York: John Wiley & Sons, Inc., 1985.

Egan, M. David. *Architectural Acoustics*. New York: McGraw-Hill Book Company, 1988.

Evans, Benjamin H. *Daylight in Architecture*. Architectural Record Books. New York: McGraw-Hill Book Company, 1981.

Fairey, Philip, Robin Vieira, and Alp Kerestecioglu. "Dessicant Enhanced Nocturnal Radiation: A New Passive Cooling Concept." In *Proceedings of the 10th National Passive Solar Conference, the Solar Energy & Utilities Conference, The Daylighting Applications Conference, The Building with the Sun Conference*, edited by Alexander T. Wilson and William Glennie, 271–276. Boulder: American Solar Energy Society, Inc., 1985.

Fitch, James Marston. "Experiential Context of the Aesthetic Process." *Journal of Archtiectural Education* 41 no. 2 (1988): 4–9.

Flynn, John E., and Arthur W. Segil. *Architectural Interior Systems: Lighting, Air Conditioning, Acoustics*. New York: Van Nostrand Reinhold Company, 1970.

Hubka, Thomas C. *Big House, Little House, Back House, Barn: The Connected Farm Buildings of New England*. Hanover, NH: University Press of New England, 1984.

International Conference of Building Officials. *Uniform Building Code*, 1991 ed. Whittier, CA: International Conference of Building Officials, 1991.

Kaufman, John E. (ed.). *IES Lighting Handbook: The Standard Lighting Guide*, 5th ed. New York: Illuminating Engineering Society, 1972.

Knowles, Ralph L. *Energy and Form: An Ecological Approach to Urban Growth*. Cambridge, MA: The MIT Press, 1974.

BIBLIOGRAPHY

Labs, Kenneth, and Keith Harrington. "A Comparison of Ground and Above-Ground Climates for Identifying Appropriate Cooling Strategies." *Passive Solar Journal* 1 no. 1 (1982): 4–11.

Lechner, Norbert. *Heating, Cooling, Lighting: Design Methods for Architects*. New York: John Wiley & Sons, 1991.

Libbey-Owens-Ford Company. *Sun Angle Calculator*. Toledo: Libbey-Owens-Ford Company, 1974.

McHarg, Ian L. *Design with Nature*. Philadelphia: Natural History Press, 1969. Reprint. Garden City, NY: Doubleday & Company, Inc., 1971.

Mark, Robert. *Light, Wind, and Structure: The Mystery of the Master Builders*. New liberal arts series. Cambridge, MA: MIT Press, 1990.

Millet, Marietta S., and James R. Bedrick. "Manual: Graphic Daylighting Design Method." U.S. Department of Energy through Lawrence Berkeley Laboratory, University of California, 1980. Typescript.

Milne, Murray. *Residential Water Conservation*. California Water Resources Center Report No. 35. Davis, CA: University of California, Davis, 1976.

Moore, Fuller. *Concepts and Practice of Architectural Daylighting*. New York: Van Nostrand Reinhold Company, Inc., 1985.

Nabokov, Peter, and Robert Easton. *Native American Architecture*. New York and Oxford: Oxford University Press, 1989.

Northwest Power Planning Council. *Staff Briefing Paper on Commercial Lighting*. Portland, OR: Northwest Power Planning Council, 1987.

Olgyay, Victor. *Design with Climate: Bioclimatic Approach to Architectural Regionalism*. Princeton: Princeton Univ. Press, 1963.

Ramsey, Charles George. *Ramsey/Sleeper Architectural Graphic Standards*. Student ed. abridged from 7th ed., ed. Stephen A. Kliment, 7th ed., ed. Robert T. Packard. New York: John Wiley & Sons, Inc., 1989.

Rapoport, Amos. *History and Precedent in Environmental Design*. New York: Plenum Press, 1990.

———. *House Form and Culture*. Foundations of Cultural Geography Series. Englewood Cliffs, NJ: Prentice-Hall, Inc., 1969.

Schiler, Marc (ed.). *Simulating Daylight with Architectural Models*. Sponsored by U.S. Department of Energy, Daylighting Network of North America, Southern California Daylighting Council, and University of Southern California, 1986.

Spielvogel, Lawrence G. "How and Why Buildings Use Energy." In Watson, Donald, *Energy Conservation through Building Design*, pp.52–75. An Architectural Record Book. New York: McGraw-Hill Book Company, 1979.

BIBLIOGRAPHY

Stein, Benjamin, and John S. Reynolds. *Mechanical and Electrical Equipment for Buildings*. 8th ed. New York: John Wiley & Sons, Inc. 1991.

Stevens, Peter S. *Patterns in Nature*. An Atlantic Monthly Press Book. Boston: Little, Brown and Company, 1974.

Stoner, Carol Hupping (ed.). *Goodbye to the Flush Toilet: Water-Saving Alternatives to Cesspools, Septic Tanks, and Sewers*. Emmaus, PA: Rodale Press, 1977.

Sweet's Group. *Sweet's Catalog File: Products for General Building & Renovation*. New York: McGraw-Hill Information Services Company, 1990.

Thompson, D'Arcy Wentworth. *On Growth and Form*. Vol. 1. Cambridge, England: The University Press, 1952.

U.S. Department of Commerce. Environmental Science Service Administration. Environmental Data Service. *Climatic Atlas of the United States*. Washington: U.S. Government Printing Office, 1968.

U.S. National Oceanic and Atmospheric Administration, Environmental Data and Information Service, National Climatic Center. *Local Climatological Data: Annual Summary with Comparative Data*. Asheville, NC: U.S. Department of Commerce, 1990.

Vale, Brenda, and Robert Vale. *The Autonomous House: Design and Planning for Self-Sufficiency*. New York: Universe Books, 1975.

Venturi, Robert. *Complexity and Contradiction in Architecture*, rev. ed. The Museum of Modern Art Papers on Architecture. New York: Museum of Modern Art, 1977.

Zevi, Bruno. "Architecture." In *Encyclopedia of World Art*, Vol. 1, pp.626–694. New York: McGraw-Hill Book Publishers, Inc., 1959.

BIBLIOGRAPHY

Software Sources

See *MEEB*, Appendix H, pp.1534–1539, for descriptions of these and other programs.

ADM–DOE. ADM Associates, Inc.; 3299 Ramos Circle; Sacramento, CA 95827.

CALPAS–3. Berkeley Solar Group; 454 Santa Clara Avenue; Oakland, CA 94610.

Energy Scheming (Beta Version). G. Z. Brown; Department of Architecture; University of Oregon; Eugene, OR 97403.

DOE–2. National Energy Software Center; Argonne National Laboratory; 9700 South Cass Avenue; Argonne, IL 60439.

F–Chart. F–Chart Software; 4406 Fox Bluff Road; Middleton, WI 53562.

MICRO–DOE. Arcosoft International Inc.; 9745 East Hampden Avenue, Suite 230; Denver, CO 80231.

SOLAR–5. Murray Milne; Graduate School of Architecture; UCLA; 405 Hilgard Avenue; Los Angeles, CA 90024-1467.

WATT SUN–5. Washington State Energy Office; 809 Legion Way SE; Olympia, WA 98504-1211.